PARTICLE AND PARTICLE SYSTEMS CHARACTERIZATION

PARTICLE AND PARTICLE SYSTEMS CHARACTERIZATION
Small-Angle Scattering (SAS) Applications

Dr. Wilfried Gille

Martin Luther University of Halle-Wittenberg

CRC Press
Taylor & Francis Group
Boca Raton London New York

CRC Press is an imprint of the
Taylor & Francis Group, an **informa** business

CRC Press
Taylor & Francis Group
6000 Broken Sound Parkway NW, Suite 300
Boca Raton, FL 33487-2742

First issued in paperback 2020

Version Date: 20130923

ISBN 13: 978-0-367-57625-7 (pbk)
ISBN 13: 978-1-4665-8177-7 (hbk)

Visit the Taylor & Francis Web site at
http://www.taylorandfrancis.com

and the CRC Press Web site at
http://www.crcpress.com

Contents

Preface

The aim of this book is to demonstrate basic knowledge of the application of small-angle scattering (SAS) for the characterization of physical and chemical materials in various fields to beginners and advanced scientists in a comprehensive way. The subject is divided into 10 chapters consisting of 336 pages, including 164 figures and 237 references.

SAS is a technique of spatial density measurement; however, it does not produce an image or hologram. A monochromatic neutron, X-ray or light beam directed at a sample material containing small zones of non-uniform density allows a scattering effect to be recorded. This phenomenon is based on the interference property of electromagnetic waves.

The designation *particle* is a significant one in this book. Particles are ensembles of many atoms, which possess a (nearly) homogeneous composition. Therefore, the general characterization of homogeneous regions of a heterogeneous body is referred to as *particle characterization* for short. SAS can detect (yields) geometric parameters describing ensembles of small particles down to the nanometer region, i.e., of microscopic regions possessing a density different from that of their surroundings. SAS does not depend in any way on the internal crystal structure of the particle and is produced by the particle as a whole.

In contrast to the electron microscope, SAS experiments do not require a special sample preparation procedure, are essentially non-destructive and can be directly applied for liquid and solid samples.

While an electron microscope yields direct real-space images, SAS provides data in *reciprocal space*, which yields a scattering pattern. Compared with a typical sample micrograph of well-selected magnification, the information content of an interference intensity can be interpreted in many ways. This interference intensity is called the *scattering intensity* or *scattering pattern* of the sample for short. An interpretation of such data always requires a model as well as initial assumptions about the sample material: For a certain order range (i.e., on a certain length scale interval), the spatial density of many materials can be approximated by a two-phase model. Such a model implies two different densities. For the density contrast of particles, possessing a constant density inside and a surrounding matrix, there exists a variety of geometric approaches and applications, modeling sizes, shapes and geometric arrangements. In the ideal case, both the size and shape of the particles involved in a model can be determined from the scattering intensity expressed as a function of the scattering angle.

To do this, integral transformations are used (i.e., procedures of numerical mathematics) to transform the recorded scattering pattern to real-space functions. These functions describe a point-to-point distribution inside a particle or between two particles, which run under the name *structure functions*. Structure functions describe sets. The most important structure function defined in the field of *crystallography* and crystal structures is the *Patterson function*. In the field of SAS, the *SAS density autocorrelation function* (also referred to as *correlation function*) defined by Debye and Bueche in 1949 is the most important.

SAS has developed into a standard method in the field of materials science since the first applications of X-ray scattering in the 1930s. This book in particular is geared to scientists who use SAS for studying tightly packed particle ensembles using elements of stochastic geometry. Models developed in this mathematical field possess manifold applications for a large variety of typical heterogeneous materials, which can be effectively approximated by a two-phase approximation. Polycrystalline metals and alloys, micropowders (absorbers, color pigments, ceramic powders), ceramics and porous materials (especially controlled porous glasses, porous silica, colloids, membranes and nuclear filter materials), random composites, special construction materials and a large number of polymers (polyethylene, polypropylene, etc., as well as biopolymers and molecules discussed in the field of molecular biology, and bacteriophages) are typical materials.

The following 10 chapters give an overview of particle and particle systems characterization via SAS for two-phase particle ensembles. After explaining the SAS experiment and typical scattering patterns, Chapter 1 introduces integral transformations, which interrelate the scattering pattern with structure functions. Regions of heterogeneity exist on very different length scales, starting from 1 000 nm and greater down to the *atomic short order range*, and usually ending at (1...2) nm. The *order range L* is introduced, which is a length parameter frequently used in this book (see mainly the second part of Chapter 1, followed by applications in other chapters).

Chord length distribution densities of selected geometric bodies are summarized in Chapters 2–3, starting from the *cone* as an instructive geometric figure. These analytic results are the key interpreting of the second derivative of the sample correlation function. Operational experiences concerning both – details of analytical calculation and the interpretation of experimental curves – are explained to the reader. Particle volume fraction and *particle to particle interference* are discussed in Chapter 4. Fundamental set models of stochastic geometry, the *Boolean model* and the *Dead Leaves model*, related structure functions and scattering patterns are explained in Chapters 5 and 6. In Chapter 7, one of the questions dealt with is the following: "When does a given set of *fragment particles* have its origin in a *Dead Leaves model*?" A puzzle fitting function is defined. Different kinds of puzzles are mooted.

Models for the analysis of the particle volume fraction are described in Chapter 8. Chapter 9 deals with a general concept of chord length distributions for random two-phase systems. Chapter 10 is a summary of exercises with typical examples. All the chapters build on each other.

Not all sections or subsections appear in the table of contents; however, a large collection of *key word terms* (index) is included. Essential definitions have been abbreviated, e.g., *correlation function* (CF), *chord length distribution density* (CLDD), *puzzle cell* (PC), *linear simulation model* (LSM), etc. The symbols were chosen according to the typical symbols in the scientific literature.

In order to introduce geometric models and illustrate their application, many figures have been incorporated. Essential information is inserted in the long figure captions. This way, the main text is not encumbered. Fig. 4.20, (*SAS sphere figures*) and Fig. 4.19 are remarkable. Even experts will find Eqs. (4.64) and Eqs. (9.23) valuable for solving problems of particle description in materials science via SAS.

The principles set out in *Mathematica* (Wolfram, 1990–2013) [234] were continuously applied (i.e., text, models, programs, calculations, data evaluation, producing figures, exporting the EPS files and the basic LaTeXcode). In some places, the *Mathematica* code of a formula is included. Suitable options for typical data evaluation functions have been inserted, which include *(Non-Liner)Fit, SequenceLimit, NLimit* and *NIntegrate*. For a final check of the formulas in the printed out pages, these formulas were reinserted into a computer algebra program and checked again. Nevertheless, if the careful reader should detect misprints or even an incorrect representation of connections, the author would appreciate suggestions and corrections.

The extensive advice of Dietrich Stoyan was very helpful for the success of this project. Over the years, the author had many helpful discussions with him about the connection between set models in the field of stochastic geometry and their applications in SAS. Furthermore, I thank the SAS specialist Günter Schulze for introducing me to laboratory work in the 1970s. The cooperation with this expert has formed my knowledge about data evaluation and its limits. Gregor Damaschun († 2009) and his team encouraged me in numerous discussions to publish my results in papers and conference contributions.

The author wishes his readers a successful application of the ideas set forth in this book.

Halle/Saale, May 2013.

Wilfried Gille
Martin Luther University Halle-Wittenberg
Germany
Inst. of Physics, Hoher Weg 8, D-06120 Halle

About the Author

The author was born in Halle/Saale in 1953. He obtained his school–leaving certificate in a special class with a concentration on mathematics and physics at Martin Luther University of Halle in 1970 and began his studies of physics the same year at the Halle university. His special field was solid state physics in the team of Prof. Brümmer (diploma with the thesis: *"X-ray emission spectra of vanadium and titanium oxides"* in 1977).

In 1979 he was confronted with small-angle scattering (SAS) experiments. It is a nonsatisfactory objective fact that the whole field of SAS (experiments and data evaluation) has been developed *independently* of the mathematical results achieved in the field of stochastic geometry. It was a part of the author's life to see the connections between both these sides and try to transform and introduce known mathematical results into the field of SAS and come to application in materials science. In this way, in 1983 he got his doctorate with the thesis *"Stereological characterization of microparticles by use of SAS – applications in metal- and polymer physics."* In 1995 he wrote the postdoctoral thesis *"The concept of chord length distributions for reaching information from experimental small-angle scattering curves."* Since then, the author has usefully cooperated with mathematicians and crystallographers in laboratories in Eastern Europe, especially in Poland, Armenia and the Balkan States. He maintains his physical fitness by sports (bike trips, jogging, skiing).

Since 1977, he has been a scientific assistant in the field of physics (first at the college at Merseburg and now at the Martin Luther University of Halle), published a sequence of papers and gave lectures at numerous conferences. The main subject of this work is the intimate connection between solid state physics, data evaluation of scattering experiments and numerical mathematics, applying models of stochastic geometry. The author has given lectures in the field of materials science, theory and practice of SAS and in some special fields of numerical mathematics and computer algebra to students of natural sciences.

1

Scattering experiment and structure functions; particles and the correlation function of small-angle scattering

There is a close interrelation between physical scattering experiments and the so-called *second order characteristics* defined in stochastic geometry. In small-angle scattering (SAS), these functions, which are a bit modified sometimes, go by the name of *SAS structure functions*. In Chapter 1, the basic structure functions of SAS are introduced (Guinier & Fournét, 1955) [143]. Later on, the connections to the analogous functions defined in stochastic geometry are discussed (Stoyan et al., 1987, 1995) [210, 212].

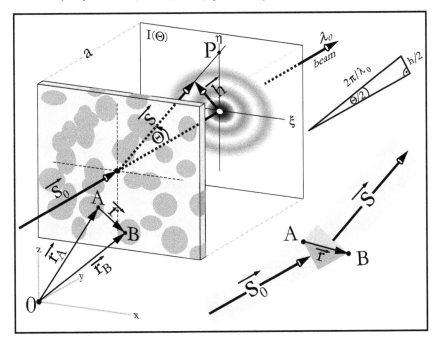

An X-ray or neutron beam of wavelength $\lambda = \lambda_0$ is scattered by an isotropic sample. This results in an *isotropic scattering pattern* $I(\theta)$. The intensity distribution $I(\theta)$ is called the *scattering curve* of the sample (see [143], the next page and Section 1.1). This sample is a *Boolean model* (see Chapter 5).

On this page, essential information about sample and scattering patterns is introduced, which includes the following: Anisotropic samples yield an anisotropic scattering. The scattering pattern notations $I(\mathbf{h})$ or $I(\mathbf{q})$ are usual and the pattern is recorded in the registration plane (ξ, η). The argument variable denotes the scattering vector $\mathbf{h}$ or $\mathbf{q}$. In cases of anisotropic samples, $I(\mathbf{h})$ depends on the sample orientation relative to the (ξ, η) plane.

The following theory holds true for the elastic scattering of neutrons, X-rays and synchrotron scattering. In this case, the relation $|\mathbf{s}| = |\mathbf{s_0}| = s = s_0$ is fulfilled. The angle $\theta/2$ exists in a right triangle with a leg of length $h/2$ and a hypotenuse of $s = s_0 = 2\pi/\lambda_0$. From $\sin(\theta/2) = (h/2)/(2\pi/\lambda_0)$, the magnitude of the scattering vector $|\mathbf{h}| = h = 4\pi/\lambda_0 \cdot \sin(\theta/2)$ results. The unit of h is a reciprocal length, which is a physical value in reciprocal space. Thus, the data points $I_k(h_k)$ represent information in reciprocal space. A relatively small angle θ and a big λ_0 correspond to a very small h. Conversely, a bigger θ like $\theta = \pi/20$ and a smaller λ_0 lead to a bigger h, e.g., $h = 3$ nm^{-1}.

Only bigger particle diameters of some nanometers can be related to the model of a nearly homogeneous particle. The investigation of bigger particles requires a bigger wavelength and smaller scattering angles. This is the origin of the term *Small-Angle Scattering*. For example, a typical Bonse-Hart type ultra-small-angle X-ray scattering instrument, using higher than 20 KeV X-rays, makes it possible to operate in the h-interval 10^{-4} nm$^{-1} < h < 10$ nm^{-1}.

Chapter 1 is organized as follows: First, the connection between the density distribution of the sample and the corresponding scattering pattern is explained by applying the principles of wave propagation and interference (Section 1.1). An important initial connection results, which is shown in Eq. (1.6). Expressions for the amplitude $A(\mathbf{h})$ of the scattered wave are derived, followed by examples for the scattering pattern of single particles in the isotropic case. Having developed a feeling for the averaging over the particle orientation in the single particle case, the reader is confronted with a more general expression for $I(\mathbf{h})$. In Section 1.1.2, the step from $A(\mathbf{h})$ to $I(\mathbf{h})$ is handled by use of the convolution theorem. Then, the model of a two-phase particle ensemble is considered. The parameters *density difference* and *density fluctuation* are introduced, which lead to the *real-space structure functions* $Z(\mathbf{r})$ and $\gamma(\mathbf{r})$. The distances r between two random points A and B, $r = \overline{AB}$, are of importance. Further specialization leads to the connection between the sample correlation function $\gamma(r)$ and $I(h)$ for isotropic two-phase particle ensembles. Manifold verification and applications of the relationship $I(h) \leftrightarrow \gamma(r)$ are discussed. The significant part of chapter 1 deals with the *order range parameter* L and the data evaluation for a specific L (Fig. 1.11).

In most practical cases, a scattering pattern described by Eq. (1.6) is not directly recorded in the registration plane. Rather, corrections of the intensity signal are indispensable. At a minimum, this includes collimation corrections, denoted by "geometrical dismearing" and corrections of the wavelength spectrum of radiation applied. An explanation of these procedures is beyond the scope of this book.

1.1 Elastic scattering of a plane wave by a thin sample

Interference is the foundation of all scattering techniques for (electromagnetic) waves. For this, basic definitions have been introduced in the initial figure of the chapter: The directions of the unit vectors s_0 and s define the propagation direction of the incident and scattered waves. These directions agree with those of the corresponding Poynting vectors.

The registration plane is arranged parallel to the sample at a distance a. The point of observation P is the hitting point of a straight line of direction s with the registration plane. The vectors s_0 and s define the *scattering angle* θ, $\theta = \arccos[s_0 \cdot s/(s_0 s)]$, and the scattering vector h, $h = s - s_0$. The vector h is nearly parallel to the registration plane, where the scattering pattern $I(h)$ is recorded by a physical device. Here, a registration matrix detects intensities at many points of observation P_i. The intensity possesses a dimension of $energy/(time \cdot surface\ area) = [\text{Nm}/(\text{s} \cdot \text{m}^2)] = [\text{W}/\text{m}^2]$. However, most laboratories are far from a precise, direct intensity measurement in P. Nevertheless, about eight independent structure parameters can be detected from a *relative* measurement of $I(h)$ in SAS (see Chapter 9).

The sample and wave propagation are described in the fixed x, y, z coordinate system with the origin 0. This point, which is directly positioned on the anterior sample surface, position vector $\mathbf{0}$, is the reference point for the propagation of the incident wave in a time interval $0 \le t < \infty$. The coordinate y runs parallel to s_0.

A plane wave of frequency f and phase velocity c_p propagates in the y direction and is scattered by the sample. Let $\omega = 2\pi f = 2\pi c_p/\lambda_0$ be the circular frequency of the wave and $\Delta\varphi = \Delta\varphi(P)$ the phase angle in P. The incident and scattered wave are one-dimensional waves by nature. Points belonging to the anterior sample surface plane $y = 0$, $-\infty < x, z < \infty$, are characterized by the oscillation $A(t) = A_0 \exp[i(\omega t)]$. The amplitude of the incident wave can be described by $A(y, t) = A_0 \cdot \exp[i(\omega t - (2\pi/\lambda_0)y)]$. Neglecting the factor of polarization $A_0 = A_0(\theta)$ for $\theta \to 0$, the incident wave is written $A(y, t) \sim \exp[i(\omega t - (2\pi/\lambda_0) \cdot y)]$.

The amplitude $A(P)$ of the elastically scattered wave in point P

Many elementary waves, starting in different scattering centers (different sample points A, B, ...) interfere in P. The positions A and B are given by the position vectors r_A and r_B, respectively. A huge number of scattering centers are independently distributed in the sample, i.e., many real-space vectors r_A, r_B, ... have to be considered. For the following, the phenomenon of multiple scattering is disregarded, i.e., points A, B, ... contribute to the scattering one time. Polarization factors can be neglected since only relatively small θ, where about $0 < \theta < \pi/20$, are analyzed.

Head waves, starting in the points 0 and A, interfere in point P. The contribution of other sample points like B is considered later in Eq. (1.5). Denoting the distance from the origin 0 to point P by $r_0 = \overline{0P}$, the amplitude of the scattered wave in point P is

$$A(P) \sim \exp[i(\omega t - (2\pi/\lambda_0) \cdot \overline{0P} + \Delta\varphi)] = \exp[i(\omega t - (2\pi/\lambda_0) \cdot r_0)] \times \exp[i\Delta\varphi].$$
$$(1.1)$$

The phase angle $\Delta\varphi$ in P will be traced back to the parameters $\mathbf{r_A}$, r_0, r_P, $\mathbf{s_0}$ and $\mathbf{s}$ by geometric considerations (see also Fig. 10.1). In the end, the surprisingly simple result will be that $\Delta\varphi$ is independent of the lengths r_0 and r_p to an extremely precise degree [see Eqs. (1.4) to (1.6)].

Now, the phase angle $\Delta\varphi(P)$ in P is analyzed (for details see Section 10.1). Starting from points 0 and A, point P is reached by two paths of different path lengths: On the one hand, there is the direct path of length $r_0 = \overline{0P}$, i.e., the shortest connection between point 0 to point P, which is denoted as *path length₁*. On the other hand, the plane wave has to "travel" from the starting plane first to point A and then from point A to point P, which means that *path length₂* $= \mathbf{r_A} \cdot \mathbf{s_0} + \overline{AP}$. In order to detect the phase angle involved in Eq. (1.1) with

$$\Delta\varphi = \Delta\varphi(P) = \frac{2\pi}{\lambda_0} \cdot (path\ length_2 - path\ length_1) = \frac{2\pi}{\lambda_0} \cdot (\mathbf{r_A} \cdot \mathbf{s_0} + r_P - r_0),$$
$$(1.2)$$

an analysis of the triangle $0PA$ of side lengths $r_0 = \overline{0P}$, $r_p = \overline{AP}$ and $r_A = \overline{0A}$ (not included in initial figure; however, see Fig. 10.1 in exercise Chapter 10) is phased. Both the side lengths $\overline{0P}$ and $\overline{AP}$ of the triangle are much bigger than the length of side r_A, i.e., sides $\overline{0P}$ and $\overline{AP}$ are nearly parallel. Typical experimental conditions are $r_A \approx 20$ nm and $r_0 \approx r_P > 0.2$ m. Let α be the angle $A0P$ (i.e., the angle between r_A and r_0). The cosine theorem yields $r_p = \sqrt{r_0{}^2 + r_A{}^2 - 2r_0 r_A \cdot \cos(\alpha)} \equiv r_0\sqrt{1 + r_A{}^2/r_0{}^2 - 2r_A/r_0 \cdot \cos(\alpha)}$. Since $r_A/r_0 \ll 1$, the square root can be avoided by the approximation

$$r_P = r_0 - r_A \cdot \cos(\alpha) + \tfrac{1}{2}r_0\sin^2(\alpha) \cdot \left(\tfrac{r_A}{r_0}\right)^2 + \tfrac{1}{2}r_0\cos(\alpha)\sin^2(\alpha) \cdot \left(\tfrac{r_A}{r_0}\right)^3 +$$
$$\tfrac{r_0}{16}(5\cos(2\alpha) + 3)\sin^2(\alpha)\left(\tfrac{r_A}{r_0}\right)^4 + \tfrac{r_0}{32}(9\cos(\alpha) + 7\cos(3\alpha))\sin^2(\alpha)\left(\tfrac{r_A}{r_0}\right)^5$$
$$+ O\left[\left(\tfrac{r_A}{r_0}\right)^6\right].$$
$$(1.3)$$

All the terms of Eq. (1.3) are applied in the field of *dynamic interference theory*. Additionally, this more general approach takes into account the absorption of the wave inside the sample, multiple scattering and deviations of λ_0 from the vacuum wavelength, i.e., a varying refractive index, which is a bit bigger than 1.

Neglecting all of these effects and especially the power terms $...(r_A/r_0)^4$, $(r_A/r_0)^3$ and $(r_A/r_0)^2$ and also taking into account $\alpha = \arccos(\mathbf{r_A} \cdot \mathbf{s}/r_A)$ (where α is the angle between the vector $\mathbf{r_A}$ and the unit vector $\mathbf{s}$), Eqs. (1.2) and (1.3) result in the simple, nearly perfect approximation

$$r_P \approx r_0 - r_A \cdot \cos(\alpha) = r_0 - \mathbf{r_A} \cdot \mathbf{s}. \tag{1.4}$$

The triangles on the right in the starting figure of this chapter illustrate the parallel components of the vector $\mathbf{r}$ in the directions of $\mathbf{s_0}$ and $\mathbf{s}$, respectively.

By using the length r_p, which is approximately defined by Eq. (1.4), the phase angle in point P [see Eq. (1.2)] is written $\Delta\varphi(P) = (2\pi/\lambda_0) \cdot (\mathbf{r_A} \cdot \mathbf{s_0} + r_0 - \mathbf{r_A} \cdot \mathbf{s} - r_0) \approx (2\pi/\lambda_0) \cdot [\mathbf{r_A} \cdot \mathbf{s_0} - \mathbf{r_A} \cdot \mathbf{s}] = (2\pi/\lambda_0) \cdot [\mathbf{r_A} \cdot (\mathbf{s_0} - \mathbf{s})]$. Hence, based on Eq. (1.4), $\Delta\varphi(P)$ *does not at all depend* on the length $r_0 = \overline{0P}$. This fact essentially simplifies the approach. In all the following parts of this book, $\Delta\varphi(P) = -(2\pi/\lambda_0)\mathbf{r_A} \cdot \mathbf{h}$ is applied. Thus, the scattered wave [see Eq. (1.1)] is fixed by the product $A(P) = A(\mathbf{h}) \sim \exp[i(\omega t - (2\pi/\lambda_0) \cdot r_0)] \times \exp[-i(\mathbf{r_A} \cdot \mathbf{h})]$.

These considerations restrict to the one point A, which is defined by the vector $\mathbf{r_A}$ relative to the origin. Taking into account two different sample points A and B, and introducing the two different densities $\rho(\mathbf{r_A})$ and $\rho(\mathbf{r_B})$ at these points, the *density averaged amplitude* $A(P)$ of the scattered wave in the point of observation P is

$$A(\mathbf{P}) = A(\mathbf{h}) = A(\mathbf{h}, \varrho(\mathbf{r_A}), \varrho(\mathbf{r_B})) \sim \exp[i(\omega t - (2\pi/\lambda_0) \cdot r_0)] \times$$
$$\left(\frac{\varrho(\mathbf{r_A})}{\varrho(\mathbf{r_A}) + \varrho(\mathbf{r_B})} \exp[-i(\mathbf{r_A} \cdot \mathbf{h})] + \frac{\varrho(\mathbf{r_B})}{\varrho(\mathbf{r_A}) + \varrho(\mathbf{r_B})} \exp[-i(\mathbf{r_B} \cdot \mathbf{h})] \right). \tag{1.5}$$

For a typical SAS experiment, all prefactors in Eq. (1.5) – including the one for time and frequency – are unimportant. These factors are not interrelated with the specific geometric information about the density distribution in the sample. Hence, the time-dependent term taken into account with the frequency of the incident wave is no longer considered. A detection (or even an analysis) of the time dependence would require significant effort and complicated measurements in the field of quantum electronics.

After neglecting all prefactors and not fixing the vector $\mathbf{r_A}$ at one point A, but integrating it over the whole irradiated sample volume V_i and introducing a normalized density $\rho(\mathbf{r_A})$, the wave amplitude and the intensity $I(\mathbf{h})$ in point P from Eq. (1.5) is

$$A(\mathbf{h}) \sim \int_{V_i} \varrho(\mathbf{r_A}) \exp[-i(\mathbf{r_A} \cdot \mathbf{h})]dV_i, \quad I(\mathbf{h}) \sim \left| \int_{V_i} \varrho(\mathbf{r_A}) \exp[-i(\mathbf{r_A} \cdot \mathbf{h})]dV_i \right|^2. \tag{1.6}$$

Equations (1.6) represent a volume average for a constant $\mathbf{h}$. Clearly, $\mathbf{r_A}$ is a radius vector, but not a vector between two random points A and B. This difference is key for understanding the SAS structure functions. For these functions, the argument variable $\mathbf{r}$ is reserved, where $r = |\mathbf{r}|$ denotes a random distance like $r = \overline{AB}$ between any two points positioned inside the sample. In stochastic geometry, the distance between random points is frequently denoted just by h [206, 231]. Such contradicting symbols can lead to difficulties for beginners and experts as well.

Simplification of $A(\mathbf{h})$ and $I(\mathbf{h}) \sim \overline{A(\mathbf{h}) \cdot A(\mathbf{h})} = \overline{A(\mathbf{h})}^2$ in special cases

In the following, the initial conclusions from Eqs. (1.6) will be drawn. Up to this point, no special assumptions about the function $\rho(\mathbf{r_A})$ have been made.

Equations (1.6) hold true for any sample material of a certain spatial density distribution $\rho(\mathbf{r_A})$. If $\rho(\mathbf{r_A}) = const$ in the whole interradiated volume V_i, $A(\mathbf{h}) = \delta(\mathbf{h}-\mathbf{0}) = \delta(\mathbf{h})$ results, i.e., the amplitude of the scattered wave of a huge particle only differs from zero near a very small region $\mathbf{h} \to \mathbf{0}$. Let $\overline{\rho}$ be the mean density in V_i. SAS "lives" on a density fluctuation $\eta(\mathbf{r_A}) = \rho(\mathbf{r_A})-\overline{\rho}$ inside the irradiated sample volume. For the function $\eta(\mathbf{r_A})$, many special cases and special symmetries are possible. Each special case possesses practical relevance.

The representation in the next subsection will investigate some normalized functions $I(h)$ for single *centro-symmetric moving particles*, i.e., for single particles which possess a center of symmetry and which rotate randomly around this center. In this way, all spatial particle orientations can be reached. It is assumed that each orientation exists with the same probability, i.e., there is no favored spatial particle orientation. For such cases, $dV_i = 4\pi r_A{}^2 dr_A$, $\rho(\mathbf{r_A}) = \overline{\rho}(r_A)$ and $\overline{\exp[-i(\mathbf{r_A} \cdot \mathbf{h})]} = \sin(hr_A)/(hr_A)$ (for details see Section 1.1.2). Thus, from Eq. (1.6) for a radial symmetric particle, $A(h) \to \overline{A(\mathbf{h})} \sim \int_0^\infty \overline{\rho}(r_A) \cdot \sin(hr_A)/(hr_A) \cdot 4\pi r_A{}^2 dr_A$ holds.

In the special case of a spherically symmetric density distribution, $\rho(\mathbf{r_A}) = \rho(r_A)$ holds. Rotation of such an inhomogeneous "particle" does not influence the amplitude $A(\mathbf{h})$ in P. Hence, the scattering pattern $I(\mathbf{h}) = I(h) \sim \overline{A^2(h)}$ is even still proportional to the square of the averaged term $A(\mathbf{h})$, i.e., $I(h) \sim [\overline{A(h)}]^2 = \overline{A^2(h)}$.

The function $\rho(\mathbf{r_A})$ yields $I(\mathbf{h})$ of a single particle. Based on this, a direct interpretation of the behavior of a scattering pattern $I(h)$ will be explained in Section 1.1.1 for some centro-symmetric particles. The calculation method from the geometric particle shape to the corresponding function $I(h)$ is analyzed for the general case of a barbell possessing *four* shape parameters. In contrast to this, Platonic bodies possess one shape parameter, i.e., the edge length a. As experience has shown, this fact does not necessarily simplify the determination of a scattering pattern $I(h)$.

After these special results, *real-space structure functions* are introduced. This approach (see Sections 1.1.2 and 1.1.3) relates $I(\mathbf{h})$ with structure functions. The foundations of particle description via real-space structure functions are also explained.

In summary, Eq. (1.6) will be further modified. The *Patterson function* and SAS *correlation function* will be defined and the approach of an isotropic two-phase sample will be introduced. These simplifying steps alone will represent points of connection to the field of stochastic geometry.

1.1.1 Guinier approximation and Kaya's scattering patterns

Following is an explanation of the analysis of the function $I(h)$ for special cases. The first method used to analyze the size and shape of microparticles was the small-angle scattering of X-rays.* Guinier and Fournét introduced the radius of gyration R_g (frequently called the *Guinier radius*) into a small-angle scattering technique [142, 143]. For homogeneous particles, the length parameter R_g is defined by the volume integral over the volume region V of the centro-symmetric particle (or particle agglomerate), $R_g^2 = 1/V_0 \cdot \int_V r^2 dV$. Regardless of the particle shape, the second term of the Taylor series of the scattering pattern at the origin is a quadratic one, which is expressed as $I(h) = 1 - R_g^2 \cdot h^2/3 + \ldots$ In this light, Fig. 1.1 explains the universal approach for the variety of $I(h, L_c, r_c, R_s, \pm h_s)$ of capped cylinders [160, 161].† For any combination of the parameters L_c, r_c, R_s and $h_s = \pm\sqrt{R_s^2 - r_c^2}$, which are involved in the approach, analytic expressions for the R_g^2-volume integral and the corresponding scattering pattern $I(h)$ are known (see Table 1.1). The following formulas include an orientation angle θ of the cylinder axis of the barbell, where $\theta = 0$ denotes the cylinder axis.

TABLE 1.1
Exact scattering pattern $I(h)$ and Guinier approximation

exact scattering pattern	Guinier approximation
$I(h) = \dfrac{\int_0^{\frac{\pi}{2}} A(h,\theta)^2 \sin(\theta)d\theta}{\int_0^{\frac{\pi}{2}} A(0,\theta)^2 \sin(\theta)d\theta}$ $A(h)$ [see Eq. (1.7)]	$I(h) \approx \exp\left[-\dfrac{R_g^2 h^2}{3}\right] = 1 - \dfrac{R_g^2 h^2}{3} + \ldots$ R_g [see Eq. (1.8)]

The variables $A(h) = A(h, \theta)$ and R_g are defined by

$$A(h, \theta) = 4\pi R^3 \cdot \int_{-\frac{h_s}{R}}^{1} \left[\cos[h \cdot (h_s + \tfrac{L_c}{2} + R_s t) \cos(\theta)] \times \right.$$
$$\left. (1 - t^2) \cdot \frac{J_1\left[h \cdot R_s \sin(\theta)\sqrt{1-t^2}\right]}{h \cdot R_s \sin(\theta) \cdot \sqrt{1-t^2}} \right] dt + \quad (1.7)$$
$$\pi L_c r^2 \cdot \frac{\sin[h \cdot L_c \cos(\theta)/2]}{h \cdot L_c \cos(\theta)/2} \cdot \frac{2J_1[h \cdot r_c \sin(\theta)]}{h \cdot r_c \sin(\theta)}$$

and

$$R_g^2 = \frac{1}{6L_c r_c^2 - 4(h_s - 2R_s)(h_s + R_s)^2} \left[\frac{24}{5} R_s^5 + \right.$$
$$3(4h_s + L_c)R_s^4 + 2(2h_s + L_c)^2 R_s^3 + 3h_s L_c(2h_s + L_c)R_s^2 + \quad (1.8)$$
$$\left. \tfrac{1}{5}h_s^3[4h_s^2 - 5L_c(h_s + L_c)] + \tfrac{1}{2}L_c r_c^2(L_c^2 + 6r_c^2) \right].$$

---

*In 1895, Wilhelm Conrad Roentgen discovered X-rays. This tool introduced the possibility of microparticle sizing. The electron microscope was developed later.
†The actual notations have been slightly modified from those found in Kaya's papers.

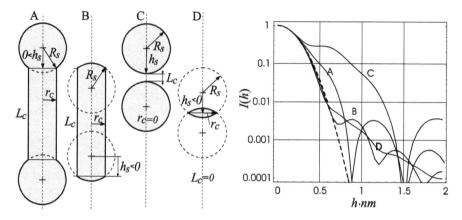

FIGURE 1.1

Definition of size/shape parameters for a cylinder with spherical end caps and four typical scattering patterns $I(h)$.

A variety of geometric shapes can be handled by varying the parameters L_c, r_c, R_s and $h_s = \pm\sqrt{R_s{}^2 - r_c{}^2}$. The scattering patterns of 4 elementary geometric figures are normalized, $I(0) = 1$ and trimmed to $R_g = const$, $R_g = 6.25$.

Curve A: $L_c = 2$, $r_c = 3$, $h_s = 4$, $R_s = 5$; curve B: $L_c = 7.88$, $r_c = 6$, $h_s = 0$, $R_s = 6$; curve C: $L_c = 5.62$, $r_c = 0$, $h_s = 3$, $R_s = 3$; curve D: $L_c = 0$, $r_c = 9.73$, $h_s = -5.62$, $R_s = 11.24$.

The comparison of the scattering patterns $I_{A,B,C,D}(h)$ (solid line) with the Guinier approximation (dashed line) demonstrates the meaning of R_g. The interpretation of this parameter requires the selection of a well-suited particle shape model. One specific length R_g can be puzzled out for different particle shapes. A (averaged) length parameter cannot define the shape of a particle.

Based on L_2 norms, estimations for the average value R_g result from a linearizing transformation, i.e., via a plot $\ln(I)$ vs. h^2. This task can be handled by the *Mathematica* module L2xy[data] [117]. This algorithm involves a specialized L_2 approximation, where the result is invariant with respect to the exchange of the axes of the coordinate system. This property is significant because both the variables I and h involve the experimental errors Δh and ΔI, mostly in the same order of magnitude, which is typically $\Delta h/h \approx \Delta I/I \approx$ (1–5)%. This special approach is based on the starting equation for an x, y coordinate system and a data set (x_i, y_i) with weights (σ_x, σ_y) or for exchanged data (y_i, x_i) with weights (σ_y, σ_x), which is expressed by

$$\sum_{i=1}^{n} \frac{[y_i - (a \cdot x_i + b)]^2}{\sigma_y{}^2 + a^2 \cdot \sigma_x{}^2} \rightarrow Min, \quad (-\infty < a < \infty, \ -\infty < b < \infty). \qquad (1.9)$$

The approach in Eq. (1.9) leads to (at least) two solutions $a = a_{1,2}$ and

$b = b_{1,2}$ for $y = a \cdot x + b$. The simplest way to find the global minimum is to insert these values and check the resulting deviations (see the last lines of the following module). There exist analytic expressions for the estimation of the errors of the parameters in terms of the data given, i.e., $a = a_{1,2} \pm \Delta a_{1,2}$ and $b = b_{1,2} \pm \Delta b_{1,2}$. These expressions, which are not given here, involve extensive interval splittings and are more complicated than those for a and b. Here is a *Mathematica* module for the L_2 approximation according to Eq. (1.9).[‡]

```
L2xy[data_]:=Module[{sigmax=1, sigmay=1, n, i, Xi, Yi, Xquer, Yquer, xi, yi,
                Alphax, Alphay,p,B,a1,a2,b1,b2,suma1,suma2},
( (*Special L2_approximation of a list "data" given with y=a*x+b *)
n = Length[data];
Xi = Table[data[[i, 1]], {i, 1, n}];
Yi = Table[data[[i, 2]], {i, 1, n}];
Xquer = Apply[Plus, Xi]/n;
Yquer = Apply[Plus, Yi]/n;
xi = Xi - Xquer;
yi = Yi - Yquer;
Alphax =(xi.yi)/(xi.xi);
Alphay =(yi.yi)/(xi.yi);
p = sigmay/sigmax;
B =(p^2/Alphax -Alphay)/2;
a1 =-B + Sqrt[B^2+p^2]; (* solution 1 *)
a2 =-B - Sqrt[B^2+p^2]; (* solution 2 *)
b1 = Yquer - a1*Xquer;
b2 = Yquer - a2*Xquer;
(* Checking L2 for a1 and a2 *)
suma1 = Apply[Plus,(yi - a1*xi)^2]/(sigmay^2+a1^2*sigmax^2);
suma2 = Apply[Plus,(yi - a2*xi)^2]/(sigmay^2+a2^2*sigmax^2);
{a = Which[suma1 < suma2, a1, True, a2],
 b = Which[suma1 < suma2, b1, True, b2]} )];
```

For a data list with $n = 3$ points, given as $data = \{\{0,0\}, \{1,2\}, \{2,1\}\}$, this algorithm [see Eq. (1.9)] yields the plausible result $y = x$, which contradicts most of the commonly used approximation programs operating with the L_2 norm.

Hence, a data evaluation strategy of a scattering experiment could consist of specific formulas for scattering patterns for a well-selected sequence of particle shapes. One of the most universal ideas in this field is Kaya's *barbell cylinder formula.*

Such investigations always involve a sensitive point: the orientation(s) of the particle(s) with respect to the plane of registration. The following pages explain the spatial averaging approach in the case of *isotropic uniform random orientation* (IUR) of the particles. This is considered in the equations that have already been discussed.

[‡]This algorithm was explained at the Weierstrass Institute for Applied Analysis and Stochastics in Berlin at the "Interdisziplinäres Kolloquium zur Anwendung von *Mathematica* in den Naturwissenschaften" on November 20, 2009. There are other approaches (Stoyan, 1979) [209].

Averaging of $I(\mathbf{h})$ with respect to particle orientation

After a discussion of the special example of a *barbell particle*, Eq. (1.6) is specialized for any single homogeneous particle shape. The approach will be explained for the *Platonic bodies*. Let (h_x, h_y, h_z) be the components of the scattering vector $\mathbf{h}$. The end position of the vector $\mathbf{r_A}$ pinpoints all points belonging to the particle. Stretching only the volume integral in Eq. (1.6) over the *one particle volume*

$$I(h) = \overline{I(h)} \sim \overline{A(\mathbf{h})^2} = \overline{\left[\int e^{-i\,\mathbf{h}\cdot\mathbf{r_A}}\, d^3\mathbf{r_A} \right]^2}. \qquad (1.10)$$

The "overlined parts" involved in Eq. (1.10) represent averaging with respect to particle orientation. In order to handle this average for a certain *distribution law of orientation*, the scattering amplitude of the particle $A(h_x, h_y, h_z)$ can be traced back to two angles of orientation, i.e., φ and α. These angles describe the particle orientation in a fixed x, y, z system. The polar angle φ stretches from the x axis in the positive direction, i.e., the x axis means $\varphi = 0$. The angle α stretches from the z axis in a positive direction, i.e., the z axis means $\alpha = 0$. The scattering amplitude is written $A(h, \varphi, \alpha)$, where $h = \sqrt{h_x^2 + h_y^2 + h_z^2}$ denotes the amount of $\mathbf{h}$ in terms of its components. In a certain case of multiple orientation, $I(h) = \overline{I(\mathbf{h})}$ results from the averaged square $\overline{[A(h, \varphi, \alpha)]^2}$, taking into account each orientation with the corresponding probabilities. The distribution laws of the random variables φ and α define these probabilities.

For bodies without a center of symmetry, both *real* and *imaginary terms* must be considered for averaging [see Eq. (1.10)]. This is the case with a tetrahedron. For the other Platonic bodies, the "real" transformation $A(\mathbf{h}) \sim \int \exp[i\,\mathbf{h}\cdot\mathbf{r_A}]d^3\mathbf{r_A} \rightarrow \int \cos(\mathbf{h}\cdot\mathbf{r_A})d^3\mathbf{r_A}$ includes all the information. This simplifies the analysis. With respect to Fig. 1.2, the simplification $A(\mathbf{h}) \sim \int \cos(\mathbf{h}\cdot\mathbf{r_A})d^3\mathbf{r_A}$ is valid for all Platonic solids, except for the tetrahedron [166].

In detail, the averaging procedure consists in the following: The distribution densities of the random angles φ and α are denoted by $f_\varphi(\varphi)$ and $f_\alpha(\alpha)$, respectively. The normalizations $\int_0^{2\pi} f_\varphi(\varphi)d\varphi = 1$ and $\int_0^{2\pi} f_\alpha(\alpha)d\alpha = 1$ hold. The mean value $\overline{A(h)^2}$ results from the integral

$$I(h) = \overline{A(h)^2} = \frac{\int_0^{2\pi} \int_0^{2\pi} A^2(h_x(\varphi,\alpha,h), h_y(\varphi,\alpha,h), h_z(\varphi,\alpha,h)) \cdot f_\alpha(\alpha) f_\varphi(\varphi) d\varphi d\alpha}{\int_0^{2\pi} \int_0^{2\pi} f_\alpha(\alpha) f_\varphi(\varphi) d\varphi d\alpha},$$

$$\qquad (1.11)$$

$$I(h) = \overline{A(h)^2} = \frac{\int_0^{2\pi} \int_0^{2\pi} A^2(h, \alpha, \varphi) \cdot f_\alpha(\alpha) \cdot f_\varphi(\varphi) d\varphi d\alpha}{\int_0^{2\pi} \int_0^{2\pi} f_\alpha(\alpha) f_\varphi(\varphi) d\varphi d\alpha}.$$

For averaging, Eqs. (1.11) involve two integrations. Usually, these types of integrals cannot be handled analytically. However, the integration limits simplify by adapting to the symmetry of the specific body (see the case of a hexahedron in the following section).

Scattering pattern of the hexahedron in the isotropic case

The *isotropic case* means an Isotropic Uniform Random (IUR) orientation of the solid. The scattering pattern results from averaging over all the orientations possible. In stochastic geometry, the abbreviation IUR orientation is commonly used (see also IUR chords in Chapter 2). This is the case with an isotropic sample.

In the following, IUR averaging is exemplified in detail for the rectangular parallelepiped. With Eq. (1.10), for the edge lengths $2a$, $2b$, $2c$, the scattering amplitude for a fixed $\mathbf{h}$ direction is written in terms of the components h_x, h_y and h_z as

$$A(a, b, c, h_x, h_y, h_z) = \frac{\sin(ah_x)}{ah_x} \cdot \frac{\sin(bh_y)}{bh_y} \cdot \frac{\sin(ch_z)}{ch_z}. \tag{1.12}$$

Actually, $f_\alpha(\alpha) = \sin(\alpha)$, $0 \le \alpha \le \pi/2$, and $f_\varphi(\varphi) = 2/\pi$, $0 \le \varphi \le \pi/2$. Hence, Eq. (1.11) is written as follows:

$$I(h) = \overline{A^2(h)} = \frac{\int_0^{\pi/2} \int_0^{\pi/2} A^2[h_x(\varphi,\alpha,h), h_y(\varphi,\alpha,h), h_z(\varphi,\alpha,h)] \cdot [\sin(\alpha) \cdot 1] d\varphi d\alpha}{\int_0^{\pi/2} \int_0^{\pi/2} [\sin(\alpha) \cdot 1] d\varphi d\alpha}, \tag{1.13}$$

$$I(0) = \qquad 1.$$

The components h_x, h_y, h_z are: $h_x = h \cdot \cos(\varphi)\sin(\alpha)$, $h_y = h \cdot \sin(\varphi)\sin(\alpha)$, $h_z = h \cdot \cos(\alpha)$. With these components and Eqs. (1.12) and (1.13), a representation of the IUR scattering pattern $I(h)$ of the parallelepiped results. Two steps of integration are required, as shown by

$$I(h) = \frac{\int_0^{\pi/2} \int_0^{\pi/2} \left[\frac{\sin(a \cdot h \cdot \cos(\varphi)\sin(\alpha))}{a \cdot h \cdot \cos(\varphi)\sin(\alpha)} \cdot \frac{\sin(b \cdot h \cdot \sin(\varphi)\sin(\alpha))}{b \cdot h \cdot \sin(\varphi)\sin(\alpha)} \cdot \frac{\sin(c \cdot h \cdot \cos(\alpha))}{c \cdot h \cdot \cos(\alpha)} \right]^2 \cdot [\sin(\alpha) \cdot 1] d\varphi d\alpha}{\int_0^{\pi/2} \int_0^{\pi/2} [\sin(\alpha) \cdot 1] d\varphi d\alpha},$$

$$I(0) = 1. \tag{1.14}$$

For a hexahedron, $2a = 2b = 2c$; therefore Eq. (1.14) simplifies to

$$I(h) = \frac{2}{\pi} \int_0^{\pi/2} \int_0^{\pi/2} \left[\frac{\sin(ah\cos(\varphi)\sin(\alpha))}{ah\cos(\varphi)\sin(\alpha)} \cdot \frac{\sin(ah\sin(\varphi)\sin(\alpha))}{ah\sin(\varphi)\sin(\alpha)} \cdot \frac{\sin(ah\cos(\alpha))}{ah \cdot \cos(\alpha)} \right]^2 \times$$

$$[\sin(\alpha)] d\varphi \, d\alpha. \tag{1.15}$$

See a program for the numerical integration of Eq. (1.15) below.

The radius of gyration R_g of a hexahedron of edge length $2a$ is $R_g(2a) = 2a/2 = a$. For $0 \le h \cdot R_g < 1$, the function $I_{approx}(h) = \exp(-h^2 R_g^2/3) = \exp(-h^2 a^2/3)$ is an approximation of $I(h)$. The last line of the following program includes a sensitive check of the scattering pattern via Porod's invariant [see Eqs. (1.37) and (1.38)].

```
(* scattering pattern of a hexahedron of edge length A = 2a  *)
i[h_, a_]:=(RG = (2a/2);
Which[0 <= h*a < 1, Exp[-h^2*RG^2/3],True,
2/Pi*NIntegrate[((Sin[a h*Cos[Phi]*Sin[Alpha]])/(a h*Cos[Phi]*Sin[Alpha]))^2*
                ((Sin[a h*Sin[Phi]*Sin[Alpha]])/(a h*Sin[Phi]*Sin[Alpha]))^2*
                ((Sin[a h        *Cos[Alpha]])/(a        *Cos[Alpha]))^2*
                Sin[Alpha], {Alpha,0,Pi/2},{Phi,0,Pi/2}]]
       );

Plot[i[h, 3],{h,0, 2},AxesLabel ->{"h", "I(h)"},PlotLabel -> "edge length=6"]
(* edge length=2 for the invariant *)
invariant=2^3*NIntegrate[h^2*i[h, 1],{h, 0, Infinity}, PrecisionGoal -> 4]
```

In scattering theory, the particle volume V is a significant parameter. Sometimes it is advantageous to compare $I(h, a)$ with the scattering pattern $I_{iso}(h, R)$ of the so-called *isovolumic sphere*. For a (single) hexahedron of edge length $2a$, the radius R follows from $(2a)^3 = 4\pi R^3/3$. A plot, comparing the scattering pattern of the *isovolumic sphere* with that of a hexahedron, is inserted into Fig. 1.2. If $0 \le h \cdot R < 4$, the differences remain surprisingly small. The part of the curve near the origin, $I(0) = 1$, can be approximated by a parabola as $I(h) = 1 - R_g{}^2 h^2/3$ where $R_g = (2a)/2 = a$ (see Table 1.1 and Fig. 1.1). Actually, this approximation only involves the one structure parameter a. For an analytical summary of all Platonic solids, see (Xin Li et al., 2011) [166]). Besides a detailed explanation of all analytic terms $I(h, a)$, specific references for each solid are discussed.

More generally, the volumes of all Platonic solids can be formulated in terms of the edge length a by $V = const. \times a^3$, where *const.* is a specific constant (for the icosahedron $V \approx 2.18 \cdot a^3$; for the dodecahedron $V \approx 7.66 \cdot a^3$) (see Fig. 1.2). The *concept of the isovolumic sphere* of radius R can be applied. For a solid volume $V(a)$ in terms of the edge length a, the isovolumic sphere radius R results from $V(a) \equiv 4\pi R^3/3$. The bigger the edge number of the Platonic solid is, the smaller the deviations between $I(h, a)$ and $I_{iso}(h, R)$ are. From the tetrahedron to the dodecahedron, the scattering pattern increasingly approaches that of a sphere. The smallest deviations are expected for the dodecahedron. Nevertheless, a typical SAS experiment is so precise that these deviations are easy to recognize. Of course, the parameter R_g is not sensitive enough for this comparison.

These examples show that already *the direct investigation of the scattering pattern of the single homogeneous particle* via Eq. (1.10) includes several surprisingly interesting aspects.

The next sections will introduce another strategy of particle description. *Structure functions* will be defined and analyzed, for example, the SAS *correlation function* $\gamma(r)$. However, as far as the author knows, analytic expressions of $\gamma(r)$ [see Eq. (1.31) and preceding equations] or the *chord length distribution* (Chapter 2) for the icosahedron and the dodecahedron are still unknown.

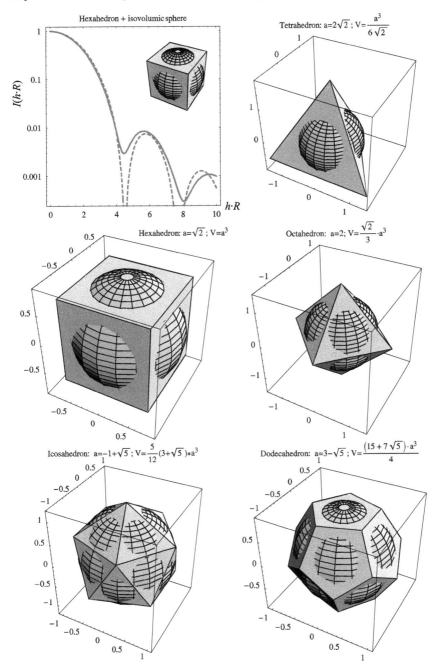

FIGURE 1.2

Platonic solids are useful models in SAS. The scattering pattern $I(hR)$ of a cube (solid line) and the isovolumic sphere (dashed line) are compared.

1.1.2 Scattering intensity in terms of structure functions

The inverse Fourier transformation of Eq. (1.6) connects a local density $\rho(\mathbf{r_A})$ with the resulting amplitude of the scattered wave as depending on the scattering vector $\mathbf{h} = \mathbf{s} - \mathbf{s_0}$. The origin of the vector $\mathbf{r_A}$ is arbitrary since the length $r_0 = \overline{OP}$ is exclusively involved in the time-dependent term of Eq. (1.5) and not involved at all in Eqs. (1.6). These are an essential result in diffraction theory, but only an initial step in the field of SAS. It describes the case of an anisotropic sample in a general manner. Only in exceptional cases does a model for materials research resort to this kind of general relation right away.

The interrelationship between scattering pattern, Patterson function $P(r)$ and SAS correlation function $\gamma(r)$ will now be explained. In SAS, there exist specialized approaches for evaluating a scattering pattern. In this context, the analysis of *point-to-point-distances*, represented by $r = \overline{AB}$, is very useful. Let the real-space convolution square of the density $\varrho = \varrho(\mathbf{r_A})$, stretched over the whole irradiated volume V_i, be denoted by $\int \varrho(\mathbf{r_A}) \cdot \varrho(\mathbf{r_B}) dV_i = \int \varrho(\mathbf{r_A}) \cdot \varrho(\mathbf{r_A}+\mathbf{r}) dV_i = \int \varrho(\mathbf{r_B}-\mathbf{r}) \cdot \varrho(\mathbf{r_B}) dV_i = [\varrho(\mathbf{r_A}) * \varrho(\mathbf{r_A})](\mathbf{r})$. Introducing the symbol F for a three-dimensional Fourier transformation, the scattered amplitude [see Eqs. (1.6)], is written $A(\mathbf{h}) = F[\varrho(\mathbf{r_A})](\mathbf{h})$.

The scattering pattern $I(\mathbf{h})$, recordable at a point of observation in the registration plane, is proportional to the product $A(\mathbf{h}) \cdot A(\mathbf{h}) = F[\varrho](\mathbf{h}) \cdot F[\varrho](\mathbf{h}) = [F[\varrho](\mathbf{h})]^2$. The product can be traced back to the real-space convolution term $[\varrho(\mathbf{r_A}) * \varrho(\mathbf{r_A})](\mathbf{r})$. Points A and B can be exchanged. Since the product of two Fourier transforms equals the Fourier transformation of its convolution, $[F[\varrho](\mathbf{h})]^2 = F[\varrho(\mathbf{r_A}) * \varrho(\mathbf{r_A})](\mathbf{h})$, the scattering intensity

$$I(\mathbf{h}) \sim F[\varrho(\mathbf{r_A}) * \varrho(\mathbf{r_A})](\mathbf{h}) =$$
$$F[\int_{V_i} \varrho(\mathbf{r_A})\varrho(\mathbf{r_A} - \mathbf{r})dV_i] = \int_{V_i} \left[\int_{V_i} \varrho(\mathbf{r_A})\varrho(\mathbf{r_A} - \mathbf{r})dV_i \right] e^{-i\,\mathbf{h}\cdot\mathbf{r}}dV_i$$
$$(1.16)$$

results. The integration variable in the inner integral of Eq. (1.16) is $\mathbf{r_A}$. This position vector samples the volume V_L, $V_L \ll V_i$. Integration with respect to $\mathbf{r_A}$ yields a function of $\mathbf{r}$, which is finally transformed by the exponential term $e^{-i\cdot\mathbf{h}\cdot\mathbf{r}}dV_i$. The integration region of the outer volume integral is the irradiated volume V_i. If the density function $\rho(\mathbf{r_A})$ is known, Eq. (1.16) allows $I(\mathbf{h})$ to be detected for any fixed $\mathbf{h}$, i.e., to determine the intensity in P.

For real crystals (i.e., in basic crystallography), the function term $P(\mathbf{r}) = \int \varrho(\mathbf{r_A}) \cdot \varrho(\mathbf{r_A} - \mathbf{r})d^3\mathbf{r_A}$ involved in Eq. (1.16) is defined as the *Patterson function*, (Patterson, 1935) [187]. Starting with a certain crystal structure based on the well-defined term $\varrho(\mathbf{r_A})$, $P(\mathbf{r})$ results.

Modification of the Patterson function for two-phase particle ensembles of volume fraction c

The approach by Patterson is not suitable for describing ensembles of homogeneous particles. However, Eq. (1.16) can be adapted to the two-phase

approximation. The goal of this approach is the characterization of homogeneous particles (phase 1, nearly constant density ϱ_1) embedded in a matrix
region (phase 2, nearly constant density ϱ_2). This means that the average
density in the surroundings of the particles for a certain order range L is denoted by $\varrho(\mathbf{r_A}) = \varrho_2$. Phase 2 surrounds the particles arranged in a typical
volume V_L, as $V_L \ll V_i$, which the experimenter is interested in. In this case,
spatial density difference $\Delta\varrho(\mathbf{r_A})$, defined by $\Delta\varrho(\mathbf{r_A}) = \varrho(\mathbf{r_A}) - \varrho_2$, is a non-
negative function, which contrasts with the basic density level, ϱ_2. Let c and
$1 - c$ be the volume fractions of the phases 1 and 2, respectively. The roles of
ϱ_1 and ϱ_2 can be exchanged. Actually, $\varrho_2 < \varrho_1$ is assumed, which means that
the volume V_L involves particles (and no holes).

The mean *density difference* $\overline{\Delta\varrho}$ is defined by the volume integral $\overline{\Delta\varrho} = (1/V_L) \cdot \int \Delta\rho(\mathbf{r_A})dV_L = c \cdot \varrho_1$. Furthermore, the mean *squared density difference* $\overline{\Delta\varrho^2} = (1/V_L) \cdot \int [\Delta\rho(\mathbf{r_A})]^2 dV_L = c \cdot \rho_1^2$ follows. Hence, the volume
fractions c and $1-c$ can be expressed by $c = [\overline{\Delta\rho}]^2/\overline{\Delta\rho^2}, 1-c = 1-[\overline{\Delta\rho}]^2/\overline{\Delta\rho^2}$.
This relationship will be used later.

The specialized Eq. (1.16) is written in terms of $\Delta\varrho(\mathbf{r_A})$ and ϱ_2 as

$$I(\mathbf{h}) \sim \int_{V_i} \left(\int_{V_L} [\Delta\varrho(\mathbf{r_A}) + \varrho_2][\Delta\varrho(\mathbf{r_A} - \mathbf{r}) + \varrho_2]dV_L \right) e^{-i\,\mathbf{h}\cdot\mathbf{r}}dV_i. \quad (1.17)$$

The outer integral transformation identifies a certain region size, which is
much smaller than V_i in the variable $\mathbf{r}$. Formally expanding the product in
the inner integral yields four terms. To analyze these terms with respect to
SAS, it is important that the maximum amount of $\mathbf{r}$ is much smaller than the
diameter of the irradiated volume V_i, i.e., the region of final integration dV_i.
Furthermore, the Fourier transformation of a constant term leads to terms
that are proportional to $\delta(\mathbf{r} - \mathbf{0})$. Thus, only the term $\int_{V_L} \Delta\varrho(\mathbf{r_A}) \cdot \Delta\varrho(\mathbf{r_A} - \mathbf{r})dV_L$ involved in the inner integrand *survives* the final integration dV_i with
respect to the variable $\mathbf{r}$. Introducing a normalization factor $\overline{\Delta\varrho^2} \cdot V_L$ to the
denominator, Eq. (1.17) reduces to

$$I(\mathbf{h}) \sim \int_{V_i} \left[\frac{\int_{V_L}[\Delta\varrho(\mathbf{r_A})\Delta\varrho(\mathbf{r_A} - \mathbf{r})]dV_L}{\overline{\Delta\varrho^2} \cdot V_L} \right] e^{-i\,\mathbf{hr}}dV_i \sim \int_{V_i} Z(\mathbf{r})e^{-i\,\mathbf{h}\cdot\mathbf{r}}dV_i.$$

$$(1.18)$$

Equation (1.18) traces $I(\mathbf{h})$ back to the self-convolution of the density dif-
ference $\Delta\varrho(\mathbf{r_A})$. The non-negative, normalized function $Z(\mathbf{r})$, $Z(\mathbf{0}) = 1$, is
referred to as the "Belegungsfunktion" (Porod, 1951) [191]. The standard
translation is *function of occupancy*.

In the following, Eq. (1.18) is adapted to the experimental conditions much
better by introducing the *density fluctuation* $\eta(\mathbf{r_A}) = \Delta\varrho(\mathbf{r_A}) - \overline{\Delta\rho}$. The
function $\eta(\mathbf{r_A})$ oscillates around zero like an alternating current. For the
quadratic mean $\overline{\eta^2}$ of $\eta(\mathbf{r_A})$ (related to a volume V_L), the well-known relation
$\overline{\eta^2} = (\varrho_1 - \varrho_2)^2 \cdot c(1 - c)$ results. The roles of ϱ_1 and ϱ_2 can be exchanged.

Inserting $\overline{\Delta\varrho} = c\varrho_1$, the function $Z(\mathbf{r})$ involved in Eq. (1.18) is

$$
\begin{aligned}
Z(\mathbf{r}) &= \frac{\int_{V_L} [\eta(\mathbf{r_A})+\overline{\Delta\varrho}]\cdot[\eta(\mathbf{r_A}-\mathbf{r})+\overline{\Delta\varrho}]dV_L}{\overline{\Delta\varrho^2}\cdot V_L}, \\
&= \frac{\int_{V_L} [\eta(\mathbf{r_A})+c\varrho_1]\cdot[\eta(\mathbf{r_A}-\mathbf{r})+c\varrho_1]dV_L}{\overline{\Delta\varrho^2}\cdot V_L}, \\
&= \frac{\int_{V_L} \eta(\mathbf{r_A})\eta(\mathbf{r_A}-\mathbf{r})+c\varrho_1[\eta(\mathbf{r_A})+\eta(\mathbf{r_A}-\mathbf{r})]+c^2\varrho_1{}^2 dV_L}{\overline{\Delta\varrho^2}\cdot V_L}.
\end{aligned}
$$
(1.19)

With $\int_{V_L}[\eta(\mathbf{r_A}) + \eta(\mathbf{r_A} - \mathbf{r})]dV_L = 0$ and partly inserting $\overline{\Delta\rho^2} = c\rho_1{}^2$, the function $Z(\mathbf{r})$ in Eq. (1.19) can be dissected into two parts, a *variable part* + a *constant part*,

$$
Z \equiv \frac{\int_{V_L} \eta(\mathbf{r_A})\eta(\mathbf{r_A} - \mathbf{r}) + c^2\varrho_1{}^2 dV_L}{\overline{\Delta\varrho^2}\cdot V_L} = \frac{\int_{V_L}\eta(\mathbf{r_A})\eta(\mathbf{r_A}-\mathbf{r})dV_L}{\overline{\Delta\varrho^2}V_L} + \frac{(c\varrho_1)^2 V_L}{c\varrho_1{}^2 V_L}.
$$
(1.20)

The constant part (i.e., the last part) of Eq. (1.20) equals c. The variable part is a bit more complicated to handle. Since $\overline{\eta^2} = \overline{(\Delta\rho(\mathbf{r_A}) - \overline{\varrho})^2} = \overline{\Delta\varrho^2} - [\overline{\Delta\varrho}]^2$, the denominator of the variable part is written in terms of η^2 and c as

$$
\frac{1}{\overline{\Delta\varrho^2}} = \frac{1}{\overline{\Delta\varrho^2}}\cdot 1 = \frac{1}{\overline{\Delta\varrho^2}}\cdot\frac{\overline{\Delta\varrho^2} - [\overline{\Delta\varrho}]^2}{\overline{\eta^2}} = \frac{1-c}{\overline{\eta^2}}.
$$
(1.21)

Hence, from Eqs. (1.20) and (1.21) the simplified connection

$$
Z(\mathbf{r}) = (1-c)\cdot\frac{\int_{V_L}\eta(\mathbf{r_A})\eta(\mathbf{r_A}-\mathbf{r})dV_L}{\overline{\eta^2}\cdot V_L} + c, \quad Z(\mathbf{r}) = (1-c)\cdot\gamma(\mathbf{r}) + c \quad (1.22)
$$

results. The dimensionless function $\gamma(\mathbf{r})$ in Eq. (1.22) represents the SAS correlation function (CF) (Debye & Bueche, 1949) [27], which is defined by

$$
\gamma(\mathbf{r}) = \frac{1}{V_L}\cdot\frac{\int_{V_L}\eta(\mathbf{r_A})\eta(\mathbf{r_A}-\mathbf{r})dV_L}{\overline{\eta^2}}, \quad \gamma(\mathbf{0}) = 1, \quad \gamma(\mathbf{L}) = 0.
$$
(1.23)

By use of the function $\gamma(\mathbf{r})$, Eq. (1.18) is written for all c, $0 < c < 1$,

$$
I(\mathbf{h}) \sim \int_{V_i}[(1-c)\gamma(\mathbf{r}) + c]\cdot e^{-i\mathbf{h}\cdot\mathbf{r}}dV_i \sim \int_{V_i}\gamma(\mathbf{r})\cdot e^{-i\mathbf{h}\cdot\mathbf{r}}dV_i.
$$
(1.24)

Thus, the scattering intensity $I(\mathbf{h})$ results via Fourier transformation of $Z(\mathbf{r})$ via Eq. (1.18), or in terms of $\gamma(\mathbf{r})$ via Eq. (1.24). The function $\gamma(\mathbf{r})$ is better suited for practical application. Operating with $\gamma(\mathbf{r})$ in Eq. (1.24), the volume fraction is an unimportant parameter. Care must be taken with truncation errors, which are inherent in the back-transformations, i.e., from an experimental function $I(\mathbf{h})$ to $\gamma(\mathbf{r})$.

Different techniques of normalization for $I(\mathbf{h})$ are possible. The simplest is $|I(\mathbf{0})| = 1$, which requires a division by the term $\int_{V_i}\gamma(\mathbf{r})dV_i$. It should be stressed that Eq. (1.24) seems to be directly connected with Eqs. (1.6). This is not correct because the variables $\mathbf{r_A}$ and $\mathbf{r}$ differ from each other.

Orientation of single particles and isotropic particle ensembles

A specialization of Eq. (1.23) for a single homogeneous particle possessing selected orientations can be made. For this, $L = L_0$, i.e., $c \to 0$ and $0 \leq r \leq L_0$. Two (random) orientation angles describe the spatial orientation of a particle or particle agglomerate. A weighted averaging over all existing spatial orientations of the particle must be performed as shown by

$$\gamma_0(\mathbf{r}) = \frac{1}{V_0} \cdot \frac{\overline{\int_{V_0} \eta(\mathbf{r_A})\eta(\mathbf{r_A} - \mathbf{r})dV_0}}{\overline{\eta^2}}, \quad \gamma_0(\mathbf{0}) = 1, \ \gamma_0(\mathbf{L_0}) = 0. \quad (1.25)$$

The case that is most applied in practice is the IUR orientation of the particle. For the two homogeneous bodies cone and pyramid, this is illustrated in Figs. 2.1 and 2.14. Analytic expressions of IUR-isotropized CFs were evaluated for many bodies, e.g., sphere, hemisphere, tetrahedron and parallelepiped (see [191, 104, 91, 19, 60]). Chapter 2 starts with the evaluation of the function $\gamma_0(r)$ of a right circular cone. This example clearly demonstrates that the procedure of averaging with respect to particle orientation is complicated. CFs are simple to use but complicated to determine.

In the case of isotropic particle ensembles, the formulas partly simplify. Equation (1.18) or Eq. (1.24) fixes the scattering pattern $I(\mathbf{h})$ of an anisotropic sample. With isotropic samples for $\theta = \theta_0$, symmetrical scattering rings of constant intensity $I(h_0) = I(h(\theta_0)) = const.$ result in the registration plane. The scattering pattern contains an axis of revolution, which coincides with that of the primary beam (see Figs. 1.1 and 1.2). Terms involving the scalar product $\mathbf{h} \cdot \mathbf{r}$ and the spatial volume element dV_i simplify. Let φ, $0 \leq \varphi \leq \pi$, be the angle between the vectors $\mathbf{h}$ and $\mathbf{r}$, $\varphi = \arccos[\mathbf{h}\cdot\mathbf{r}/(hr)]$. In a total isotropic case, φ is a random variable characterized by the distribution density $\sin(\varphi)/2$. The averages of the terms $\overline{\cos(\mathbf{h} \cdot \mathbf{r})} = \overline{\cos[h \cdot r \cos(\varphi)]}$ and $\overline{\sin(\mathbf{h} \cdot \mathbf{r})} = \overline{\sin[h \cdot r \cos(\varphi)]}$ are

$$\int_0^\pi \cos[hr\cos(\varphi)] \cdot \frac{\sin(\varphi)}{2}d\varphi = \frac{\sin(hr)}{hr}, \quad \int_0^\pi \sin[hr\cos(\varphi)] \cdot \frac{\sin(\varphi)}{2}d\varphi = 0. \quad (1.26)$$

The sine terms resulting from Euler's relation disappear. The anisotropic vector term $\cos(\mathbf{h} \cdot \mathbf{r})$ is substituted by the magnitude term $\sin(hr)/(hr)$. By use of Eqs. (1.26) and the isotropic volume element $dV_i \to 4\pi r^2 dr$, Eqs. (1.16), (1.18) and (1.24) simplify. Now, Eq. (1.16) is written

$$I(\mathbf{h}) \to I(h) \sim \int_{r=0}^{r=L} \overline{\int_{V_L} \rho(\mathbf{r_A})\rho(\mathbf{r_A} - \mathbf{r})dV_L} \cdot \frac{\sin(hr)}{hr} \cdot 4\pi r^2 dr. \quad (1.27)$$

For a certain particle ensemble, a geometric model has to describe the frequency of a certain length r for all the $\mathbf{r} = \overline{\mathbf{AB}}$ possible. For example, models can be inserted that involve and operate with the distribution law(s) of the possible distances r inside and between homogeneous regions of the material(s) (P. Debye & A. M. Bueche, 1949) [27].

For particle ensembles possessing two homogeneous phases, Eq. (1.27) simplifies. The whole set of relations from Eqs. (1.18) to Eq. (1.24) can be adapted to the case of an isotropic sample. The scattering intensity $I(h)$ results in terms of the density fluctuation $\eta(\mathbf{r_A})$ of the isotropic two-phase model [see Eq. (1.23)]

$$I(h) \sim \int_{Vi} \left(\overline{\int_{V_L} \eta(\mathbf{r_A})\eta(\mathbf{r_A} - \mathbf{r})dV_L} \right) \frac{\sin(hr)}{hr} \cdot 4\pi r^2 dr. \qquad (1.28)$$

From this, with the prefactor $\overline{\eta^2} = (\rho_1 - \rho_2)^2 \cdot c(1 - c)$, a normalized representation of the scattering pattern follows:

$$I(h) = (\rho_1 - \rho_2)^2 \cdot c(1 - c) \cdot \int_0^L \gamma(r) \frac{\sin(hr)}{hr} \cdot 4\pi r^2 dr. \qquad (1.29)$$

In the limiting case $\rho_1 = \rho_2$ or in the case $c \to 0$, the scattering intensity disappears, $I \to 0$. Furthermore, by analogy with the *Patterson function* $P(\mathbf{r})$, (Patterson, 1935) [187], for a homogeneous particle of volume V_0 (or for a particle agglomerate as well), possessing the isotropized CF $\gamma_0(r)$ [see Eq. (1.17)], the distance distribution density $p_0(r)$ is defined by

$$p_0(r) = \frac{4\pi r^2 \gamma_0(r)}{V_0} = \frac{4\pi r^2}{V_0} \frac{\overline{\int \eta(\mathbf{r_A})\eta(\mathbf{r_A} - \mathbf{r})dV_0}}{\int \eta(\mathbf{r_A})^2 dV_0}, \quad p_0(r) \sim r^2\gamma_0(r). \qquad (1.30)$$

Equations (1.27)–(1.30) describe the case of isotropic SAS of homogeneous particles in terms of the CF $\gamma(r) = \gamma(r, L)$. CFs will be used in all the chapters. A typical problem discussed in Chapter 4 considers the distribution density $p_0(r)$ of random distances r between points A and B belonging to two unit spheres, touching each other.

In the last equations, the vectors $\mathbf{r_A}$, $\mathbf{r_B}$ and $\mathbf{r} = \mathbf{r_B} - \mathbf{r_A}$ are applied. In fact, $\mathbf{r_A}$ is the position vector of point A and $\mathbf{r}$ is the vector from sample point A to sample point B. In this light, it does not make sense to write $\gamma(\mathbf{r_A})$, $Z(\mathbf{r_A})$ or $p_0(r_A)$. Furthermore, a denotation $i(r)$ for a particle indicator function is also useless.

Sometimes, such superficialities exist in publications about scattering theory and SAS. These contradictions, which are frequently tolerated by experts, should be avoided. Any newcomer would be happy to understand that the meaning(s) of the variable r in the terms $A(h) = \int \rho(r) \exp[-ihr]dV$ and $p_0(r) = 4\pi r^2 \gamma_0(r)/V_0$ are different.

Nevertheless, writing the term $i(\mathbf{r})$ for an indicator function means that $\mathbf{r}$ is a position vector [it would be better to write $i(\mathbf{r_A})$ instead]. However, writing the same variable r in $\gamma(r)$ means that r denotes the distance between two correlated points. The latter is the meaning of r in the following sections.

1.1.3   Particle description via real-space structure functions

The approaches of Section 1.1.1 *directly* interpret the behavior of the signal function, i.e., of the scattering intensity $I(h)$. Based on an a priori particle shape assumption, real-space parameters are detected by tricky data operations in reciprocal space. Contrary to this, based on the Fourier transformation of $I(h)$ (see Section 1.1.2), other appropriate methods of data evaluation are available [218, 219, 129, 20].

The approach dealt with in the following chapters does not start from any a priori particle shape assumption. Of course, the assumption of particles, possibly possessing a random size *and* shape, is still a useful idealization. A microparticle object can be studied in different order ranges.[§] To begin the study, the size interval of interest has to be selected by specifying an order range L. In the two-phase approximation, an indicator function $i(\mathbf{r_A})$ can be introduced. Taking values 1 (inside a homogeneous particle) or 0 (outside a homogeneous particle), $i(\mathbf{r_A})$ defines regions of different sizes, shapes and densities in the material. Via L_0 and $i(\mathbf{r_A})$, a spatial ensemble of (possibly differently shaped) particles is described in real space. Via $i(\mathbf{r_A})$, particle ensembles can be considered as sets. Basic structure functions defined in SAS and in the field of stochastic geometry are as follows:

- an isotropized normalized density auto-correlation function $\gamma(r, L_0)$
- an isotropized linear erosion $P(r, L_0)$
- an isotropized chord length distribution density $A(r, L_0)$
- an isotropized distance distribution density $p(r, L_0)$
- an isotropized set covariance $C(r, L_0)$

These functions, sometimes called *second order characteristics*, are related to geometric parameters in real space. The success of the two-phase approach largely hinges on the existence of geometric models, which yield analytic expressions for $\gamma(r, L_0)$ ... $C(r, L_0)$. See the following chapters.

There exist very useful techniques of image analysis, (Ohser & Schladitz, 2009) [185]. Compared with these techniques, which begin with a micrograph or 3D image, the interpretation of a structure function is more complicated and does not seem to be relevant for practice. However, operating with scattering techniques and structure functions has enormous advantages: The sample preparation is simple. Most SAS experiments are nondestructive and the signal function is averaged over a huge sample volume. On the other hand, microscope sample preparation can modify (or even destroy) a material. In contrast to image analysis, *a single scattering experiment* simultaneously records information about millions of nanometric particles.

[§]This consideration is of importance in many fields. This was emphasized in a paper by Sahian [202], who studied the structure of the observed field of an astronomic object via the set covariance $C(r, L_0)$. For an atomic length scale, i.e., $L < 0.2$ nm, a two-phase indicator function $i(\mathbf{r_A})$ is too rough an approximation because it cannot reflect continuously changing macroscopic densities.

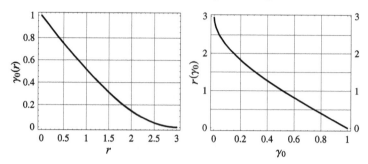

FIGURE 1.3
SAS CF of a sphere of diameter $d = 3$ (left) and its inverse function (right).
While $\gamma_0(r, d)$ represents a geometric probability, $0 \leq \gamma_0(r, d) \leq 1$, the function $r(\gamma_0, d)$, $0 \leq r \leq L_0 = d$, detects the length r in terms of a known (geometric) probability γ_0.

Properties of the correlation function $\gamma_0(r, d)$ of a single sphere

The auto-correlation function [abbreviated as CF for correlation function] of a single sphere of diameter d is written $\gamma_0(r, d) = 1 - 3r/(2d) + r^3/(2d^3)$; $0 \leq r < d$, else $\gamma_0(r, d) = 0$. For several applications in SAS, it is useful to know the inverse function of $\gamma_0(r)$, i.e., the function $r = r(\gamma_0)$. A real-valued analytic expression is

$$r(\gamma_0) = r(\gamma_0, d) = \frac{d \cdot (1 - \gamma_0)}{\frac{1}{2} + \cos\left(\frac{\pi - \arccos\left[2(1-\gamma_0)^2 - 1\right]}{3}\right)}, \quad 0 \leq r \leq d = L_0. \quad (1.31)$$

Equation (1.31) is exemplified for the case $L_0 = d = 3$ in Fig. 1.3.

There is an interesting application of Eq. (1.31). Starting from the CF γ_0 of a (homogeneous) body, the corresponding CF γ_ρ of the parallel body can be constructed. The set operation $X_\rho = X_0 \oplus B(0, \rho)$ is explained in the textbooks by (Serra, 1982) [205], (Stoyan & Stoyan, 1992) [211] and (Stoyan et al., 1995) [212]. In the spatial case, the structure element $B(0, \rho)$ denotes a sphere of radius ρ, $0 \leq \rho < \infty$. Both the sets X_0 and X_ρ are described by a specific correlation function. The simplest case is that of the CF $\gamma_0(r, d)$ of an *initial sphere* to the CF of the *parallel sphere*, $\gamma_\rho(r) = \gamma_0(r, d + 2\rho)$. This transformation can be handled by simultaneously plotting the CFs $\gamma_0(r, d)$ and $\gamma_0(r, 2 \cdot \rho)$ in the same coordinate system. The function $\gamma_\rho(r)$ is then obtained by formally adding (shifting) both curves parallel to the r-axis. This step can be proved by use of the inverse function terms $r_1 = r_1(\gamma_0)$ and $r_2 = r_2(\gamma_\rho)$. Altogether, the simple sum of two inverse functions, $r = r_1 + r_2$, already yields the CF of the parallel sphere.

This approach can be extended to all convex smooth bodies. The transformation of an initial ellipsoid X_0, $x^2/a^2 + y^2/b^2 + z^2/c^2 = 1$, to the cor-

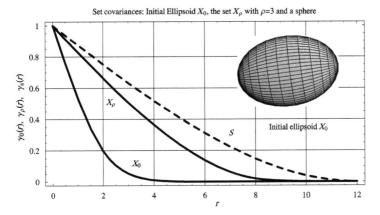

FIGURE 1.4
SAS correlation functions $\gamma_0(r)$, $\gamma_\rho(r)$ and $\gamma_S(r)$ of three compact sets [usually called (micro)particles].
X_0: Ellipsoid $[a = 1, b = 2, c = 3]$ of largest diameter $L_0 = 2c = 6$. This body has a volume $V_0 = 25.13$ and a surface area $S_0 = 48.88$.
X_ρ: The corresponding parallel set has the largest diameter $L_\rho = 12$.
S: The dashed line represents the CF of a sphere of diameter $d = 12$.
The CF includes the information about the volume and surface area of a particle (see the following sections). Furthermore, the chord length distribution densities (CLDDs) of these particles can be derived in terms of the CFs (see Chapters 2–9).

responding body X_ρ was the subject of intense investigations by the author. An example is illustrated in Fig. 1.4. This and similar questions of particle description are connected with a specific length parameter: the *mean caliper diameter* $\overline{b}$ and the (mean) integral $\overline{M}$ of the mean *surface curvature* $M = 1/(4\pi) \cdot \int (1/R_1 + 1/R_2)dS$ of a particle. It holds $\overline{M} = 2\pi \cdot \overline{b}$. The (mean) integral of the mean surface curvature for an ellipsoid is given by the *surface integral of the first kind*:

$$\overline{M}(a, b, c) =$$

$$\int \frac{abc \cdot [2c^2 - 2(a^2 - b^2)\cos(2\theta)\sin^2(\varphi) + 3(a^2 + b^2) + (a^2 + b^2 - 2c^2)\cos(2\varphi)]}{8 \cdot [a^2 b^2 \cos^2(\varphi) + c^2 \sin^2(\varphi)(b^2 \cos^2(\theta) + a^2 \sin^2(\theta))]^{3/2}} dS ,$$

(1.32)

where dS denotes the surface element of the ellipsoid and $x = a\cos(\theta)\sin(\varphi)$, $y = b\sin(\theta)\sin(\varphi)$, $z = c\cos(\varphi)$, $0 \leq \varphi \leq \pi$, $0 \leq \theta \leq 2\pi$ are used. The integral on the right hand side of Eq. (1.32) can be traced back to standard functions of mathematical physics. In summary, $\overline{b}(1, 2, 3) = \overline{M}(1, 2, 3)/(2\pi) = 4.20246...$ results.

1.2 SAS structure functions and scattering intensity

The sample scattering intensity $I(h)$ is recorded by a scattering experiment and interrelated with the SAS structure functions. These are the correlation function $\gamma(r, L)$, the distance distribution density function $p(r, L)$ and the chord length distribution density $A_\mu(r, L)$ of the particle(s) [66]. There exist common properties between $A_\mu(r, L)$ and $p(r, L)$ for spherical symmetrical particles. See the limiting cases *spherical shell* and *spherical half-shell* [106]. Essential papers in this field were published by O. Glatter [133].

1.2.1 Scattering pattern, SAS correlation function and chord length distribution density (CLDD)

In classical textbooks, general principles of diffraction experiments are summarized, for example, in (Guinier & Fournét, 1955) [143] and (Glatter & Kratky) [129], (Feigin and Svergun, 1987) [36, p. 25]. There are many synonymous formulas connecting experimental functions with theoretical ones, (Glatter, 1977, 1979, 1980) [130, 131, 132], (Müller & Glatter, 1982) [178], (Feigin & Svergun, 1987), [36], (Müller et al., 1983) and [179]. Each formula, integral transformation and numerical procedure is closely interrelated with a special experimental situation involving its own respective restrictions. Searching for a *general integral transformation technique* that connects the reciprocal space (i.e., diffraction/scattering pattern) with real-space structure functions for many different cases is a time-intensive task.

Isotropic sets are considered. The scattering pattern is normalized (always possible for a limited particle volume) to $I(0) = 1$ and interrelated with the pair distribution density $p(r)$ of the particle (maximum diameter L_0) or a particle ensemble (order range L) via

$$I(h) = \int_0^{(L,L_0)} p(r) \cdot \frac{\sin(hr)}{hr} dr = \frac{1}{v_c(L, L_0)} \int_0^{(L,L_0)} 4\pi r^2 \gamma(r) \cdot \frac{\sin(hr)}{hr} dr.$$

(1.33)

In the case of independent single particles (usually referred to as "in the single particle case"), the denotation *characteristic volume*, $v_c(L) = \int_0^L 4\pi r^2 \gamma(r) dr$, simplifies to the single particle volume V_0. It seems paradoxical that the functions $p(r)$ and $\gamma(r)$ in Eq. (1.33) resulting from $I(h)$ depend on L. This fact is still neglected here. For details, see Section 1.4 and especially Fig. 1.11.

There is a large number of different numerical and analytical approaches for handling the inverse transformation of Eq. (1.33). At least two facts make this step complicated:

1. The data always involves at least *random errors* Δh_k and ΔI_k.
2. A *collimation correction procedure* and other corrections are indispensable.

One field-tested approach is Glatter's indirect Fourier transformation [138]. In this broad field, a large variety of computer programs have been developed. See (Fritz et al., 2000) [40], (Feigin & Svergun, 1987) [36, p. 45] and others [26, 51]. A special approach is presented in Section 1.4.

Regardless of the origin of $I(h)$, twice integrating Eq. (1.33) by parts yields

$$I(h) = \frac{4\pi}{v_c(L, L_0)} \int_0^{(L, L_0)} \gamma''(r) \cdot \frac{2 - 2\cos(hr) - hr\sin(hr)}{h^4} dr, \quad I(0) = 1.$$
$$(1.34)$$

Equations (1.33) and (1.34) allow several conclusions. See topics 1–6 below:

1. Composition of normalized scattering patterns, $I_k(0) = 1$, of (widely) separated particles of volume V_k: The "scattering power" of a particle sensitively depends on its volume. The direct superimposition of scattering patterns $I_1(h)$, $I_2(h)$,... of particles with volumes V_1, V_2, ... [or of the corresponding CFs $\gamma_1(r)$, $\gamma_2(r)$,...] is of importance for diluted and quasi-diluted particle ensembles. Special cases are discussed in Chapters 4, 7 and 8. Neglecting any particle-to-particle interference, the *summarized* scattering pattern $I(h)$ and the *summarized* CF $\gamma(r)$ are the volume averaged sums

$$I(h) = \frac{I_1(h) \cdot V_1^2 + I_2(h) \cdot V_2^2}{V_1^2 + V_2^2}, \quad \gamma(r) = \frac{V_1 \cdot \gamma_1(r) + V_2 \cdot \gamma_2(r)}{V_1 + V_2}. \quad (1.35)$$

At the origin, the scattering patterns $I_1(h, d_1)$ and $I_2(h, d_2)$ of two spheres of diameters d_1, d_2 are interrelated by $I_1(0, d_1)/I_2(0, d_2) = d_1^6/d_2^6$.

2. The invariant of SAS: For the normalized (dimensionless) scattering pattern, $[I(0) = 1$ and $v_c(L) = 4\pi \int_0^L r^2\gamma(r)dr]$, the relation

$$inv = v_c(L) \cdot \int_0^\infty h^2 I(h)dh = 2\pi^2 \quad (1.36)$$

holds true, i.e., regardless of the background of the scattering pattern.

A proof directly interrelates Eq. (1.34) and Eq. (1.36). The invariant is represented by

$$inv = 4\pi \int_0^\infty \int_0^L \frac{\gamma''(r) \cdot [2 - 2\cos(hr) - hr\sin(hr)]}{h^2} dr dh. \quad (1.37)$$

The order of integration of the double integral can be changed. It holds that $\int_0^\infty \frac{2 - 2\cos(hr) - hr\sin(hr)}{h^2} dh = \frac{\pi r}{2}$. The remaining integral part simplifies by use of the properties $\int_0^L r \cdot \gamma''(r)dr = 1$, $\gamma(L) = 0$, $\gamma'(L) = 0$ and $\gamma''(L) = 0$. Hence,

$$inv = 4\pi \int_0^\infty \gamma''(r) \cdot \left(\frac{\pi r}{2}\right) dr = 2\pi^2 \int_0^L r \cdot \gamma''(r)dr = 2\pi^2. \quad (1.38)$$

The invariant integral Eq. (1.36) is useful for checking model calculations of scattering patterns (see the following chapters).

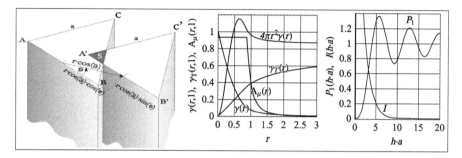

FIGURE 1.5
The particle model (left) (here a triangular rod) and the scattering intensity
curve (right). The function $\gamma(r)$ results from the overlapping volume of the
original rod and its r-translate [72]. The real-space structure functions $\gamma(r,a)$,
$4\pi r^2 \gamma(r,a)$, $\gamma_T(r,a)$ and $A_\mu(r,a) = a\sqrt{3}/3 \cdot \gamma''(r)$ result. The scattering
intensity $I(h)$ (here $I(0) \to +\infty$), and finally $P_1(h \cdot a)$ follow.

**3. Analytic representations of $\gamma(r)$ are more complicated than
those of $\gamma''(r)$:** Alternatively, from the second derivative of the CF $\gamma''(r)$ the
formula $\gamma(r) = 1 + \gamma'(0) \cdot r + \int_0^r (r-t) \cdot \gamma''(t)dt$ results, where $\gamma'(0)$ is connected
with integral parameters of the particle ensemble. Usually, the analytic repre-
sentation of a model function $\gamma''(r)$ is shorter and simpler than that of the CF.
As shown in examples [151, 60], the step $\gamma''(r) \to \gamma(r)$ leads to more complicated integrals and does not simplify the representation of the model function.

4. Normalization of the scattering pattern: In certain limiting
cases, a normalization $I(0) \to 1$ is unfeasible. This is the case with very long
cylinders. Here $I(0) \to \infty$; see (Porod, 1951, 1952) [191, 192]. Frequently,
transformed representations of $I(h)$, like $\log(I(h))$, $h \cdot I(h)$, $h^2 \cdot I(h)$ and
$h^4 \cdot I(h)$, are discussed. The latter, which is called the *Porod plot* [191], is the
most important. The *normalized Porod plot* $P_1(h)$,

$$P_1(h) = \frac{\pi \cdot h^4 \cdot I(h)}{4|\gamma'(0)| \int_0^\infty h^2 I(h)dh} = 1 - \frac{1}{2} \int_0^{L,L_0} \frac{\gamma''(r)}{|\gamma'(0)|}[2\cos(hr) + hr\sin(hr)]dr,$$

$$(1.39)$$

analyzes the scattering behavior at large h values. Porod's length parameter
l_p is an abbreviation of the typical length $1/|\gamma'(0)|$, $l_p = 1/|\gamma'(0)|$. Studies
have been done on the function $P_1(h)$ [14, 15]. Basically, $P_1(h)$ involves a
more or less oscillating behavior around $P_1 = 1$ (see the right of Fig. 1.5).
Such special plots can effectively compare experiment and models.

5. Experimental experience and ill-posed problems: The Eqs. (1.33)
to (1.39) are fundamental for performing the step from a theoretical real-space
structure function to the (non-negative) scattering intensity. Trial and error

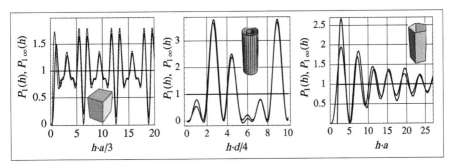

FIGURE 1.6
Typical behavior of $P_1(h)$: The solid lines denote the exact curves, whereas
the dashed lines denote simple asymptotic approximations. Left: cuboid with
edges $a = 3$, $b = 4$, $c = 5$ [60]. Center: hollow cylinder $d = 4$, $d_i = 2$,
$H \to \infty$ [72, p. 330]. Right: rod with pentagonal right section $a = 1$ [116].

methods, which are frequently applied in practice, allow a comparison of models and experiments. Certainly, the inversion of Eqs. (1.33) and (1.39) is a
sophisticated matter. In any case, special assumptions are required for performing an ill-posed integral transformation. An indispensable assumption is
the concept of a fixed range order $L = L_0$. Experimental experience is hardly
translatable into a compact formula or a computer program.

   6. **Parameter estimation from $P_1(h)$ – linear approximation:** Structure parameters can be estimated from $P_1(r)$ (see examples in Fig. 1.6). There
exist approximations consisting of simple terms [see Fig. 1.6 and Eqs. (1.40)–
(1.41)]. For a hollow cylinder with inner diameter $d_i \to 2$, outer diameter
$d \to 4$ and height $H \to 3$,

$$P_{1cyl}(h, d, d_i) \approx 1 - \tfrac{1}{2}\cos[h(d - d_i)] + \sin\left[h\left(\tfrac{d-d_i}{2} + d_i\right)\right] - \tfrac{d}{d+d_i}\sin(hd) - \tfrac{d_i}{d+d_i} \cdot \sin(hd_i)$$

$$(1.40)$$

follows. Even without the addition of a Kirste-Porod term, the approximation
is excellent. In the parallelepiped case, [edges a, b, c; surface area S; volume
V (see Section 2.3)],

$$P_1(h) \approx 1 - \frac{12}{\pi S} \cdot \frac{1}{h^2} - \left[\frac{2bc}{S} \cdot \cos(ah) + \frac{2ac}{S} \cdot \cos(bh) + \frac{2ab}{S} \cdot \cos(ch)\right]. \quad (1.41)$$

For cylindrical particles, the particle shape can be reconstructed operating
with such $P_1(h)$ representations, (Ciccariello, 2002) [18].
   In most cases of polydisperse particle ensembles, the informative oscillations
are almost completely smoothed out. Then, for large h, $P_1(h)$ reduces to an
almost constant $P_1 \to 1$. In just this case, $P_1(h) = const. = 1$ is a well-known
essential law of scattering theory (Porod, 1951) [191].

1.2.2   Indication of homogeneous particles by $i(\mathbf{r_A})$

In materials science, the denomination *particle*, i.e., part of a whole ensemble of objects, is important. For writing down a formula indicating homogeneous particles, it is useful to introduce an indicator function $i(\mathbf{r_A})$ (Serra, 1982) [205, p. 128, p. 272, p. 425, p. 486], (Serra, 1988) [206, p. 257]. From $i(\mathbf{r_A})$, a black and white figure of the particle(s) results. Evidently, there does not exist a *super apparatus*, which directly records $i(\mathbf{r_A})$ based on scattering data. Other concepts, which are closer to experimental possibilities, have been introduced. Intermediate steps involving computer programs connect experiment and particle indicator function. In this process, the direct relationship *experiment ⇒ particle indication* is lost.

There exist at least two concepts for particle indication from scattering experiments. A particle indication succeeds when operating with the sum of the volume elements of the particle (1) or with its border position in space (2). Concept (1) is connected with the distance distribution density $p(r)$ of the particle (see Fig. 1.7 and the left of Fig. 1.8). Concept (2) is interrelated with the chord length distribution density $A_\mu(r)$ (CLDD) of the particle (see Fig. 1.9). There exist common properties and differences between these structure functions. The properties of $p(r)$ and $A_\mu(r)$ have been carefully investigated. For convex particles, the CLDD moments are interrelated with those of the distance distribution (Damaschun & Pürschel, 1969) [25].

1. Particles may be considered and analyzed as the sum of their volume elements dV_i. The distance distribution density $p(r)$ was carefully investigated by O. Glatter [129]. This function analyzes the distances r_i between all points belonging to the particle volume. The connection between γ and p for a typical three-dimensional particle and for particle arrangements is explained in Fig. 1.8. The particle stretches along its largest dimension from endpoint E_1 to point E_2, $L_0 = \overline{E_1 E_2}$. The function $p(r)$ involves information about the curvature of the particle at these endpoints. This can be studied in terms of the scattering pattern of the particle (Gille, 1999) [59].

   Both the volume elements dV_1 and dV_2 are independent of each other and evenly distributed in the particle volume V_0. A distribution function $F(r) = \int_{(inside)} 4\pi r^2 \gamma(r)dr/V_0$ results. Thus, $p(r)$ and $\gamma(r)$ are interrelated, $F'(r) = p(r) = 4\pi r^2 \gamma(r)/V_0$. The frequency of the distances between the volume elements of the particle are summarized by the density function $p(r)$. In classical SAS, the distribution density of the random variable r, the function $p(r)$, is referred to as the "distance distribution." This can lead to contradictions in transforming the formulas and results of SAS to the field of stochastic geometry.

2. A smooth particle can be defined by use of a well-defined border with a finite surface area (see Fig. 1.8). A CLDD $A_\mu(r)$ reflects the behavior of the particle border. It is not influenced by the inner parts of

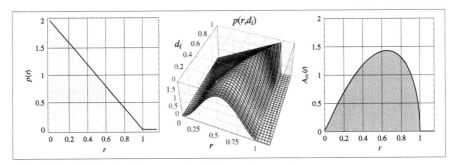

FIGURE 1.7
The function $p(r)$. Left: rod of length $L = 1$; center: hollow sphere with outer diameter $d = 1$ (varying inner diameter d_i in $0 \leq d_i \leq d$); right: SHS of diameter $d = 1$. The SHS-CLDD $A_{cc}(r)$ involves a much simpler representation than that of a hemisphere [61, 103, 104]. The function $p(r)$ corresponds to $A_{cc}(r)$.

a particle. Particle borders are interfaces between two phases of the material. For a particle, a function $A_\mu(r)$ is more shape sensitive than the corresponding $p(r)$. Merely the analysis of the interfaces of a particle leads to an asymptotic expansion of the scattering pattern for big scattering vectors, $P_{1\infty}(h)$, which is sometimes an excellent approximation in an unexpectedly wide h region, (Ciccariello, 1991) [14]. There exist fully developed theories (see Lu & Torquato, 1993 [169] and Mering & Tchoubar, 1968 [182]) for interpreting the CLDD *of* and *between* random tightly packed particles. In the simplest case for the particle volume fraction c, a first order mean value is clearly defined by the mean chord length l_1 of the particle phase and the mean chord length m_1 of the intermediate space (Fig. 1.9, Section 1.3).

The CLDD of a rod of length L is $A(r) = \delta(r-L)$. The connection to $p(r, L)$ is clearly illustrated in Fig. 1.7. The function $p(r)$ consists of two linear parts and involves a singularity at $r = L$. The hollow sphere case involves the limiting case of a spherical shell. In this limiting case, $p(r)$ and the IUR CLDD $A_\mu(r)$ agree (this is the linear case, where the singularity marks the outer diameter). Additionally, a CLDD plot of a spherical half shell (SHS), i.e., $A_\mu(r) = A_{cc}(r)$, is explained. The functions in Fig. 1.7 are normalized. The function $p(r)$ differs from the SAS CF.

In summary, a function $p(r)$ is less sensitively characterized by the particle shape than $A_\mu(r)$ because a CLDD provides a suitable description of the particle border (see Chapter 2).

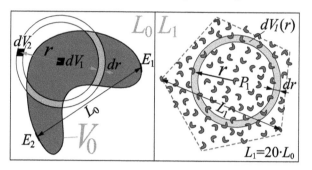

FIGURE 1.8

The functions $p(r)$ for L_0 and $g(r)$ for $L_1 = 20 \cdot L_0$ (illustration in the plane). Left: order range L_0: Random distances r inside a homogeneous particle of volume V_0 can be described by the distance distribution function $F(r)$ and by the distance distribution density $p(r) = F'(r)$. The random variable r is the distance between the centers of two volume elements dV_1 and dV_2.

Right: order range L_1: Let P_1 denote an arbitrary mini-particle of largest diameter L_0. The term $g(r)$ is the average number of mini-particles (points), positioned between the spherical shells of diameters r and $r + dr$.

Particles of diameter L_0 indicate particles of diameter L_1

An essential aspect of particle detection by second-order characteristics is shown in Fig. 1.8 (right). A regular pentagon is defined by a point process of constant intensity (point density) λ_0 ending at the interface of the pentagon. The particle border is not clearly defined. There exists a distribution law of the distance r inside the "huge" particle, which is described by a function $p(r_i)$. The distances r_i are particle-to-particle distances inside a geometric figure [64]. The smaller λ_0 is, the less $p(r_i)$ of the ensemble of mini-particles (volumes dV_1, dV_2, ...) is influenced by the all-enveloping pentagon.

There is also another aspect. Since the mini-particles are small, their CLDD $A_\mu(r)$ cannot be investigated unless all the volume elements are magnified by a size factor k. The reader will recognize $k \approx 20$, because the mini-particles dV_i (which might have an arbitrary shape) are 20 times smaller than the pentagon. Care must be taken in defining *a particle size under investigation*. Such a length scale definition can always be performed by introducing the well-defined order ranges L_i a priori: For an investigation of the huge pentagon, a well-suited order range is L_1, which is a multiple of a certain length L_0. In contrast, the investigation of a typical, single mini-particle of volume V_0 requires $L = L_0$. Thus, $k = L_1/L_0$. Any intermixing of L_0 and L_1 leads to contradictions. This separation of *huge* and *mini* is reflected by the upper integration limits in Eqs. (1.33) and (1.39). The specific *range intervals* $L_1 < L < L_2$ can be picked out by subtracting two integrals from the upper integration limits L_2 and L_1.

Two definitions of the pair correlation function $g(r)$ and examples

Let a particle ensemble be given (see the mini-particles inside the pentagon in Fig. 1.8). The function $g(r)$ describes all existing distances from particle center to particle center. A value $g(r_0)$ does not have the meaning of a probability. Instead, $0 \leq g(r) < \infty$ is possible. Compared with the functions $\gamma(r)$ or $p(r)$, $g(r)$ does not characterize the particle; $g(r)$ describes an isotropic point field, i.e., point-to-point distances.

Let $\lambda = $ *number of points/volume* be the intensity of a homogeneous point process, i.e., λ is constant inside the pentagon. By neglecting border effects,¶ *any positive* test window thrown at random into the point ensemble can be used to estimate λ. Now, two independent definitions of the function $g(r)$ are explained. The second one, based on a technique for estimating λ, is more popular in physics. The first approach defines $g(r)$ in a more general manner.

Approach 1: Let dV_1 and dV_2 be two infinitesimally small volumes, the centers of which are separated by a distance r. Further, $Prob(r)$ denotes the probability that both the volumes dV_1 and dV_2 contain one point of the point field. Consequently, $g(r)$ is defined as a factor of proportionality for a fixed r via $Prob(r) = g(r) \cdot \lambda dV_1 \cdot \lambda dV_2$, i.e., $g(r) = Prob(r)/(\lambda^2 \cdot dV_1 dV_2)$.

Approach 2: The idea is to insert a test window of a special shape for estimating λ. Let the window be the volume region between two spherical shells of radii r and $r + dr$ (see Fig. 1.8). Thus, the mean number of points dN_{mean} expected in a volume element $dV = 4\pi r^2 dr$ is $dN_{mean} = \lambda \cdot 4\pi r^2 dr$. In order to find out the frequency of a selected point-to-point distance r for n (particle center) points, it is useful to introduce the ratios $g_i = dN_i/dN_{mean}$, e.g., for the first point $g_1 = dN_1/dN_{mean}$, for the second point $g_2 = dN_2/dN_{mean}$ and so on. The mean value of all the g_i named by $g(r)$ results in $g(r) = \frac{1}{n} \cdot \sum_{i=1}^{i=n} g_i(r)$. Thus, $g(r = const)$ is the mean number of points, situated in a fixed distance $r = const$ from a center point. Another formulation is $g(r) = \overline{dN}/(\lambda \cdot 4\pi r^2 dr)$. The mean value $\overline{dN}$ is derived by repeating this procedure for all n actual points $\overline{dN} = \frac{1}{n} \cdot \sum_{i=1}^{i=n} dN_i$.

SAS experiments investigate particles like the pentagon, e.g., $L_1 =$(2–500) nm (see Fig. 1.8). Wide-angle scattering (WAS) experiments are restricted to operating with much smaller distances r, e.g., $r < 1$ nm. These are the typical distances between atoms, which usually involve an ordered arrangement that can be described well by a function $g(r) = g(r, L_1)$ up to a certain upper limit $0 < r < r_{max} = L_1$, see (Hermann, 1991) [150, p. 3], (Stoyan & Stoyan,

---

¶In fact, the shape of the pentagon is reflected (up to a certain degree) in the function $g(r)$. There exists the (plane) correlation function $\beta_5(r, L_1)$ of the pentagonal shape. Thus, the function $g(r, L_1)$ can be corrected based on $\beta_5(r, L_1)$. In a first approximation, $g := g/\gamma_5$ leads to an improved g.

1992) [211, p. 276]. The functions $\gamma(r)$ and $g(r)$ describe homogeneously filled particles and point-to-point distances, respectively. Perfect crystals of a large size are uncommon. In many cases, the perfect order of the atoms is already lost for the relatively small length r, $r_{max} < r$. The connection $g(\infty) = 1$ reflects any law of orderless distances. This limiting value never occurs in a perfect crystal. In reality, $g(L) = 1$ is an approximation.

Figure 1.7 involves three examples of the function $p(r)$, consisting of a rod, hollow sphere and spherical half shell (SHS). The thinner the hollow sphere, the more the difference between distance distribution density and chord length distribution density disappears [106]. For a spherical shell, $p(r) \equiv A_\mu(r) = 2r/d^2$, if $0 < r < d$, results. Hence, in certain limiting cases, both the distribution densities are similar and can indicate single particles, particle agglomerates and random particle ensembles. In a certain sense, $p(r)$ is the more general of both functions. For a particle arrangement with particles of extremely small volume fraction (e.g., cosmic objects in astronomy), the particle-to-particle pair correlation function $g(r)$ is not so far away from the $p(r)$ of the ensemble. In this case, the volume elements dV_1 and dV_2 (see Fig. 1.8) are two selected particles of the extremely diluted ensemble.

1.3   Chord length distributions and SAS

The interrelations between the scattering pattern $I(h)$ of an ensemble of hard particles and the *chord length distribution density* (CLDD) have been extensively investigated. Several types of chords are used. Fixing random chords relative to a particle can be handled differently. Isotropic uniform random chords (IUR chords, which are sometimes called μ-chords) are associated with many experimental situations. This type of averaging random chord length is illustrated in Fig. 1.9. A certain order range L of a material is considered. There exist connections between chord lengths along a long straight line, which intersect a part of the sample material and the particle volume fraction.

The random chord segments l in phase 1 and m in phase 2 are interrelated with the particle volume fraction c (see Chapter 8). Let N, $N \to \infty$ be the number of homogeneous hard particles, which are randomly arranged in the plane/space. Let V_t be the total volume (the volume inside a large sphere/circle of diameter L). In the interest of simplicity, it is assumed that each particle possesses a fixed volume V_1 and a fixed surface area S_1. Regardless of the particle shape, the following can be concluded: Per definition, $c = NV_1/V_t$. The volume of the connected region of the material is $(1-c)V_t$. The connection between c, the mean chord length of the single particle $l_1 = 4V_1/S_1$ and the mean chord length of the connected region between

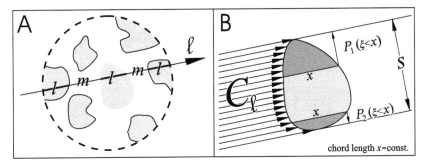

FIGURE 1.9
Chord length $l = x$ and m of a particle ensemble and x of a single particle—
illustration for the plane case for a fixed field of straight lines C_l.
A: Testline ℓ intersects many particles, generating two CLDDs, $\varphi(l)$ and $f(m)$.
B: Analysis of the single particle for a fixed chord direction. The ratio s_i/s,
perpendicular to C_l, defines the distribution function of the random chord
length x, $F(x)$. Per definition $F(x) = P(\xi < x)$, where $P(\xi < x) = P_1(\xi < x) + P_2(\xi < x)$. In the spatial case, the lengths s_i are substituted by the
projection areas S_i. For IUR chords (μ-chords), all directions of C_l exist with
the same probability.

the particles m_1 is

$$\frac{1}{l_p} = \frac{1}{c \cdot m_1} = \frac{1}{(1-c) \cdot l_1} = \frac{1}{l_1} + \frac{1}{m_1}. \tag{1.42}$$

Here, l_p is Porod's length parameter [see Eq. (1.39)]. Equation (1.42) (see
Chapter 9) has been frequently ascertained in materials research and SAS
experiments. Eliminating m_1 yields the Rosiwal theorem (Rosiwal, 1898) [197]
and agrees with the extension of the Cauchy theorem discussed by Mazzolo
et al. [175]. The mean chord length m_1, given by the ratio

$$m_1 = \frac{4(V_t - NV_1)}{NS_1} = \frac{4V_t(1-c)}{NS_1}, \tag{1.43}$$

is a universal structure parameter. Two limiting cases, $m_1 \to \infty$ and $m_1 \to 0$,
are of importance. The first case denotes a so-called infinitely diluted particle
ensemble. The latter case is interrelated with particles, the shapes of which
originate from puzzle fragments that fit together (see Chapter 7).

A two-phase material involves an *inner* and an *outer surface area*. Typically, the inner surface area is much bigger than the outer surface area (see the
dashed particle borders in Fig. 1.9). Outer sample interfaces can be neglected
if N is sufficiently large. Taking into account $V_t/N = V_1/c$, it can be recognized that the contents of Eq. (1.42) and Eq. (1.43) agree. The application of
the Cauchy theorem for the non-convex region (sometimes called matrix region) is in agreement with the fundamentals of SAS (see Porod [191, 192, 194]).

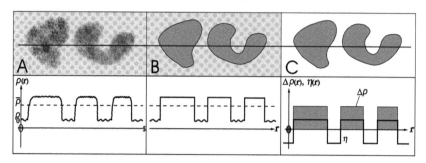

FIGURE 1.10

From reality to a model: Characteristic regions embedded in a matrix phase (two-phase density approximation).

A: Realistic, inhomogeneous regions of mean density $\bar{\varrho}$ and the inhomogeneous matrix phase of mean density ϱ_0. There is a particle border region of a certain thickness.

B: Smoothing out the particle inhomogeneities leads to idealized homogeneous particles (without a border region).

C: Idealized homogeneous particles embedded in a zero density matrix, $\varrho_0 = 0$. The latter assumption can be made without any restrictions. Instead of the density $\varrho(r)$, the fluctuation $\eta(r) = \varrho(r) - \bar{\varrho}$ and the density difference $\Delta\varrho(r) = \varrho(r) - 0$ are investigated. These functions fix $\gamma(r)$ and $Z(r)$.

1.3.1 Sample density, particle models and structure functions

The following is exclusively based on an isotropic sample.

The model of an isotropic three-dimensional sample is closely interrelated to the one-dimensional approach, which will be explained in the following. The volume averaged integration dV must be substituted by a one-dimensional integration dx (see also [102]). A straight line possessing a length bigger than L in direction $\mathbf{r} = \overline{AB}$ alternatively intersects the particle and matrix phase (Fig. 1.9). This simple approach is a special case of the three-dimensional analysis (see Section 1.1.2). As a result, the connection between sample CF $\gamma(r)$, particle volume fraction c and the function of occupancy $Z(r)$ [191] will be verified. Figure 1.10 illustrates the approach in detail. Density fluctuations and density differences are analyzed as depending on r. Realistic sample conditions, i.e., phases that are not completely homogeneous and blurred particle borders, are explained. Further idealization leads directly to the *linear simulation model* discussed in Chapter 8.

Along a fixed direction, the smoothed density $\rho(r)$ splits into two values. These are the density in the particle ρ_1 (phase 1) possessing a volume fraction c and the density $\rho_0 = const.$ outside the particles, i.e., that of the matrix phase 2. For the considered volume (length) described by L, there exists a mean density $\bar{\rho}$ resulting from $\bar{\rho} = c \cdot \rho_1 + (1-c) \cdot \rho_0$. Based on $\bar{\rho}$, the density fluctuation $\eta(r) = \rho(r) - \bar{\rho}$ follows. Furthermore, the non-negative density difference is defined by $\Delta\rho(r) = \rho(r) - \rho_0$.

Without any restriction, $\rho_0 = 0$ can be assumed (Fig. 1.10). Following from this is $\bar{\rho} = c \cdot \rho_1$, $\eta(r) = \rho(r) - c \cdot \rho_1$, $\Delta\rho(r) \equiv \rho(r)$ and $\Delta\rho(r) = \eta(r) + \bar{\rho}$. Thus, from $\eta(r)^2 = (\Delta\rho(r) - \bar{\rho})^2$, the average $\overline{\eta^2} = \overline{(\Delta\rho(r) - \bar{\rho})^2} = \overline{(\Delta\rho)}^2 - \bar{\rho}^2$ results. Furthermore, the particle indicator function $i(r)$ results from the transformation $i(r) = \Delta\rho(r)/\Delta\rho(0)$. Therefore, $i(r) \equiv 1$ inside a particle and $i(r) \equiv 0$ outside.

The mean convolution square of the (electron) density is $\overline{\rho^2} = c \cdot [\rho_1 - \bar{\rho}]^2 + (1-c) \cdot [\rho_0 - \bar{\rho}]^2$. This yields $\overline{\rho^2} = (\rho_1 - \rho_0)^2 \cdot c(1-c) = (\Delta\rho)^2 \cdot c(1-c)$. Usually, this term is a prefactor to the scattering intensity if $I(h)$ is not normalized to unity in the origin. Clearly, in the limiting cases $\Delta\rho \equiv 0$, $c \equiv 0$ or $c \to 1$, no scattering intensity appears, $I(h) \to 0$.

In the isotropic case, the SAS correlation function $\gamma(r)$ introduced by Debye [27] (see also [28]) and the function of occupancy $Z(r)$ introduced by Porod [191] are defined as restricting to any selected direction $\vec{r} = \vec{r}(x, y, z)$. For $y = 0$ and $z = 0$ (i.e., along the x-direction and for a fixed L),

$$\gamma(r) = \frac{\int_0^L \eta(x)\eta(x+r)dx}{\int_0^L [\eta(x)]^2 dx} = \frac{\int_0^L \eta(x)\eta(x+r)dx}{L \cdot \overline{\eta^2}},$$

$$Z(r) = \frac{\int_0^L \Delta\rho(x)\Delta\rho(x+r)dx}{\int_0^L [\Delta\rho(x)]^2 dx} = \frac{\int_0^L \Delta\rho(x)\Delta\rho(x+r)dx}{L \cdot \overline{\Delta\rho^2}} \tag{1.44}$$

results. Based on $\Delta\rho(x) = \eta(x) + \bar{\rho}$ (see the introduced restriction $\rho_0 \equiv 0$), a simple representation for $Z(r)$ is obtained since the parametric integral $\int_0^L [\eta(x) + \eta(x+r)] \cdot \bar{\rho}dx$ disappears for all r:

$$Z(r) = \frac{\int_0^L [\eta(x) + \bar{\rho}] \cdot [\eta(x+r) + \bar{\rho}]dx}{L \cdot \overline{\Delta\rho^2}} = \frac{\int_0^L \eta(x) \cdot \eta(x+r)dx}{L \cdot \overline{\Delta\rho^2}} + \frac{L \cdot (\overline{\Delta\rho})^2}{L \cdot \overline{\Delta\rho^2}} \tag{1.45}$$

Equation (1.45) restricts to the direction x. It is closely interrelated with Eqs. (1.20) to (1.22). Directly incorporating the volume fraction c, $c = (\overline{\Delta\rho})^2 / \overline{\Delta\rho^2}$ and inserting the identity $[(\overline{\Delta\rho})^2 - \bar{\rho}^2]/\overline{\eta^2} \equiv 1$, one of the denominator terms in Eq. (1.45) can be substituted with $1/\overline{\Delta\rho^2} = (1-c)/\overline{\eta^2}$. By doing this, similar to the three-dimensional case already handled in Section 1.1, Eqs. (1.44) and (1.45) give the relationship between $Z(r)$ and $\gamma(r)$:

$$Z(r) = (1-c) \cdot \gamma(r) + c, \quad \gamma(r) = \frac{Z(r) - c}{1-c}, \quad (0 \le c < 1). \tag{1.46}$$

Equation (1.46) is a typical starting point for model calculations with ensembles of homogeneous particles in the isotropic case. Of course, Eq. (1.46) is an approximation, if the number of particles N is small, such as for $N < 100$. The multiple averaging procedures and the definition of c and L require averaging over millions of particles.

1.4   SAS structure functions for a fixed order range L

In the following sections and chapters, the order range L is the essential parameter. Simply said, to extract a certain length L means to *select a typical particle size (or object size)* when investigating a sample material. The integral transformation, which gives $\gamma(r, L)$ in terms of an all-including scattering experiment $I(h)$, will be explained. Figure 1.11 illustrates the order range concept in detail. Typically, an *all-including scattering pattern* involves information *about several order ranges* $L_1, L_2, L_3....$ A clever experimenter is aware of this. Several different L_i should be introduced beforehand, namely those, which are of interest.

For an unmistakably clear scattering pattern I_L the denotation $I_L(h, L)$ must be introduced. In this approach, *all-including case* simply means $L \to \infty$, where the usual isotropized kernel function $KP(h, \infty, s) = sin(sr)/(sr)$ is used for the Fourier transformation. In this limiting case, the scattering pattern will remain unchanged, i.e., no smoothing at all is taken into consideration. Hence, for a relatively big L, the identity $I_\infty = I_L(h, \infty) \equiv I(h)$ holds true. However, the data evaluation will be totally transparent if certain selected L_i split $I_L(h, L_i)$ into certain meaningful cases.

Data evaluation of a scattering pattern for specific $L = L_i$

Let the selection $L = L_1$ be made beforehand. The scattering experiment and data evaluation are planned and performed for investigating the specific order range L. The following approach *inserts this information* about *the specific $L = L_1$* into the data evaluation. In the end, a specifically smoothed scattering pattern and the corresponding autocorrelation function $\gamma(r, L_1)$ will result from a parametric integral involving L. For detecting an "unmistakably clear" scattering pattern $I_L = I_L(h, L)$, two idealizations are given to start with.

1. A quasi-diluted particle ensemble is assumed (see Chapter 4).
2. An *all-including scattering pattern* $I(h)$ is analyzed for the case of an unlimited h-interval.

Altogether, the following derivations, performed step by step, represent a clear concept of data evaluation, which is essential for Chapters 4–9. Based on Fig. 1.11, the derivation starts with the classical, well-known pair of transformations between $\gamma(r)$ and $I(h)$. Equation (1.47) does not include L yet.

$$\gamma(r) = \frac{1}{2\pi^2} \int_0^\infty h^2 I(h) \frac{\sin(h \cdot r)}{h \cdot r} dh, \qquad (1.47)$$

$$I(h) = \int_0^\infty 4\pi r^2 \gamma(r) \frac{\sin(h \cdot r)}{h \cdot r} dr,$$

$$I(h) = \int_0^L 4\pi r^2 \gamma(r) \frac{\sin(h \cdot r)}{h \cdot r} dr. \qquad (1.48)$$

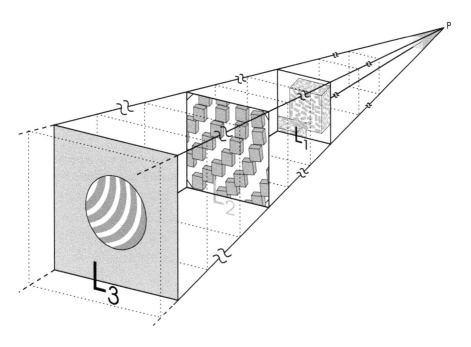

FIGURE 1.11

The order range concept.

A consideration of three different order ranges L_1, L_2, L_3, i.e., the choice of specific lengths L_i, leads to different geometric constellations. Three different geometric objects (sub-structures) must be described by appropriate real-space parameters. Consequently, different geometric models are required for a relevant order range L_1 or L_2 or L_3.

Big spheres come into sight if $L = L_3$. However, these huge particles are composed of smaller elements.

By setting $L = L_2$, nearly parallel chains can be seen. However, a bigger magnification gives even more details.

If $L = L_1$, the originating "boxes" come into sight.

These three order ranges must be inspected and interpreted separately, i.e., one after the other, not simultaneously. A specific micrograph possesses a specific magnification, i.e., the magnification controls and selects the order range shown by an image.

In SAS, the length parameter L is directly involved in an integral transformation as a free parameter. This parameter controls the behavior of the resulting structure functions. Consequently, the meaning of a term $\gamma(r)$ is not clear at all, without knowing the corresponding L. However, after fixing a specific L, the different terms $\gamma(r, L_1)$, $\gamma(r, L_2)$, $\gamma(r, L_3)$ ..., precisely reflect the sample geometry down to a particle size of about (1–2) nm.

Equations (1.48) operate with an idealized, all-including scattering pattern. The normalization of $\gamma(r)$ and $I(h)$ is initially neglected. It will be considered in detail later.

After changing the denotation by writing s instead of h, the function γ is written as follows:

$$\gamma(r, L) = \gamma(r, L) = \frac{1}{2\pi^2} \cdot \int_0^\infty s^2 I(s) \cdot \frac{\sin(s \cdot r)}{s \cdot r} ds. \tag{1.49}$$

Equation (1.49) will now be inserted into the second version of Eqs. (1.48). In this approach, this step introduces the variable L into the equations. A function $I_L(h, L)$ may be derived from the all-including scattering pattern $I(h)$ by

$$I_L(h, L) = \int_0^L 4\pi r^2 \cdot \left(\frac{1}{2\pi^2} \cdot \int_0^\infty s^2 I(s) \frac{\sin(s \cdot r)}{s \cdot r} ds \right) \cdot \frac{\sin(h \cdot r)}{h \cdot r} dr$$

$$= \frac{2}{\pi} \cdot \int_0^L r^2 \cdot \left(\int_0^\infty s^2 I(s) \frac{\sin(s \cdot r)}{s \cdot r} ds \right) \cdot \frac{\sin(h \cdot r)}{h \cdot r} dr. \tag{1.50}$$

Based on this step, the influence of the order range L on the intensity $I_L(h, L)$ can be analyzed. An instructive numerical check is suppressed.

However, there exist useful simplifying manipulations of Eq. (1.50). In the beginning, the sequence of integrations is changed. At first, the r integration is handled with

$$I(h) = \frac{2}{\pi} \cdot \int_0^\infty \int_0^L \left(r^2 \cdot s^2 \cdot \frac{\sin(s \cdot r)}{s \cdot r} \cdot \frac{\sin(h \cdot r)}{h \cdot r} \right) dr \cdot I(s) ds. \tag{1.51}$$

The inner integral in Eq. (1.51),

$$K(s, L, h) = \int_0^L \left(r^2 \cdot s^2 \cdot \frac{\sin(s \cdot r)}{s \cdot r} \cdot \frac{\sin(h \cdot r)}{h \cdot r} \right) dr, \tag{1.52}$$

is independent of any experimental data and can be simplified. The function $K(s, L, h)$ is a sum of two sine terms. Finally, the specific scattering pattern $I_L(h, L)$ is written

$$I_L(h, L) = \frac{2}{\pi} \cdot \int_0^\infty I(s) \cdot K(s, L, h) ds, \text{ with}$$

$$K(s, L, h) = s \left(\frac{\sin[L(h-s)]}{2h(h-s)} - \frac{\sin[L(h+s)]}{2h(h+s)} \right). \tag{1.53}$$

The integral transformation, Eqs. (1.53), allows a *controlled smoothing* of the *all–including experimental function* $I(s) = I(h)$ (see theright in Fig. 1.12). On the right, seven curves are plotted corresponding to different L_i.

However, Eqs. (1.53) are by far not the end of the *modified controlled data manipulation* (Gille, 2010) [120]. The modified scattering pattern I_L is the starting point for determining the set of all classical structure functions in terms of L, especially $\gamma(r, L)$ (see Eq. (1.47) and Section 1.4.1).

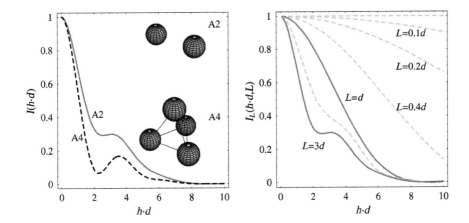

FIGURE 1.12
Analysis of the smoothed intensity $I_s(h, L) \equiv I_L(h, L)$ in terms of the *"all–including"* intensity $I(h)$ and L via Eqs. (1.53) for a two-sphere aggregate A2. The distance between two sphere centers is $s = d = const.$
Left: The *"all–including"* isotropic scattering patterns $I(h \cdot d)$ for two spheres (A2, solid line), and four spheres in a tetrahedral constitution (A4, dashed line). In case A4, the spheres are centered in the tips of a tetrahedron of edge length s. The aggregates [see Chapter 4, Eq. (4.59)] involve two order ranges; $L = d$ and $L = s + d$ holds.
The A2 case is analyzed further on the right. The functions $I_g(h \cdot d, L) = I_L(h \cdot d, L)$ represent smoothed scattering patterns. The smaller L is, the smoother the curves are. Beginning with the top, the lengths $L = 0.01d$, $L = 0.1d$, $L = 0.2d$, $L = 0.4d$, $L = d$, $L = 2d$ and $L \geq 3d$ are inserted. The dashed lines on the right are unimportant for practice. However, there exist many samples investigated by SAS which involve more than two order ranges L_i. Any positive L can be inserted into Eqs. (1.53). In the limiting case $L \to 0$, $I_L = const = 1$ follows (see the upper curve for $L = 0.01d$). The case $L = L_0 = d = 1$ exactly selects a single sphere. For any L with the property $3d = 3 \cdot L_0 < L$, the scattering pattern $I(h \cdot d)$ from the left can be reproduced. Formulas for $I(h \cdot d)$ are explained in detail in Chapter 4. The functions are based on $\gamma(r, s, d)$ of a pair of (two) spheres A, B. The case A4 is described by the CF $\gamma(r, s, d) = \gamma_0(r, d) + 3 \cdot \gamma_{AB}(r, s, d)$, where $\gamma_0(r, d) = 1 - 3r/(2d) + r^3/(2d^3)$ if $0 \leq r < d$, else $\gamma_0(r, d) = 0$. Here, the term $\gamma_{AB}(r, s, d)$ is defined by

$$s - d \leq r \leq d + s, \ \gamma_{AB}(r, s, d) = \frac{d^5 - 5(r-s)^2 d^3 + [5d^2 - (r-s)^2] \cdot |r-s|^3}{20rsd^3};$$

$$-\infty < r < s - d, \ d + s < r < \infty, \ \gamma_{AB}(r, s, d) \equiv 0.$$

1.4.1   Correlation function in terms of the intensity $I_L(h, L)$

Inserting $I_L(h, L)$ [instead of the usually applied $I(h)$ given by Eq. (1.48)] into Eq. (1.47) yields a compact representation of the CF in terms of L,

$$\gamma(r, L) = \frac{1}{2\pi^2} \cdot \int_0^\infty h^2 \cdot I_L(h, L) \cdot \frac{\sin(h \cdot r)}{h \cdot r} dh,$$

$$\gamma(r, L) = \frac{1}{2\pi^2} \cdot \int_0^\infty h^2 \cdot \left(\frac{2}{\pi} \cdot \int_0^\infty I(s) \cdot K(s, L, h) ds \right) \cdot \frac{\sin(h \cdot r)}{h \cdot r} dh. \tag{1.54}$$

Hence, with the $K(s, L, h)$ term from Eq. (1.52)

$$\gamma(r, L) =$$
$$\frac{1}{2\pi^2} \int_0^\infty h^2 \cdot \left[\frac{2}{\pi} \int_0^\infty I(s) s \left(\frac{\sin[L(h-s)]}{2h(h-s)} - \frac{\sin[L(h+s)]}{2h(h+s)} \right) ds \right] \cdot \frac{\sin(h \cdot r)}{h \cdot r} dh \tag{1.55}$$

follows. Considering the h integration in detail yields

$$\gamma(r, L) =$$
$$\frac{1}{2\pi^2} \cdot \frac{2}{\pi} \int_0^\infty \left(\int_0^\infty h^2 s \left[\frac{\sin[L(h-s)]}{2h(h-s)} - \frac{\sin[L(h+s)]}{2h(h+s)} \right] \frac{\sin(h \cdot r)}{h \cdot r} dh \right) \cdot I(s) ds. \tag{1.56}$$

The inner integral does not depend on $I(s)$. By inserting the denotation $g_0(s, L, r)$ instead,

$$g_0(s, L, r) = \int_0^\infty h^2 \cdot s \left[\frac{\sin[L(h - s)]}{2h(h - s)} - \frac{\sin[L(h + s)]}{2h(h + s)} \right] \cdot \frac{\sin(h \cdot r)}{h \cdot r} dh, \tag{1.57}$$

$$\gamma(r, L) = \left(\frac{1}{2\pi^2} \right) \cdot \frac{2}{\pi} \int_0^\infty I(s) \cdot g_0(s, L, r) ds, \quad \text{with} \tag{1.58}$$

$$g_0(s, L, r) = \frac{s}{2r} \int_0^\infty \left(\frac{\sin[L(h-s)]}{h-s} - \frac{\sin[L(h+s)]}{h+s} \right) \cdot \sin(h \cdot r) dh$$

follows. Equation (1.58) controls the analysis of the CF as depending on L. After considering the limiting value $r \to 0$, $g_0(s, L, 0) = \pi s^2/2$, the normalization $\gamma(0) = 1$ can be handled. In summary, the simplification of the integral yields the expected representation for an infinite upper integration limit,

$$g_0(s, L, r) = \begin{cases} r = 0 & : \frac{\pi}{2}s^2 \\ r < L & : \frac{\pi}{2} \cdot \frac{s\sin(sr)}{r}. \\ L < r < \infty & : 0 \end{cases} \tag{1.59}$$

Thus, for $L \to \infty$, and only in this case, $g_0(s, L, r) = (\pi/2)s^2 \sin(sr)/(sr)$.

Hence, the best-known form of the Fourier transformation

$$\gamma(r, L) = \frac{1}{2\pi^2} \int_0^\infty I(s) \frac{2}{\pi} \cdot \frac{\pi}{2} s^2 \frac{\sin(sr)}{sr} ds = \frac{1}{2\pi^2} \int_0^\infty I(s) s^2 \cdot \frac{\sin(sr)}{sr} ds \tag{1.60}$$

results. However, as the reader now understands, in light of the approach performed, at best, Eq. (1.60) is a simple approximation!

The function g_0 oscillates with a linearly increasing amplitude in terms of s (see Fig. 1.13, which investigates a case where $L = 10$, $r = 5 < L$).

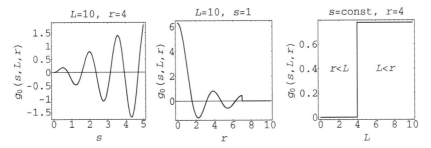

FIGURE 1.13

Illustration of the behavior of the function g_0 in terms of s, r and L.

1.4.2 Extension to the realistic experiment $I(s)$, $s < s_{max}$

With respect to experimental limitations (truncation errors) and bearing in mind the *backward extrapolation approach* (see [66]), a more general function $g(b, s, L, r)$, where the parameter b defines the upper integration limit of the integral, will be inspected [see the following Eq. (1.61)]. Thus, Eqs. (1.58) are written as follows

$$\gamma(r, L) = \quad \frac{1}{2\pi^2} \cdot \frac{2}{\pi} \int_0^b I(s) \cdot g(b, s, L, r) ds$$

$$g(b, s, L, r) = \frac{s}{2r} \int_0^b \left(\frac{\sin[L(h-s)]}{h-s} - \frac{\sin[L(h+s)]}{h+s} \right) \cdot \sin(h \cdot r) dh. \tag{1.61}$$

Equation (1.61) fixes the SAS CF $\gamma(r, L)$ in terms of the scattering pattern $I(s)$, $0 \le s < s_{max}$, $0 \le b \le s_{max}$ for a specific L. In most experimental cases, a restriction $s \le b$ is reasonable. Special *window functions*, which reduce truncations errors, are not used.

Analysis of the function $g(b, s, L, r)$

If b is relatively small, the functions g_0 and g differ completely for $L < r$. Obviously, $g(\infty, s, L, r) \equiv g_0(s, L, r)$. A numerical check for g is $g[4, 3, 2, 1] = 0.483441...$; however, contrary to this, $g_0(3, 2, 1) = 0.665$. In the following, more intrinsic properties of the g-function are investigated (see special plots). A sequence of different cases must be distinguished with $g(b, s, L, r)$. At least, $s = 0$, $r = 0$, $r = L$, ($s = b$ & $L < r$) and ($s = b$ & $r < L$) must be inspected separately. The following *Mathematica* pattern yields a 50-digit approximation of the function g by simple numerical integration.

```
gnum[b_, s_, L_, r_]:= s/(2r)*NIntegrate[(Sin[L*(h - s)]/(h - s) -
                      Sin[L*(h + s)]/(h + s)]*Sin[h*r], {h, 0, b},
                      PrecisionGoal->50, WorkingPrecision->100];

(* r < L *)  {gnum[4, 3, 2, 1], g[4, 3, 2, 1]}
(* L < r *)  {gnum[4, 3, 2, 13],g[4, 3, 2, 13]}
Out={ 0.48344139207812180304875055062833129684619632848640419700, 0.483441}
Out={-0.000093845995348593298003170216990388246450125032819137 26, -0.000093846}
```

Bigger parameters lead to highly oscillating integrands. A general numerical approach breaks down and cannot be applied in practice (relatively big CPU time). The analytic expression of $g(b, s, L, r)$ is indispensable.

The function g is important for the scattering pattern step up to the structure functions. Hence, each analytic representation, involving standard functions of mathematical physics, is important. Analytical integration of the g-integral terms in Eq. (1.61) is a complex matter. The function g can be represented in terms of classical functions of mathematical physics. The reader may find the integral tables by Ryshik and Gradstein [201] and a textbook about special functions (Mocica, 1988) [181] useful for further reading. The following *Mathematica* module incorporates a solution $g = g(b, s, L, r)$. Numerical test procedures operating with randomly selected parameters b, s, L, and r were applied in order to confirm these analytic results.

```
g = Compile[{b, s, L, r},
  Module[{A, B, C, D, E, F}, (A = Abs[L - r]*(b - s);
          B = (L + r)*(b - s);
          C = Abs[L - r](b + s);
          D = (L + r)*(b + s);
          E = (r - L)*(b - s);
          F = (L - r)*(b + s);
(* Splitting for the limiting cases *)
Which[s == 0, 0, r == 0,
s/(4*L)*(2*Cos[L(b + s)] - 2*Cos[L(b - s)] +
2*L*s*SinIntegral[L(b - s)] + 2*L*s*SinIntegral[L(b + s)]),
r == L,
 s*Cos[L*s]/(4L)*(-CosIntegral[2L(b - s)] +
   CosIntegral[2L(b + s)] + Log[(b - s)/(b + s)]) +
 s*Sin[L*s]/(4L)*(+SinIntegral[2L(b - s)] +
   SinIntegral[2L(b + s)]), s == b && L < r,
 b*Cos[b*r]/(4r)*(-CosIntegral[2b*Abs[r - L]] +
   CosIntegral[2b(L + r)] + Log[Abs[r - L]/(r + L)]) +
 b*Sin[b*r]/(4r)*(-SinIntegral[2b*Abs[r - L]] +
   SinIntegral[2b(L + r)]),
s == b && r < L,
 b*Cos[b*r]/(4r)*(-CosIntegral[2b*Abs[r - L]] +
   CosIntegral[2b(L + r)] + Log[Abs[r - L]/(r + L)]) +
 b*Sin[b*r]/(4r)*(+SinIntegral[2b*Abs[r - L]] +
   SinIntegral[2b(L + r)]), True,
+s*Cos[r*s]/(4*r)*(+CosIntegral[A] - CosIntegral[B] -
                   CosIntegral[C] + CosIntegral[D])+
 s*Sin[r*s]/(4*r)*(-SinIntegral[E] + SinIntegral[B] +
                   SinIntegral[F] + SinIntegral[D])] )] ];
(* test  g[4, 3, 2, 1] = 0.483441 *)  g[4, 3, 2, 1]
```

Differences in the behavior of $g_0(s, L)$ and $g(b, s, L, r)$

The functions g and g_0 are decisive for data evaluation. An investigation of the difference $g(s) - g_0(s)$ is important. The following figures (Figs. 1.14, 1.15

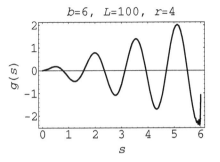
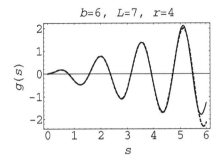

FIGURE 1.14

Similarities and differences between the functions $g_0(s, L, 4)$ (see dashed line) and $g(6, s, L, 4)$ (see solid line) as depending on L. It does not hold $g \equiv g_0$. As expected, for such relatively big L (see the case $L = 100$, left), the differences between g and g_0 remain very small in the whole s-interval $0 \le s \le b$. In contrast, there exist clear deviations between g and g_0 for $L = 7$, $s = 6$ (see right).

On the other hand, but not of direct importance for data evaluation of a scattering pattern, the behavior of g and g_0 completely differs for $b < s < \infty$.

and 1.16) demonstrate common and different properties of the functions g_0 and g. The limiting case $L \to \infty$ is included. The function g_0 is a special case of the more general and more complicated g function. Several plots compare the behavior of g_0 and g in selected cases:

1. Figure 1.14 investigates $g(s)$ for $r = 4 = const.$, $b = 6$ and two different L.
2. Figure 1.15 compares $g_0(s, L)$ with general plots of g for $r = 2$.
3. Figure 1.16 analyzes the differences between both functions.

The parametric integrals are real-valued for the all–including case $0 < s < b$. An extension of Fig. 1.14 is Fig. 1.15, which illustrates significant properties of the behavior of these functions.

Three-dimensional plots analyzing the functions g_0 and g: At first, Fig. 1.15 shows $g_0(s, L)$ for all possible cases. In contrast, a plot of g requires interval splitting. The cases $L < r$ and $r < L$ are considered separately in the lower part of Fig. 1.15.

Furthermore, Fig. 1.16 demonstrates that the functions $g(b, s, L, r)$ and $g_0 = g(\infty, s, L, r)$ are not identical (this applies to special cases as well). The analysis shows that there are always bigger or smaller differences.

The function $g(b, s, L, r)$ is tailor made for reducing truncation errors via the *Mathematica* function *SequenceLimit*. Thereby, the upper integration limit $s = b$ "works" by inserting a sequence b_i. Evidently, only those b_i for which $I(s)$ is recorded can be inserted.

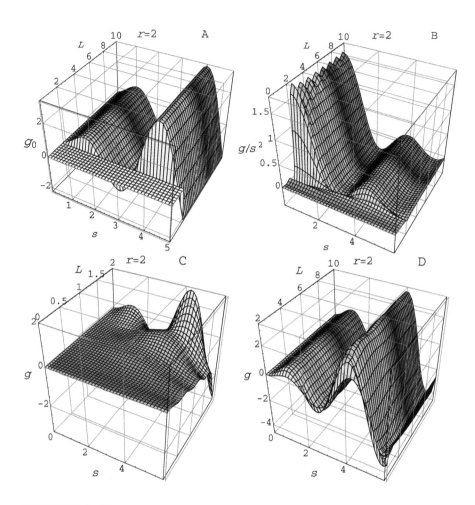

FIGURE 1.15

Behavior of the functions $g_0(b, s, r)$ and $g(b, s, L, r)$ for depicted parameters.
Upper part: Plots of $g_0(s, L, r = 2)$ (see A) and $g(b, s, L, r)/s^2$, with $b = 6$,
$r = 2$, $0 < L < 10$ (see B). The gray levels specify special L values. Both
cases $L < r$ and $r < L$ are involved in the plot. The function g_0 'abruptly'
disappears in the interval $L < r < \infty$ (see the white region). In light of the
approach performed, this demonstrates that g_0 is only an approximation.
Lower part: The function $g(b, s, L, r)$ for the cases $L \leq r$ (see C) and $r \leq L$
(see D) for the parameters $r = 2$, $b = 6$.
The gray levels used for showing a special L value are the same as in Fig. 1.15,
but g clearly differs from g_0. A white region where $g \equiv 0$ holds (compare with
the upper part), does not exist at all. The function g always oscillates around
the zero level.

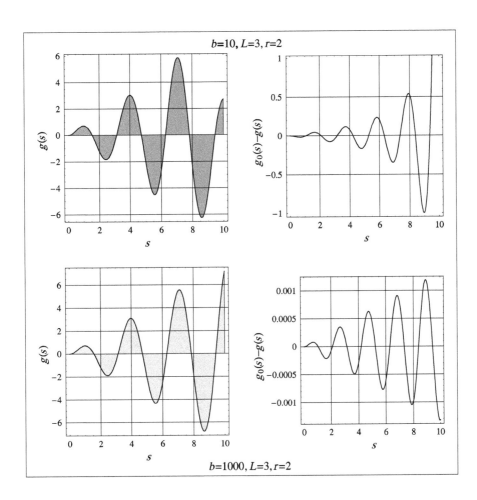

FIGURE 1.16

Analysis of the difference between the functions g and g_0 in terms of the parameter b.

Left: The exact function $g(b, s, 3, 2)$; right: The difference function for $b = 10$ and $b = 1\,000$. The maximum differences are in the order of magnitude $1/b$ for bigger b. For $b \to \infty$, the differences disappear.

1.5   Aspects of data evaluation for a specific L

As depending on a well-selected order range L, the approach explained can be used to determine approximations for smoothed scattering patterns $I_L = I_L(h, L)$ and CFs $\gamma(r, L)$. Furthermore, the whole set of SAS structure functions depends on L via $\gamma(r, L)$.

Regardless of these functions, the r-interval "available" is limited. In no case can the CF be analyzed for $0 \leq r < \infty$. Let an experimental scattering pattern $I(s)$ be given, which begins at $s = s_{min}$ and ends at $s = s_{max}$, i.e., $s_{min} < s < s_{max}$. Then, operating with

$$\gamma(r) = \frac{1}{2\pi^2} \int_0^b I(s) \cdot s^2 \cdot \frac{\sin(s \cdot r)}{s \cdot r} ds \qquad (1.62)$$

for a limited s, $0 < s < b = s_{max}$, the first zero of the function $\sin(s \cdot r) = 0$ defines a specific length r_{min}. This length $r_{min} = \pi/b$ equals the lower resolution limit of the experiment [26]. Equation (1.62) cannot be applied for $r < r_{min}$. More generally, there exists the useful length interval $r_{min} < r < r_{max}$ in real space.

The more general transformation additionally involves the parameter L and the function $g(b, s, L, r)$ [see the last subsections and Eq. (1.63)]:

$$\gamma(r, L) = \frac{1}{\pi^3} \int_0^b I(s, \infty) \cdot g(b, s, L, r) ds. \qquad (1.63)$$

Like Eq. (1.62), Eq. (1.63) is only useful for certain r-intervals $r_{min} < r < r_{max}$ and $r_{min} < L < r_{max}$. Here, the lower resolution limit r_{min} depends on the first zero of the function $g(r) = g(b, s = b, L, r)$ and on $b = s_{max}$ and L, i.e., $r_{min} = r_{min}(b, L)$. This fact is independent of the scattering pattern itself (see Fig. 1.17). Here, by analogy with a plot of a function $\sin(\pi s/s_{max})$ possessing the property $0 = \sin(\pi s_{max}/s_{max})$, the function $g(b, s, L, r)$ is plotted in terms of s for $b = s_{max} = const$, $L = const$ and $r = r_{rmin} = const$. The case $g(b, b, L, r) = 0$ corresponds to $\sin(\pi s_{max}/s_{max}) = 0$.

Extrapolating the case illustrated in Fig. 1.17, Table 1.2 and Fig. 1.18 result. The following *Mathematica* pattern $reslimrmin[h_{max}, L]$ determines r_{min} in terms of $b = s_{max} = h_{max} > 0$ and L, $L > 0$.

```
(*Fixing the resolution limit r=rmin in terms of hmax and L*)
reslimrmin[hmax_,L_]:=Module[{solution},x},
solution=FindRoot[g[hmax,hmax,L,x]==0,{x,{0,Pi/hmax}}];
solution[[1,2]] ];
{reslimrmin[0.5,1],reslimrmin[2.5,20]}
{9.00904,1.27268}
```

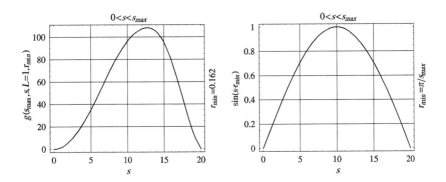

FIGURE 1.17
The resolution limit $r_{min}(b, L)$ is connected with $g(b, s, L, r)$ and $\sin(s \cdot r)$.
Left: For $L = const = 1$ nm and $b = s_{max} = const = 20$ nm^{-1}, the function
$g(s)$ has a zero at $r = r_{min} = 0.162$ nm, $g(s_{max}, s_{max}, L, r) = 0$. The zero
fixes the resolution limit for any $I(s)$ which ends at $s_{max} = 20$ nm^{-1} for $L = 1$
nm. Right: The analogous case of a simple sine function for $s_{max} = 20$ nm^{-1}.
Such an s_{max}-value case is typical of wide-angle scattering (WAS).

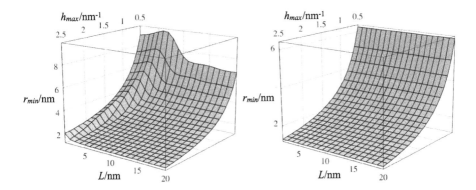

FIGURE 1.18
Lower resolution limit r_{min} in terms of the maximum amount of the scattering
vector h_{max} and L. The functions on the left and right are similar.
Left: The solution for r_{min} based on the first zero $r = r_{min}$ of the function
$g(r) = 0$. Right: The approximation $r_{min} = \pi/h_{max}$ is independent of L.
This holds better the larger $h_{max} \cdot L$ is (see Table 1.2).

TABLE 1.2

Lower resolution limit r_{min} as depending on selected h_{max} and L.

h/nm	$r_{min}(h,1)$/nm	$r_{min}(h,5)$/nm	$r_{min}(h,10)$/nm	$r_{min}(h,15)$/nm
0.5	9.01	9.54	7.47	6.96
1	4.53	3.73	3.37	3.28
1.5	3.06	2.32	2.18	2.16
2	2.33	1.68	1.62	1.61
2.5	1.91	1.32	1.29	1.27

Truncation errors can be reduced according to the concept of the *transition function* $T(h)$, $0 < h < h_{max}$ (see [66, 114, 120]). The idea is to introduce a variable upper integration limit x, $x < b = h_{max}$ of the parametric integrals in question. This results in the integration being truncated for a fixed length r, $r_{min} < r$, at $0 < x \leq b = h_{max}$. The transition function $T(x)$ defined in the finite interval $0 < x \leq h_{max}$ involves the property $T(0) = 0$. The last value $T(h_{max})$ does not automatically produce the best approximation! A better approximation can be obtained based on an analysis of the sequence of the truncated results $T(x_1), T(x_2), T(x_3), ..., T(x = h_{max})$.

There exist several statistical approaches which analyze the behavior of the function $T(x)$ in detail. Obviously, such intermediate steps which have to be performed for the whole sequence of abscissas $r_i = const.$ require much CPU time. A possible approach consists in applying the *Mathematica* function *SequenceLimit* for the sequence $[0, T(x_1), T(x_2), T(x_3), ..., T(x_{max})]$ with $x_{max} = b = h_{max}$ (see [111, 234]). Here, the method of integration is not significant at all, i.e., this can be handled by simple numerical integration in a limited h-interval by use of Simpson's rule.

Figure 1.19 shows an application for the model case of two spheres of diameter d, the centers of which are separated by a distance $s = 2d$. As this example shows, the approach is useful for determining SAS correlation functions from scattering patterns for varying L.

In the following, applications of this strategy are explained for isotropic samples, starting with the CF analysis of a copper(I) cyanide/platinum micropowder. The next example analyzes atomic distances in an AlLa 30at% alloy. After the investigation of a blend polymer sample, Porod's invariant of the L-smoothed scattering pattern $I_L(h, L)$ will be considered. Furthermore, an approach for the pre-selection of one or more lengths L_i is explained. The representation includes practical examples.

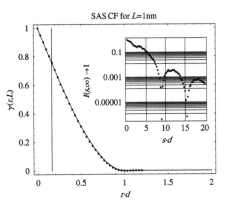

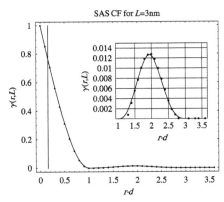

FIGURE 1.19

Determination of the SAS CF in terms of the order range from a simulated data set.

In this model case, the maximum L involved in the scattering pattern is $L = 3d$, $I(s, \infty) \equiv I(s, 3d)$. However, the single sphere case is also involved in the pattern $I(s, 3d)$. In fact, both the CFs $\gamma(r, L = d)$ and $\gamma(r, L = 3d)$ can be reproduced from one and the same scattering pattern. In principle, many L-values, e.g., $0 < r_{min} < L < r_{max} < \infty$, can be inserted here. Actually, only the two lengths $L = d$ and $L = 3d$ are inherent in the model inserted. Furthermore, a sequence of second derivatives $\gamma''(r)$ results in terms of L.

The example analyzes the step from the *all-including scattering pattern* $I(s, \infty)$ to the SAS correlation function(s) $\gamma(r, L)$ for $L = d$ and $L = 3d$. The transformation Eq. (1.63) has been applied for $n = 80$ simulated data points $I_k(s_k)$. These data points are plotted in the insert of the figure. An 8% noise term is included and $d \cdot s_{max} = 20$. The numerical results – marked by the points – are compared with the theoretical curves (solid lines).

As is to be expected, the behavior of the resulting CFs in cases $L = d$ and $L = 3d$ completely differs. Two vertical lines mark two different resolution limits $r_{min}(L_{1,2})$.

Left: Inserting $L = d$ means selecting the single particle CF from $I(s, \infty)$.

Right: In the all-including case, $3 \leq L$, the particle-to-particle interference leads to a maximum in the CF $\gamma(r, L = 3d)$. This curve part with the maximum is plotted in the insert. This CF includes information about the single spheres as well.

Application for a copper(I) cyanide/platinum micropowder; $L_1 = 200$ nm

The characterization of modified CuCN micropowders (cuprous cyanide powders) for order ranges (20–200) nm is of importance in the catalysis of polymerizations and in electroplating copper and iron. A thin powder layer 0.1 mm already represents an isotropic sample with an enormous scattering effect. The scattering pattern $I(h, L_1)$ and SAS correlation function $\gamma(r, L_1)$, $L_1 = 200$ nm, are illustrated in Fig. 1.20.

This structure function describes the morphology of the particles of the powder. This has nothing in common with the "internal atomic structure" of CuCN. For investigating the latter, the h-interval must be extended to much bigger values and relatively small h values are not important. However, for electroplating copper and iron, the internal structure represents a fixed, well-known (unimportant) parameter.

A relatively small length $L = L_0$ will be briefly considered using the example of an aluminum–lanthanum alloy.

Atomic distances in an AlLa 30at% alloy from WANS experiments; $L_0 = 1$ nm

The alloy, which was melted in the laboratory of metal physics of Martin Luther University Halle-Wittenberg (Gille et al., 1996) [56], was investigated via neutron scattering experiments in the Dubna laboratories (Dutkievicz et al., 2001) [29]. The parameter $L_0 = 1$ nm was inserted for all steps of the data evaluation (see Fig. 1.21). The pair correlation function $g(r)$ (see Chapter 4) and the radial distribution function $RDF = 4\pi r^2 g(r)$ result in $0 \leq r < L_0$ from the WAS data evaluation (Hermann, 1991) [150].[‖]

In comparing these examples for the order ranges $L_0 = 1$ nm and $L = 200$ nm, it is obvious that the denotation *volume fraction c* loses its meaning for very small L. The parameter c (for details see Chapter 8) depends on L, i.e., $c = c(L)$ (see Gille et al., 2003) [94].

In the next chapters, specific order ranges belonging to the actual models are always assumed for the description of the scattering patterns. Sometimes this fact is emphasized in detail, and other times it is not emphasized. In Chapter 2, L denotes the largest particle diameter, which is the largest chord length. In summary, all considerations are limited to the particle itself, its surface and its chord length distribution density (CLDD). Particle-to-particle distances/interactions inherent in and reflected by bigger L will be analyzed at the end of Chapter 3, as well as in Chapters 4–8.

[‖] See Section SAS and WAS of Chapter 4. The function $g(r)$ is a central function in the field of WAS. Furthermore, it is closely interrelated to the sample CF $\gamma(r)$ in the field of SAS.

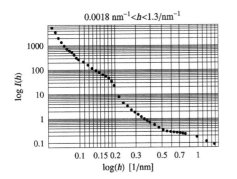

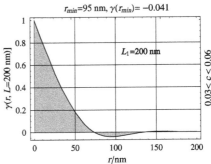

FIGURE 1.20

The isotropic scattering of the micropowder (a nearly ideal isotropic two-phase sample) was investigated for 0.0018 nm$^{-1} \leq h \leq 1.3$ nm^{-1} by use of a Kratky plant at room temperature.

After collimation correction, the resulting scattering pattern has a large dynamic (5 orders of magnitude) (see the log-log plot). The resulting CF involves a local minimum at $r = r_{min}$. Based on the relation $\gamma(r_{min}, L_1) = -c/(1-c)$, the volume fraction of the powder particles belonging to $L = L_1$ can be estimated to the value $c = 4\%$. This result nearly agrees with the value $c = 5\%$, which follows from the initial slope of the CF and the assumption of nearly spherical particles. More information about the arrangement of the particles can be found by studying the function $g(r) = \gamma''(r)/|\gamma'(0)|$ (see the following chapters). Actually, $\gamma''(r)$ is a non-negative function for $0 \leq r < 100$ nm. Let l_1 be the mean chord length of the particle phase and m_1 the one of the intermediate space between the particles [see Eqs. (1.42) and (1.43)]. This gives an approximate result of $l_1 = 50$ nm and $m_1 = 1\,000$ nm $\gg L_1$. The Rosiwal relation $l_1/m_1 = c/(1 - c)$ results in $c = 0.05$. Consequently, there should exist microparticles possessing a mean intercept length (mean chord length) of 50 nm. Therefore, for spheres starting at $l_1 = 2d/3$, a mean diameter of $d \geq 75$ nm results.

There do not exist any oscillations of the function $\gamma(r, L_1)$. This fact proves the very small order of the centers of these microparticles. This property is again related to the inserted $L = L_1$. This order range involves two typical particles and their interrelation, but does not explain the configuration inside such a particle. In order to investigate this, smaller order ranges must be inserted. If this is done, the values estimated for the volume fraction are meaningless and much bigger c values result.

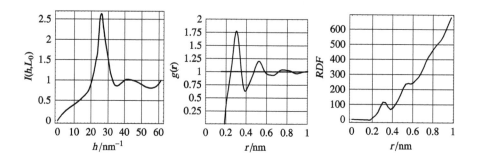

FIGURE 1.21

WANS of an amorphous AlLa 30at% alloy (see [56, 58]).

The intensity $I = I(h, L_0)$ [$n = 1300$ points] was recorded for 0.01 nm^{-1}< h < $h_{max} = 60$ nm^{-1} with a normalization of $I(h_{max}) = 1$. After correction by the factor of polarization and Compton scattering, the data was approximated by use of a Bezier-Spline function, followed by a Fourier transformation for $L = L_0 = 1$ nm. There exist about (50–60) atoms/nm^3. The functions $g(r)$ and RDF describe the (spatial) radial distribution law of the centers of the Al and La atoms. The local maximum, the first-order peak of $g(r_0)$ at $r_0 = 0.3$ nm, reflects the most typical distances, followed by the second-order peak $r_1 = 0.5$ nm. These are typical short-order peaks. The greater r is, the more the sequence of the resulting peaks approaches to the limit $g = 1$, which means it is *orderless*.

For such small L, the behavior of $I(h, L_0)$ near the origin $h = 0$ is unimportant for data interpretation/evaluation. Contrary to this, the limit h_{max} sensitively influences data evaluation. Actually, the resolution limit is $r_{min} = \pi/h_{max} = 0.05$ nm^{-1}. According to the sampling theorem of information theory, this limit means that no information about the atomic distances for $r < r_{min}$ is involved in the scattering pattern. This is in agreement with the fact that $g(r) = 0$, if $r < 0.2$ nm. The distance 0.2 nm is the shortest distance between two atom centers of the alloy.

Clearly, even smaller important distances exist inside the atoms. However, the inserted model, which is based on impenetrable atoms, differs from those models that are applicable for $L \ll L_0$. Such investigations are beyond the field of crystallography and belong to the areas of atomic physics and nuclear physics.

Analysis of a scattering pattern for $L = L_0 = 750$ nm

The scattering pattern of a blend polymer sample (polyolefin-polycarbonate) is analyzed in Fig. 1.22. Contrary to Fig. 1.21, the actual L is extremely large. The information content involved in the actual scattering pattern is very limited. Thus, it does not make any sense to vary this order range to perform data evaluations for other (especially for smaller) L_i.

Actually, $L = 750$ nm is the maximum particle diameter of the convex particles. This value was fixed from a sequence of micrographs, as well as from the investigation of the characteristic volume v_c as depending on L as given by $v_c = v_c(L)$.

No information about particle-to-particle interference is involved in this actual scattering pattern. Of course, such interferences exist for the sample material. However, for the relatively small r considered, i.e., $r < 1\,000$ nm, the actual particle ensemble is quasi-diluted (see Chapter 4).

Some remarks about the estimation of the particle volume fraction and the particle shape based on Fig. 1.22 are of interest. From the functions $\Phi(r, L)$ or $l_p \cdot \gamma''(r)$ (see the insert of Fig. 1.22) the property $\gamma''(0) > 0$ results. Under the assumption of a (very diluted) *Boolean model* (Bm) (see Chapter 5) the particle volume fraction c directly results from the structure function $\Phi(r, L)$. It holds that $c = \Phi(0, L)$. Hence, in this limiting case it follows that $c \leq 0.1$ (see the behavior of $\Phi(r)$ near the origin). The useful function $\Phi(r)$ is called the *Puzzle fitting function* (see Chapters 6 and 7). Obviously, here $\Phi(0) < 1$. The polycarbonate particles are not connected with a *Dead Leaves tessellation model* as analyzed in Chapter 7. There is no connection between the actual particle ensemble and this type of tessellation.

In general, the function $g(r) = l_p \cdot \gamma''(r)$ describes the chord lengths of a tightly packed, isotropic two-phase particle ensemble. The actual analysis is a special case. It holds $l_1 \ll m_1$. Consequently, $g(r) \approx A(r)$, i.e., the term $l_p \cdot \gamma''(r)$ is an approximation of the mean CLDD $\overline{A}(r)$ of all single particles. A detailed study of the behavior of the CLDD $\overline{A}(r) \sim \gamma''(r)$ (see the insert of Fig. 1.22) results in:

The first moment is $l_1 \approx \int_0^L r \cdot l_p \cdot \gamma''(r)dr$. On the other hand, l_1 directly results from $\gamma''(r)$ and L via $l_1 \approx \int_0^L r \cdot \gamma''(r)dr / \int_0^L \gamma''(r)dr$. The function $\overline{A}(r)$ involves a maximum at $r \approx 300$ nm. Furthermore, $\overline{A}(r) \to 0$ if 750 nm $< r < \infty$. Based on the examples of CLDD functions $A(r)$ of elementary particle shapes for μ-chords investigated in Chapter 2, this length represents the *smallest particle diameter* (for a hemisphere this is the radius, for a long cylinder this is the diameter). Hemispheric or lens-like particles are a first approximation. As $0 < \overline{A}(0+)$, the case of completely isolated ellipsoids can be excluded. This shape at most can be a second approximation. The particle ensemble includes surface singularities. All the polyolefin-polycarbonate particles are not smooth. Particle to particle overlapping or agglomeration also leads to sharp edges.

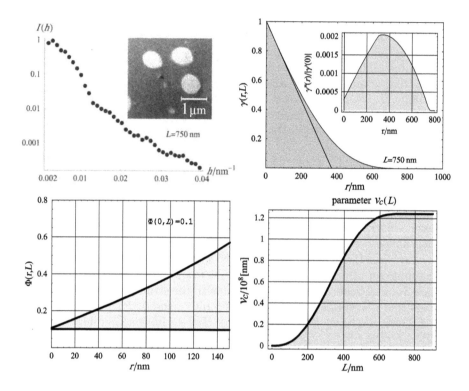

FIGURE 1.22

Analysis of a polyolefin-polycarbonate sample via SAS.

The mean particle volume is about $1.2 \cdot 10^8$ nm^3. This results from the investigation of the characteristic volume v_c in terms of L (see curve $v_c(L)$, $0 < L < 900$ nm). For Porod's length parameter, $l_p(L) = 1/|\gamma'(0+)|$, $l_p = 370$ nm results. The mean chord length l_1 (chords inside the convex particles) is about $l_1 \approx 400$ nm (see Chapter 2). In the limiting case $r \to 0$, investigation of the term $\Phi(r, L) = \gamma''(r, L)/[\gamma'(r, L)]^2$ yields $\Phi(0+, L) = 0.1$ (see Chapter 5). From this, the particle volume fraction $c = c(L)$ can be estimated to be $0.04 < c(L) < 0.1$. Hence, the mean chord length m_1 between the particles is in the order of magnitude of at least $m_1 = l_p(L)/c \approx 5\,000$ nm. This means that a scattering pattern, which should involve information about these chord lengths, must be recorded starting from $h_{min} \approx 6 \cdot 10^{-4}$ nm^{-1}.

The actual scattering pattern, in which $h_{min} = 2 \cdot 10^{-3}$ nm^{-1} holds, is far from fulfilling this condition. Hence, a real-space data evaluation must be restricted to the interval $0 \le r < 1\,500$ nm $= \pi/h_{min}$. Obviously, the inserted L parameter belongs to this interval.

1.5.1 The invariant of the smoothed scattering pattern I_L

Let the all-including scattering pattern be normalized to $I(0, \infty) = 1$. Porod's invariant of this pattern is $inv = v_c(\infty) \int_0^\infty h^2 \cdot I(h, \infty)dh = 2\pi^2$ (Glatter & Kratky, 1982) [129, 138]. The parameter L strongly influences the behavior of the L-smoothed pattern $I_s(h, L) = I_L(h, L)$ resulting from Eq. (1.64) below,

$$I_s(h, L) = \frac{2}{\pi} \int_0^\infty I(s, \infty) \cdot \left(\frac{s \sin(L(h-s))}{2h(h-s)} - \frac{s \sin(L(h+s))}{2h(h+s)} \right) ds. \quad (1.64)$$

The investigation of Porod's invariant based on Eq. (1.64) is an interesting exercise. At first sight, it is not clear whether the designation *invariant of an SAS pattern* is still correct at all in this general case. From the normalization strategy $I(0, \infty) = 1$, equation $I_s(0, L) = v_c(L)/v_c(\infty)$ follows. The invariant of the L-smoothed pattern inv_s of $I_s(h, L)$ is

$$inv_s = \int_0^\infty h^2 I_s(h, L)dh$$
$$= \int_0^\infty h^2 \left[\frac{2}{\pi} \int_0^\infty I(s, \infty) \cdot \left(\frac{s \sin(L(h-s))}{2h(h-s)} - \frac{s \sin(L(h+s))}{2h(h+s)} \right) ds \right] dh. \quad (1.65)$$

The final results, $inv_s = const$ and inv_s, are independent of L and seem to be obscure [see Eq. (1.65)]. However, after changing the sequence of integrations in Eq. (1.65) and introducing a working function $H(s, L)$,

$$inv_s = \int_0^\infty I(s, \infty) \cdot \left(\frac{2}{\pi} \int_0^\infty h^2 \left(\frac{s \sin(L(h-s))}{2h(h-s)} - \frac{s \sin(L(h+s))}{2h(h+s)} \right) dh \right) ds$$
$$inv_s = v_c(\infty) \int_0^\infty I(s, \infty) \cdot s^2 \frac{H(s,L)}{s^2} dh = \frac{2\pi^2}{v_c(L)} \quad (1.66)$$

holds true (see the explanation below).

Simplification and analysis of the term $H(s, L)/s^2$ in Eq. (1.66) yields

$$\frac{H(s,L)}{s^2} = \frac{2}{\pi} \int_0^\infty \frac{h}{s} \left(\frac{\sin(L(h-s))}{2(h-s)} - \frac{\sin(L(h+s))}{2(h+s)} \right) dh,$$
$$\frac{H(s,L)}{s^2} = \lim_{h_{max} \to \infty} \frac{2}{\pi} \cdot \int_0^{h_{max}} \frac{h}{s} \left(\frac{\sin(L(h-s))}{2(h-s)} - \frac{\sin(L(h+s))}{2(h+s)} \right) dh. \quad (1.67)$$

Here, the variable h_{max} has been introduced, where $h_{max} \to \infty$. With the antiderivative of Eq. (1.67) (last integral), the identity

$$\frac{H(s, L)}{s^2} = \lim_{h_{max} \to \infty} \left[1 - \frac{2}{\pi} \cdot \frac{\sin(Lh_{max})}{L} \cdot \frac{\sin(Ls)}{s} \right] \quad (1.68)$$

results. An intermediate check of Eq. (1.68) is the case $L \to \infty$ leading to the triviality $H(s, \infty)/s^2 = 1$. This corresponds to the *all-including scattering pattern case* $I(s, \infty) = I_s(s, \infty)$. Consequently, Eq. (1.67) simplifies to Eq. (1.68).

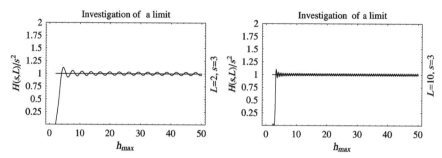

FIGURE 1.23

Properties of the parametric integral $H(s, L)/s^2$ for relatively big h_{max}.
Left: $L = 2$ and $s = 3$; Right: $L = 10$ and $s = 3$. The greater L is,
the smaller the amplitude of the oscillations around 1. In the classical case
$L \to \infty$, $H(s, \infty) = 1/s^2$ can be ascertained [see Eq. (1.68)].

Furthermore, according to Eq. (1.68), for $0 < L < \infty$ and $0 \le s < \infty$, the
limit $h_{max} \to \infty$ of the term $H(s, L)/s^2$ equals 1. Therefore, Eq. (1.67) is
written

$$\frac{H(s, L)}{s^2} = \lim_{h_{max} \to \infty} \frac{2}{\pi} \cdot \int_0^{h_{max}} \frac{h}{s} \left(\frac{\sin(L(h - s))}{2(h - s)} - \frac{\sin(L(h + s))}{2(h + s)} \right) dh \to 1.$$
(1.69)

Equation (1.69) can be ascertained by numerical integration by inserting selected parameters of L (see Fig. 1.23). Thus, Eq. (1.66) is ascertained and

$$inv_s = \int_0^\infty h^2 \cdot I_s(h, L)dh = v_c(\infty) \int_0^\infty I(s, \infty) \cdot s^2 \frac{H(s, L)}{s^2} dh,$$
$$= v_c(\infty) \int_0^\infty I(s, \infty) \cdot s^2 \cdot 1 ds = \frac{2\pi^2}{v_c(L)}$$
(1.70)

holds true. An application of Eq. (1.70) is discussed in Section 1.5.2. Table 1.3
is a summary which emphasizes the normalization. In this context, Porod's
invariant *does not depend* on L. This law can also be confirmed by numerical
integration (see Fig. 1.23).

TABLE 1.3

Porod's invariant and the relative characteristic volume.
Properties of the all-including patterns (see first line) are compared with those
of the L-smoothed patterns (see second line).

characteristic volume	normalization	invariant
$v_c(\infty) = \int_0^\infty 4\pi r^2 \gamma(r) dr$	$I(0, \infty) = 1$	$inv = \int_0^\infty h^2 I(h, \infty)dh = \frac{2\pi^2}{v_c(\infty)}$
$v_c(L) = \int_0^L 4\pi r^2 \gamma(r) dr$	$I_s(0, L) = \frac{v_c(L)}{v_c(\infty)}$	$inv_s = \int_0^\infty s^2 I_s(s, L)ds = \frac{2\pi^2}{\frac{v_c(L)}{v_c(\infty)}}$

1.5.2   How can a suitable order range L for L-smoothing be selected from an experimental scattering pattern?

The theory presented in this chapter considers the real-space analysis of an all-including scattering pattern $I(h, \infty)$. The CF and derived structure functions can be determined for any L belonging to the length interval $r_{min} < r, L < \pi/h_{min}$. This general strategy requires a great deal of CPU time, i.e., a fast computer. A sequence of many CFs should be determined.

Thus, it is a good idea to introduce an initial consideration, in order to be able to *preselect the most interesting values* for the order range parameter. The following is useful for an experimenter who is not familiar enough with a certain type of sample when starting data evaluation.

The approach proposed here is based on the analysis of the characteristic volume v_c of the sample. This parameter depends on L, i.e., $v_c(L) = \int_0^L 4\pi r^2 \gamma(r, L) dr$ (see Table 1.3). Starting with the all-including normalized intensity $I(0, \infty) = 1$, the *relative characteristic volume*

$$v_{c,rel}(L) = \frac{\int_0^L 4\pi r^2 \gamma(r, L) dr}{\int_0^\infty 4\pi r^2 \gamma(r, \infty) dr}, \quad r_{min} \leq L < r_{max} \quad (1.71)$$

is a function of L. Since $v_{c,rel}(L) = I_s(0, L)$ (see Table 1.3), the analysis can be traced back to the L-smoothed scattering pattern $I_L(s, L) = I_s(s, L)$.

Based on the transformation from $I(s, \infty)$ to $I_s(s, L)$, [see Eq.(1.51) and Eq.(1.52)], the analysis of the limit $I_s(0+, L)$ leads to the transformation

$$v_{c,rel}(L) = I_s(0+, L) = \lim_{s \to 0} I_s(s, L) = \frac{2}{\pi} \int_0^\infty I(s, \infty) \left[\frac{\sin(Ls)}{s} - L\cos(Ls) \right] ds.$$
$$(1.72)$$

The terms $I(s, \infty) = 1$ and $I_s(0+, L) = v_{c,rel}(L)$ are without a physical unit. By use of Eq. (1.72), the relative characteristic volume can be studied in terms of L (see Fig. 1.24).

Equation (1.72) does not involve terms like $s \cdot I(s, \infty)$ or $s^2 \cdot I(s, \infty)$. Hence, a given (limited) upper integration limit $s = s_{max}$ can be inserted. The resulting truncation errors will remain small. Eq. (1.72) is a modified *frequency function*. A plot $v_{c,rel}(L)$ vs. L involves characteristic features, which are useful for detecting specific L values.

Test and application of the approach

Figure 1.24 illustrates the method for a simulated scattering pattern of an aggregate of four spheres. The intensity $I(s, \infty)$ of the simulation involves a 10% noise term.

Another example for the pre-estimation of L is illustrated in Fig. 1.25. The corresponding SAS CF of this scattering pattern has already been investigated (see Fig. 1.20). Actually, the pre-estimated $L \approx 200$ nm is ascertained.

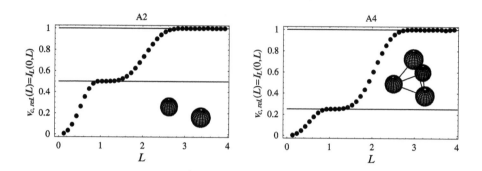

FIGURE 1.24

Relative characteristic volume $v_{c,rel}(L)$ in cases A2 (two spheres) and A4 (tetrahedral configuration): Two order ranges $L_1 = 1$ and $L_2 = 3$ are detected via $v_{c,rel}(L)$ [see Eq. (1.72)]. Actually, L_1 is the sphere diameter d, and $L_2 = s + d$ equals the largest diameter of the particle complex (see Fig. 1.12). An even larger L does not change the horizontal behavior, i.e., $v_{c,rel}(\infty) \to 1$. This corresponds to the normalization $I(0, \infty) = 1$.

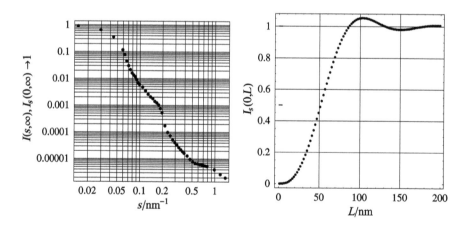

FIGURE 1.25

Scattering pattern $I(s, \infty)$ of a CuPc micropowder and the corresponding function $I_s(0, L)$ [see Eqs. (1.71) and (1.72)]. From the behavior of the *relative characteristic volume*, a typical value of $L = 200$ nm (or greater) can be estimated. This length was used in Fig. 1.20.

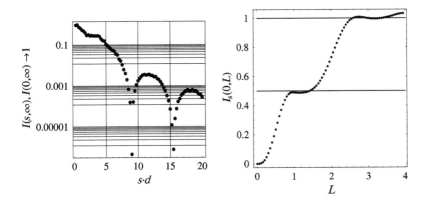

FIGURE 1.26
Looking for suitable values of the parameter L based on $v_{c,rel}(L)$.
Left: Simulation of an all-including scattering curve $I(s, \infty)$ with $I(0, \infty) = 1$
(two-spheres, diameter $d = L_1$, the centers are separated by $s = 2d$).
Right: The relative characteristic volume $v_{c,rel}(L) = I_s(0, L)$ [see Eq. (1.71)]
as depending on L from $I(s, \infty)$ [see Eq. (1.72)]. The two order ranges $L_1 = 1$
and $L_2 = 3$ can be detected. Actually, L_1 is the sphere diameter and $L_2 = s+d$
equals the largest diameter of "the particle ensemble." Inserting still larger L
does not change the horizontal behavior, i.e., $v_{c,rel}(\infty) = 1$.

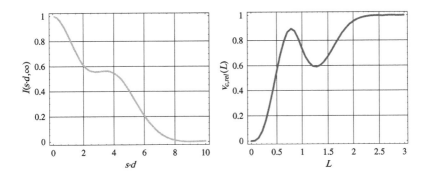

FIGURE 1.27
Linear plots of the all-including scattering pattern $I(s, \infty)$ of a DLm, constructed from spherical grains of *const.* diameter $d = 1$ and the corresponding
relative characteristic volume as depending on L given as $0 \leq L \leq 3d$.
The model can be considered to involve the two order ranges $L = L_0 = d$
and $L = L_1 = 3d$, which cannot be completely separated. There is a point of
inflection of the function $v_{c,rel}(L)$ at $L = 1$, which reflects the sphere diameter. An intermediate region $d < L < 3d$ connects L_0 with L_1. Both functions
involve information about the particle volume fraction $c = c(L)$.

The pre-estimation of the parameter L via Eq. (1.72) is too rough an approximation for detecting an exact L, e.g., $L = L_1 = (197 \pm 1)$ nm. In most samples, there exists a certain size distribution of the particles. This fact alone shows that a pre-estimation of a precise L is an illusion. In contrast to this, it is a simpler task to fix a lower limit L_0 for L as given by $L_0 < L < \infty$. From Fig. 1.25, $L_0 = 150$ nm is a secure statement.

Another example for the pre-estimation of L is explained in Chapter 8 (see Fig. 8.5), in which the scattering pattern of VYCOR glass (33% porosity) (see [86, 87]), is analyzed. This all is connected with a well-tested approach for estimating the porosity – the *linear simulation model* (see [87] and Chapter 8).

Figure 1.27 involves the very last example for the pre-estimation of L. It concerns the scattering pattern of a Dead Leaves model (DLm), which is explained in Chapter 6. The all-including scattering pattern $I(s, \infty)$ involves a shoulder stretching in $2 < s \cdot d < 4$. The SAS CF for $L = 3d$ is plotted in Chapter 8 (see Fig. 8.13). This case deals with the influence of the order range on the volume fraction c as given by $c = c(L)$. The DLm considered is not completely quasi-diluted. It involves a CF which possesses a (mainly) negative part. This explains the behavior of the function $v_{c,rel}(L)$ in this case. This example shows that the relative characteristic volume *need not be a strictly monotonously increasing function*.

The point of Chapter 1 can be summarized in two sentences: There does not exist *the one sample CF* $\gamma(r)$. However, as depending on L, there exist CFs $\gamma(r, L)$. For a well-defined L, an inversion of the integral Eq. (1.47) is entirely possible. The parametric integrals explained can be applied to handling *"many useful things"*: In this context, relevant structure models for specified lengths L_i are required (see Chapters 2–9).

CLDDs of geometric figures possessing a maximum diameter L_0 are analyzed in Chapter 2. Infinitely long cylinders, which are discussed in Chapter 3, involve the property $L_0 \to \infty$. However, characteristic diameters of such figures allow interpretations like $L_1 < L_2 < ... < \infty$. Different order ranges must be distinguished when discussing particle-to-particle interference (Chapter 4). Models of overlapping grains (Boolean models) represent an exceptional case, which is explained as follows: The disorder is so complete that the largest diameter of the largest grain fixes L of the whole ensemble of particles (Chapter 5). Chapters 6 and 7 analyze ensembles composed of very special particle shapes possessing limited L_i. An analysis of puzzle fragments always requires a fixed $L = L_0$ (see the introductory part of Chapter 7). In the limiting case $L \to 0$, even the most perfect fragments do not fit together. The particle volume fraction depends on L as given in $c = c(L)$, i.e., the parameter L belongs to the linear simulation model (Chapter 8). In addition, the formulas connecting both the chord length distributions in a two-phase particle ensemble (see Chapter 9) strictly depend on the order range inserted. If an inserted $L = L_1$ is too small, no chord lengths between two very distant objects given as $L_1 < r < L < \infty$ can be investigated. Huge objects are smoothed out inserting a small $L = L_1$.

2

Chord length distribution densities of selected elementary geometric figures

In this chapter, analytic expressions of chord length distribution densities (CLDDs) of elementary geometric figures are given, discussed and compared. Much work has been devoted to the cone case by several authors. This case is explained in detail in Section 2.1. Furthermore, the problem of automatically determining CLDDs via a computer algebra program is analyzed. The following sections involve CLDD plots calculated by *Mathematica* programs. Based on the analytic expressions, the step to the respective scattering patterns is simple.

The Pyramids of Egypt (see the Giza Pyramids in the background) belong to the largest constructions ever built. The Great Pyramid of Giza was the world's tallest building for over 2000 years, from c. 2570 BC to c. 1300 AD. Egyptian mathematicians determined the whole surface area S_0, volume V_0 and mass m_0 of this kind of pyramid. Of course, they were not able to determine the chord length distributions of their buildings.

Today, the analytic expressions of the chord length distributions of most elementary geometric bodies are known. Among the very latest results in this field is the case of a tetrahedron (Ciccariello, 2005)[19]. However, the complete analytic expression of the CLDD of a pyramid had not been known until now.

CLDDs are defined for plane and spatial geometric figures (circle, triangle, ...cube, ellipsoid etc.). The chord length r (sometimes ℓ or l is used) is the random variable. The distribution density is denoted by $f_\mu(r) = A_\mu(r) = A(r)$ (isotropic uniform random chords). The function f_μ reflects the size and shape of the figure. For a convex geometric figure, the CLDD is proportional to the second derivative of the small-angle scattering CF $\gamma(r)$. Physical apparatuses allow an automatic recording of f_μ for the purpose of partly automatic shape recognition. Scattering methods belong to the experimental techniques of materials research, which enable us to detect the CLDD of small particles. In addition, CLDDs are well-suited to describe the size of astronomic objects, e.g., via astronomic occultation experiments.

This approach exclusively reflects the "skin" of the object (particle) for a certain range order L. For that L, inside specificities are neglected and smoothed out. An estimation of size and shape of an unknown object results from comparing experimental CLDDs with theoretical models. This does not require image material (e.g., to size invisible cosmic objects in astronomy or microhomogeneities in alloys, polymers and ceramic powders). The *step* of comparing theory and experiment mainly depends on the existence of a large spectrum of model functions $f_\mu(r, a, b, c, \ldots)$ of geometric shapes described by certain parameters a, b, c, etc.

In the following sections, CLDDs of geometric figures are analyzed. After the cone case (Section 2.1), a general survey of the cases most carefully investigated is discussed. Another summarization can also be found in the textbook by Stoyan and Stoyan [212, p. 135] and related references.

2.1   The cone case–an instructive example

The functions $\gamma(r)$ and $f_\mu(r)$ of the right circular cone (height H, radius R) will be analyzed as a case in point. A simplifying substitution $v = R/H$ is introduced. Figure 2.1 exemplifies the geometry for determining the autocorrelation function (CF). Isotropic uniform random directions (IUR chords) [210] are inspected. Since the cone (like most of the examples in this chapter) belongs to the group of convex bodies, a connection between the SAS correlation function (CF) $\gamma(r)$ and the CLDD $A(r) = A_\mu(r)$ is

$$A(r) = \bar{l} \cdot \gamma''(r), \ (0 \le r \le L), \tag{2.1}$$

where L is the largest diameter of the body. Equation (2.1) has been frequently used to determine CLDDs [36, 73, 194]. The CF results in terms of the overlapping volume between the original particle and its r-translate

$$\gamma(r) = \frac{\overline{V(r, orientation)}}{V_0}, \ (0 \le r < L), \tag{2.2}$$

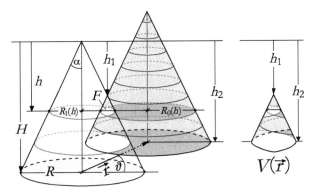

FIGURE 2.1

Analysis of the (shape) autocorrelation function of a right circular cone [63, 65].

Two cones (an original cone and its r-translate) are considered for determining the intersection volume (overlapping volume) $V(r, \vartheta, R, H)$. The symbol F denotes the intersection area of two circles of diameters $R_1 = v \cdot h$ and $R_0 = v[h + r \cdot \sin(\vartheta)]$ as depending on h, v and ϑ. The lengths $h_{1,2}$ describe the translation in the z direction.

where V_0 is the cone volume and the numerator average is the mean overlapping volume for all directions of the r-translate relative to the original, initial cone. The CF is a strictly monotonously decreasing function, which disappears for all large r values, $L < r < \infty$. Equation (2.2) is operated within the first step. The CLDD results by differentiating $\gamma(r)$.* The first CLDD moment can be traced back to V_0 and the whole surface area of the particle S_0 via $\bar{l} = l_1 = 4V_0/S_0$. Equation (2.1 and 2.2) are not approximations. Following this method, CLDDs for ellipsoids, tetrahedrons and hemispheres result [36, 52, 80].

2.1.1   Geometry of the cone case

The circular area of diameter $L = 2R$ and the rod of length $L = H$ are limiting cases involved in these studies. The cones have the surface area S_c and volume V_c and a half vertex angle α. The direction angle ϑ varies in the interval $0 \leq \vartheta \leq \pi/2$. It defines the direction of the r-translated cone (Fig. 2.1). Because of the rotational symmetry, no additional polar angle is required. The integration variable for determining the volume $V(r)$ is h

---

*There exists a sequence of overlapping cases between the original cone volume and its r-translate (see Fig. 2.1). In the following, each case is dealt with in detail as depending on r, R and H. An initial representation of the function $\gamma(r)$ results [63, 65]. Computer algebra programs are useful for an analysis of $\gamma(r)$ (see [73, 234]).

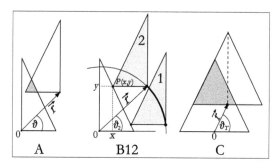

FIGURE 2.2

Analysis of overlapping cases between two cones.
A: In the basic overlapping case (a configuration as in Fig. 2.1), there exist
the limiting angles $\vartheta_{1,2}$. B12: The limiting cases of touching cones define the
limiting angles $\vartheta_{1,2}$. One or two points $P(x,y)$ may exist. C: In the case of
total overlapping, which means $\vartheta_T < \vartheta$, the analysis of $V(r)$ is trivial.

(integration limits from h_2 to h_1). The parameters h, h_1 and h_2 are the
distances from the top of the original cone up to planes parallel to the basic
plane. The radii R_0 and R_1 of the circles in a plane parallel to the basic
plane depend on h. The term $F(r, \vartheta, h)$ denotes the overlapping area of these
circles in terms of ϑ, h and r, where h is connected with all these variables.
The maximum chord length of a cone can either be $2R$ or $\sqrt{H^2 + R^2}$. Thus,
there exist two main cases for defining the functions $\gamma(r)$ and $A_\mu(r) \equiv A(r)$.
Furthermore, each of these main cases splits up into sub-cases. Altogether,
four cone types must be distinguished. Figure 2.2 includes a vivid description
of the integration limits $\vartheta_{1,2,T}$. The sums of all averaged overlapping volumes
for $r = const$ result, thereby cutting V into slates of height dh arranged
parallel to the base of the cone, which is expressed by

$$
\begin{aligned}
V_c \cdot \gamma(r) &= \int_{\vartheta(r)=\ldots}^{\vartheta(r)=\ldots} \left(\int_{h=h_1(r,\vartheta)}^{h=h_2(r,\vartheta)} F(r, \vartheta, h) dh \right) \cos(\vartheta) d\vartheta \\
&= \int_{\vartheta(r)=\ldots}^{\vartheta(r)=\ldots} V(r, \vartheta) \cos(\vartheta) d\vartheta.
\end{aligned}
\tag{2.3}
$$

The factor $\cos(\vartheta)$ controls the averaging process. In space, all positions of
the cone ghost involve the same probability. No direction is privileged [37].
The parametric integrals [Eq. (2.3)] have general integration limits, which are
not yet specified to any actual case $[r, R, H]$, via Eq. (2.3). The h integration
starts from the top toward the base. Specifications of the upper and lower
integration limits for ϑ follow based on the position of point $P(x,y)$ (Fig. 2.2,
cases B12 and total case C):

$$
\vartheta_{1,2} = \arccos \left(\frac{R}{H^2 + R^2} \cdot \left[\frac{2H^2}{r} \pm \frac{\sqrt{r^2(H^2 + R^2) - 4H^2R^2}}{r} \right] \right),
\tag{2.4}
$$

$$\vartheta_T = \arcsin\left(\frac{H}{\sqrt{R^2 + H^2}}\right) = \arcsin\left(\frac{1}{\sqrt{1 + v^2}}\right), \quad \vartheta_E = \arcsin\left(\frac{H}{r}\right). \quad (2.5)$$

The condition of real-valued $\vartheta_{1,2}$ in $0 \le r^2(H^2 + R^2) - 4H^2R^2$ is fulfilled by introducing the additional interval limit variable t in $t = 2HR/\sqrt{R^2 + H^2}$ [see Eqs. (2.19) to (2.21)]. The specifications of $h_{1,2}$ and h_{max} are written

$$h_1 = \frac{r}{2} \cdot \left[\frac{\cos(\vartheta)}{v} - \sin(\vartheta)\right], \quad h_2 = H - r \cdot \sin(\vartheta), \quad (2.6)$$

$$h_{max} = h_2 - h_1 = H - \frac{r}{2} \cdot \left[\frac{\cos(\vartheta)}{v} + \sin(\vartheta)\right], \quad (2.7)$$

(see Fig. 2.1).[†] The case $h_2 < h_1$ must be excluded. A case $h_2 = h_1$ can exist. The CF results from Eq. (2.3), which combines five integrals. The notation used, V_{symbol}, is connected with the integration limits for the direction angle ϑ given by Eqs. (2.4) and (2.5), as shown below.

$$V_{01}(r) = \int_0^{\vartheta_1} V(r, \vartheta) \cdot \cos(\vartheta) d\vartheta, \quad (2.8)$$

$$V_{2T}(r) = \int_{\vartheta_2}^{\vartheta_T} V(r, \vartheta) \cdot \cos(\vartheta) d\vartheta, \quad (2.9)$$

$$V_{0T}(r) = \int_0^{\vartheta_T} V(r, \vartheta) \cdot \cos(\vartheta) d\vartheta, \quad (2.10)$$

$$V_{TE}(r) = \int_{\vartheta_T}^{\vartheta_E} V_T(r, \vartheta) \cdot \cos(\vartheta) d\vartheta, \quad (2.11)$$

$$V_{T\pi}(r) = \int_{\vartheta_T}^{\pi/2} V_T(r, \vartheta) \cdot \cos(\vartheta) d\vartheta. \quad (2.12)$$

The terms V and V_T involved in these integrals can be traced back to a sum $F(r) = F_0(r) + F_1(r)$ [63]. These terms are

$$F_0(r) = v^2[h + r \cdot \sin(\vartheta)]^2 \cdot \arccos\left[\frac{a_0 + hb}{v[h + r \cdot \sin(\vartheta)]}\right] \\ - (a_0 + hb) \cdot \sqrt{v^2[h + r \cdot \sin(\vartheta)]^2 - (a_0 + hb)^2}, \quad (2.13)$$

$$F_1(r) = v^2 \cdot h^2 \cdot \arccos\left[\frac{a_1 - h \cdot b}{vh}\right] - (a_1 - h \cdot b) \cdot \sqrt{v^2 h^2 - (a_1 - bh)^2}. \quad (2.14)$$

Equations (2.13 and 2.14) apply the substitutions a_0, a_1 and b defined by

$$a_{0,1} = \frac{r \cdot [\cos^2(\vartheta) \pm v^2 \sin^2(\vartheta)]}{2\cos(\vartheta)}, \quad b = \frac{v^2 \cdot \sin(\vartheta)}{\cos(\vartheta)}. \quad (2.15)$$

[†]The volume of the intersection region cannot be represented for all the ϑ-values in the trivial form $V(r) = F(r, \vartheta, h_2) \cdot (h_2 - h_1)/3$, because the overlapping volume involves a non-coned surface.

The basic overlapping volumes in Eqs. (2.8)–(2.12) are obtained as shown below.

$$V(r, \vartheta) = \int_{h_1(r,\vartheta)}^{h_2(r,\vartheta)} [F_0(r, h, \vartheta) + F_1(r, h, \vartheta)]dh, \tag{2.16}$$

$$V_T(r, \vartheta) = \frac{1}{3}\pi v^2 \cdot [H - r \cdot \sin(\vartheta)]^3. \tag{2.17}$$

Equation (2.17) represents the case of a "total overlapping." Altogether, Eqs. (2.8)–(2.12), with Eqs. (2.16) and (2.17), fix the cone CF by summing up all the ϑ-intervals as depending on $r = const$. Interval splittings are useful.

Interval splitting for r

In most of the volume terms, Eqs. (2.8)–(2.12), the integration limits depend on r. It is indispensable to introduce a sequence of r-intervals for representing the overlapping integrals in Eqs. (2.8)–(2.12). The shape of the cone defined by the parameters R and H influences all the calculation steps. Four *cone types* must be distinguished, which are described below.

Type 1.1, the *low cone*, is characterized by $0 \leq H \leq R$, where $L = 2R$. The same L exists for type 1.2, where $R \leq H \leq \sqrt{3}R$, the *balanced cone*. Furthermore, type 2.1 representing the *well-balanced cone* $\sqrt{3}R \leq H \leq 2R$ and type 2.2 representing the *steep cone*, $2R \leq H < \infty$, possess the property $L = \sqrt{H^2 + R^2}$.

Inserting the special r value $t = 2R/\sqrt{1 + v^2}$ [see Eq. (2.4)], the product $V_c \cdot \gamma(r)$ results (for the four cases 1.1, 1.2 and 2.1, 2.2) by puzzling the integral terms together as follows:

$$case\ 1.1:\ V_c \cdot \gamma(r) = \begin{cases} V_{0T} + V_{T\pi}, & 0 \leq r \leq H \\ V_{0T} + V_{TE}, & H \leq r \leq \sqrt{H^2 + R^2} \\ V_{01}, & \sqrt{H^2 + R^2} \leq r \leq 2R \end{cases} \tag{2.18}$$

$$case\ 1.2:\ V_c \cdot \gamma(r) = \begin{cases} V_{0T} + V_{T\pi}, & 0 \leq r \leq H \\ V_{0T} + V_{TE}, & H \leq r \leq t \\ V_{01} + V_{2T} + V_{TE}, & t \leq r \leq \sqrt{H^2 + R^2} \\ V_{01}, & \sqrt{H^2 + R^2} \leq r \leq 2R \end{cases} \tag{2.19}$$

$$case\ 2.1:\ V_c \cdot \gamma(r) = \begin{cases} V_{0T} + V_{T\pi}, & 0 \leq r \leq t \\ V_{01} + V_{2T} + V_{T\pi}, & t \leq r \leq H \\ V_{01} + V_{2T} + V_{TE}, & H \leq r \leq 2R \\ V_{2T} + V_{TE}, & 2R \leq r \leq \sqrt{H^2 + R^2} \end{cases} \tag{2.20}$$

$$case\ 2.2:\ V_c \cdot \gamma(r) = \begin{cases} V_{0T} + V_{T\pi}, & 0 \leq r \leq t \\ V_{01} + V_{2T} + V_{T\pi}, & t \leq r \leq 2R \\ V_{2T} + V_{T\pi}, & 2R \leq r \leq H \\ V_{2T} + V_{TE}, & H \leq r \leq \sqrt{H^2 + R^2}. \end{cases} \tag{2.21}$$

Outside these r-intervals, the function $\gamma(r)$ disappears. Numerical checks with a random number generator for R and H show that these equations fulfill

$$V_c = \int_0^L 4\pi r^2 \cdot \gamma(r) dr. \qquad (2.22)$$

The final step of the project is to calculate $4V_c/S_c \cdot \gamma''(r)$ [Eq. (2.1)].

From $\gamma(r)$ to $A(r)$

Twice differentiating Eq. (2.1) can be handled numerically. Even though it takes greater effort, twice an analytical differentiation of the sum of the parametric integrals can be performed and is much more useful. The terms simplify, which is to be expected from the experiences with other geometric figures [36, 73, 3]. The differentiation of the terms $V''_{T\pi}(r)$ and $V''_{TE}(r)$ yields

$$V''_{T\pi}(r) = \frac{2\pi H v^2}{3} \cdot \left[1 - \frac{1}{(1+v^2)^{3/2}}\right] - \frac{\pi r v^2}{2} \cdot \left[1 - \frac{1}{(1+v^2)^2}\right] \qquad (2.23)$$

and

$$V''_{TE}(r) = \frac{\pi H^4 v^2}{6r^3} + \frac{\pi r v^2}{2(1+v^2)^2} - \frac{2\pi H v^2}{3(1+v^2)^{3/2}}. \qquad (2.24)$$

A detailed calculation yields parametric integrals for the three terms $V''_1(r)$, $V''_{2T}(r)$ and $V''_{0T}(r)$. Thus, the whole CLDD is fixed by a sum of at most five terms, as shown by

$$A(r) = \frac{4}{S_c}[c_{01}V''_{01}(r) + c_{2T}V''_{2T}(r) + c_{0T}V''_{0T}(r) + c_{T\pi}V''_{T\pi}(r) + \\ c_{TE}V''_{TE}(r)]. \qquad (2.25)$$

The values of the five coefficients are 0 or 1, depending on r [see Eqs. (2.18) to (2.21)]. The parametric integrals V''_{01}, V''_{2T} and V''_{0T} below result in

$$V_{01}''(r) = \int_0^{x_1(r)} I_1''(r, x) dx, \quad V_{2T}''(r) = \int_{x_2(r)}^{x_T} I_1''(r, x) dx, \\ V_{0T}''(r) = \int_0^{x_T} I_1''(r, x) dx. \qquad (2.26)$$

Equations (2.26) apply substitutions. The integration limits are defined by

$$x_T = \frac{1}{\sqrt{1+v^2}}, \quad x_{1,2} = \frac{2Hv^2 \mp \sqrt{r^2(1+v^2) - 4H^2 v^2}}{r(1+v^2)}. \qquad (2.27)$$

For I_1'' a representation

$$I_1''(r, x) = \frac{r}{2v}\left(1 - \frac{x^2}{x_T^2}\right)^{3/2} \cdot \ln\left[\frac{v[2H/r - x] + (1/x_T) \cdot \sqrt{(x_1 - x)(x_2 - x)}}{\sqrt{1 - x^2}}\right] \\ + 2v^2 x^2 \cdot (H - rx) \cdot \arccos\left[\frac{vx(rx - 2H) + (r/v) \cdot (1 - x^2)}{2(H - rx) \cdot \sqrt{1 - x^2}}\right] \\ + \frac{r \cdot x}{x_T} \cdot \sqrt{1 - \frac{x^2}{x_T^2}} \cdot \sqrt{(x_1 - x)(x_2 - x)} \qquad (2.28)$$

follows. Several synonymous versions of Eq. (2.28) are possible. However, a rigorous simplification of all these terms and expressions has not been reached.

2.1.2   Flat, balanced, well-balanced and steep cones

The CLDDs are represented by the *Mathematica* function Plot3D [234].

Since the dynamics of $A(r)$ are relatively high in this case, a complex CLDD representation requires a transformation of $A(r, R, H)$ (see Figs. 2.3 and 2.4). The transforming function $f = \tanh(x)$ possesses the property $0 \leq f < 1$ for non-negative x. For small arguments x, $f \approx x$ holds, but for larger arguments the CLDD is somewhat distorted. By use of the transformation

$$I(h) = \frac{4\pi}{V \cdot \bar{l}} \int_0^L A(r, R, H) \cdot \frac{2 - 2\cos(hr) - hr\sin(hr)}{h^4} dr, \qquad (2.29)$$

(see [36] and Eq. (1.34)), the SAS intensity $I(h) = I(h, R, H)$ results.

Analytic expression for relatively small chord lengths

Now, an analytic representation of the cone CLDD for small chord lengths is presented. A series expansion of the cone CLDD $A_\mu(r, R, H)$ for small r inside the interval $0 \leq r \leq \min\left(R, 2HR/\sqrt{H^2 + R^2}\right)$ is of interest for practical application [65].

The considerations in that paper include a computer program about the coefficients in question. The cone belongs to a certain group of shapes, for which a Taylor series of the SAS correlation function at $r = 0$ does not exist. Nevertheless,

$$A_\mu(r, R, H) = c_0 + c_1 \cdot r + c_{11} \cdot r \cdot \ln\left[\frac{r}{4R}\right] + \dots \qquad (2.30)$$

holds true. This approximation involves an absolute term, a linear term and a linear-logarithmic term. Equation (2.30) is connected with the asymptotic approximation of the scattering pattern $I(h)$, $h \to \infty$. The coefficients are defined as depending on R and H by

$$c_0 = +\frac{8\left(1 + \pi\frac{R}{H} - \frac{R}{H}\cdot\cot^{-1}\left[\frac{R}{H}\right]\right)}{3\pi H\left(\frac{R}{H} + \sqrt{1 + \frac{R^2}{H^2}}\right)}, \quad c_{11} = -\frac{3}{8R^2\sqrt{1 + \frac{R^2}{H^2}}\left(\frac{R}{H} + \sqrt{1 + \frac{R^2}{H^2}}\right)},$$

$$c_1 = -\frac{1}{2(H^2 + R^2) - 2HR\sqrt{1 + \frac{R^2}{H^2}}} + \qquad (2.31)$$

$$\frac{{}_3F_2\left(1,1,\frac{5}{2};2,4;\frac{1}{1+\frac{R^2}{H^2}}\right) - {}_2 {}_3F_2\left(1,1,\frac{3}{2};2,3;\frac{1}{1+\frac{R^2}{H^2}}\right)}{32R^2\left(1 + \frac{R^2}{H^2}\right)^{3/2}\left(\frac{R}{H} + \sqrt{1 + \frac{R^2}{H^2}}\right)}.$$

Only the coefficient c_0 defines $A_\mu(0+, R, H)$. Furthermore, Eqs. (2.30) and (2.31) are of importance for testing the concept of the *puzzle fitting function* (Gille, 2007) [108] for cone-shaped puzzle fragments. This is dealt with more fully in Chapter 7.

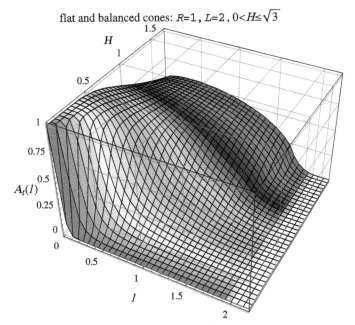

flat and balanced cones: $R=1$, $L=2$, $0<H\leq\sqrt{3}$

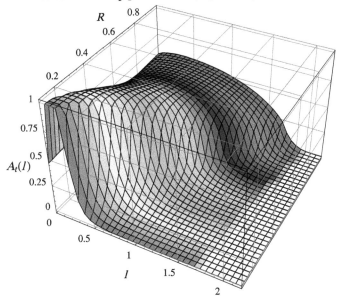

well bal. and steep 1 cones: $H=\sqrt{3}$, $H<L\leq2$, $0<R\leq1$

FIGURE 2.3

CLDD of cones: A transformation $A_t(r,R,H)=\tanh[A(r,R,H)]$ is used. Above: *Flat* and *balanced cones*, $R=1$, $L=2$, $0<H\leq\sqrt{3}$. Below: *Well-balanced* and *steep cones*, $H=const.=\sqrt{3}$, $H<L\leq2$, $0<R\leq1$.

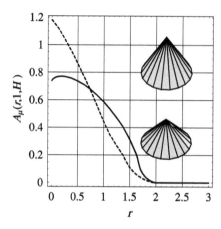
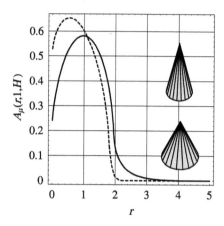

FIGURE 2.4

The function $A_\mu(r, R, H)$ of four cones with $R = 1$ and varying H.
Left: $L = 2$ from $(R = 1, H = 1)$ dashed line and $L = 2$ from $(R = 1, H = 1.5)$
solid line; right: $L = \sqrt{5}$ from $(R = 1, H = 2)$ dashed line and $L = \sqrt{17}$ from
$(R = 1, H = 4)$ solid line. In the limiting case $R/H \to 0$, $A_\mu(r) \to \delta(r)$
results.

2.1.3   Summarizing remarks about the CLDD of the cone

The cone is an elementary geometric figure. Besides [63, 65], there evidently
does not exist any paper involving a more compact analytic formula. This
CLDD is represented by a transformed 3D plot (see Fig. 2.3). The integrals
can be traced back to standard functions of mathematical physics. The expen-
diture for this is immense. This fact has been explained by Ciccariello [15, 16].
In most of the CLDD cases, it is simpler to write down the *initial integrals*.
Two reasons for these restrictions of representation are:

    1. There does not seem to be a way for avoiding r-interval splitting. The
considerations yield 4 basic cases: flat- , balanced-, well-balanced- and steep
cone. Computer experiments, which operate *without* a detailed investigation of each of these global cases, will probably fail. The computer algebra
programs that do exist are too weak for automatically detecting an interval
splitting, i.e., Eqs. (2.18)–(2.21) are indispensable.

    2. Restricting to a pure numerical analysis of the parametric integrals re-
quires effort. See the pyramidal case at the end of this chapter. In many cases,
"the big solutions" resulting from antiderivatives of relatively simple integral
representations [see Eqs. (2.18) to (2.22)] involve functions of mathematical
physics which are complicated to handle.

    It can be emphasized: Reproducible results can be obtained operating
with all versions of *Mathematica*. With big effort, parametric integrals like
[Eqs. (2.26)] can be summarized (see Filipescu et al. [37]).

2.2 Establishing and representing CLDDs

2.2.1 *Mathematica* programs for determining CLDDs?

After going through Section 2.1, the reader will probably be asking himself if computer algebra programs are powerful enough to determine simplified analytic expressions of CLDDs. This long-standing problem has been the subject of intense consideration [73, 234, 151]. The author's opinion is that computer algebra programs are not yet powerful enough to do this.

There is another question: How can a CLDD formula be written at all? It is interesting to remark that not all length parameters have to be included. This leads to a simplification in formulating a large variety of CLDDs. If possible, this will be taken into account. Obviously for the cone case, such a simplification is not feasible without restricting the universal applicability because the two independent parameters R and H exist.

Module for optimum representation of CLDDs in special cases

This question was touched upon by the author (see Section 5) in his post-doctoral thesis, "Transformation of the chord length" (Gille, 1995) [52]). In many cases, a geometric figure is fixed by use of one length parameter p, e.g., the circle by its diameter $p = d$ or the tetrahedron by an edge length $p = a$. Then, a module

$$A_\mu[r\_, p\_] := Module[\{x\}, x = r/p; f[x]/p] \qquad (2.32)$$

fixes the CLDD for arbitrary sizes of the figure. The function $f(x)$ represents the CLDD in the simple "unit length case." Inside the module, x is local. The size transformation from the "unit length case" to any other size case results from two steps, which are $x = r/p$ and $A_\mu(r, p) = f(x)/p$. Such an approach leads to simpler analytic representations. The CLDDs of two plane figures, the regular pentagon $P_5(r, a)$ and the regular hexagon $P_6(r, a)$, demonstrate this.

2.2.1.1 Regular pentagon

Let a be the edge length and $P_5(x, a)$ the CLDD. The maximum chord length is $L_5 = a \cdot (\sqrt{5} + 1)/2$. The x-intervals, where $0 \le P_5(x, a)$, are: $i_{51} = \{0 \le x \le a\}$, $i_{52} = \{a < x \le \frac{a}{2}\sqrt{5 + 2\sqrt{5}}\}$ and $i_{53} = \{\frac{a}{2}\sqrt{5 + 2\sqrt{5}} < x \le \frac{\sqrt{5}+1}{2} \cdot a\}$. The shortest distance d_5 from a corner up to the opposite side is $d_5 = \frac{a}{2}\sqrt{5 + 2\sqrt{5}} = 1.53884 \cdot a$. For $a = 1$ [see Eq. (2.32)]

$$f(x) = P_5(x, 1) = \begin{cases} i_{51} : (1 - 2\pi/5/\tan(2\pi/5))/2 \\ i_{52} : T_{52} \\ i_{53} : T_{53} \end{cases} \qquad (2.33)$$

results. The terms $T_{52}(x)$ and $T_{53}(x)$ in Eq. (2.33) were determined by the Armenian mathematicians Ohanyan & Harutyunyan (2011) [149], and are expressed as

$$T_{52}(x) = \frac{\sqrt{5+2\sqrt{5}}\,\pi - \sqrt{5}\sqrt{10+2\sqrt{5}}\,\arcsin\left[\frac{\sqrt{10+2\sqrt{5}}}{4x}\right]}{5} - \frac{\sqrt{2}\left(5+3\sqrt{5}\right)-8x^2\sqrt{3+\sqrt{5}}}{4x^2\sqrt{8x^2-\left(5+\sqrt{5}\right)}},$$

(2.34)

$$T_{53}(x) = \frac{2\sqrt{3+\sqrt{5}}}{\sqrt{8x^2-\left(5+\sqrt{5}\right)}} - \frac{2\left(2+\sqrt{5}\right)}{\sqrt{4x^2-\left(5+2\sqrt{5}\right)}} + \frac{25+11\sqrt{5}}{2x^2\sqrt{8\left(3-\sqrt{5}\right)x^2-2\left(5+\sqrt{5}\right)}}$$
$$- \frac{2\sqrt{5}}{2x^2\sqrt{\left(3\sqrt{5}-7\right)\left(5+\sqrt{5}-8x^2\right)}} - \sqrt{2+\tfrac{2}{\sqrt{5}}}\,\mathrm{arccsc}\left[\sqrt{2-\tfrac{2x}{\sqrt{5}}}\right]$$
$$+ \tfrac{\sqrt{5}}{5}\sqrt{1+\tfrac{2}{\sqrt{5}}}\left(\pi - 2\sqrt{5}\,\mathrm{arcsec}\left[2\sqrt{1-\tfrac{2x}{\sqrt{5}}}\right]\right).$$

(2.35)

Harutyunyan and Ohanyan also studied the CLDD of a regular hexagon [148].

Analytic expressions for the planar convex n-sided polygon with perimeter u and inner angles β_i, $i = 1(1)n$ are known. The CLDDs $P_n(x, \beta_i, u)$ start near the origin with a constant value of

$$P(0+, \beta_i, u) = 1/(2u) \cdot \sum_{i=1}^{n}\left[1 - (\pi - \beta_i) \cdot \beta_i\right].$$

2.2.1.2 Regular hexagon

Here, $L_6 = 2a$ results. The x-intervals, where $P_6(x, a)$ differs from 0, are $i_{61} = \{0, a\}$, $i_{62} = \{a, a\sqrt{3}\}$ and $i_{63} = \{a\sqrt{3}, 2a\}$.

The distance d_6 between two parallel sides is $d_6 = a\sqrt{3}$. For $a = 1$ [see Eq. (2.32)], the CLDD $P_6(x, a)$,

$$P_6(x) = \begin{cases} i_{61} : (1 - 2\pi/6/\tan(2\pi/6))/2 \\ i_{62} : \frac{\sqrt{4x^2-3}}{2x^2} + \frac{2}{\sqrt{3}}\left(\arccos\left[\frac{\sqrt{3}}{2x}\right] - \frac{\pi}{4}\right) \\ i_{63} : \frac{6-x^2}{x^2\sqrt{x^2-3}} - \frac{1}{2} + \frac{1}{\sqrt{3}}\left(\frac{\pi}{6} - \arccos\left[\frac{\sqrt{3}}{x}\right]\right) \end{cases}$$

(2.36)

results. From Eq. (2.36), the first moment $\bar{l}_{1,6} = \pi S/u = \pi a\sqrt{3}/4$ results and $P_6(x, a)$ possesses a pole at $x = \sqrt{3} \cdot a$.

2.3 Parallelepiped and limiting cases

The CLDDs of parallelepipeds with edges a, b, c and the SAS CF $\gamma(r, a, b, c)$ are known for all limiting cases possible [60]. By use of the abbreviations $L = \sqrt{a^2 + b^2 + c^2}$, $S = 2(ab + cb + ac)$, $f(t, r) = (2r^2 + t^2)\sqrt{r^2 - t^2}\big/3$, $s_a = \sqrt{b^2 + c^2}$, $s_b = \sqrt{a^2 + c^2}$ and $s_c = \sqrt{a^2 + b^2}$, eight r-intervals, where the CLDD differs from zero, $i_1, i_2, \ldots i_8$, can be introduced: $i_1 = \{0 < r < a\}$, $i_2 = \{a < r < b\}$, $i_3 = \{b < r < (s_c + c - |s_c - c|)/2\}$, $i_4 = \{s_c \le r < c\}$, $i_5 = \{c < r \le s_c\}$, $i_6 = \{(c + s_c + |c - s_c|)/2 < r \le s_b\}$, $i_7 = \{s_b \le r \le s_a\}$ and $i_8 = \{s_a \le r \le L\}$. Altogether, the CLDD $A_\mu(r) \equiv A_\mu(r, a, b, c)$ can be formulated by use of the following terms T_i as

$$A_\mu(r) = \frac{8}{\pi S}\{i_1 : T_1,\ i_2 : T_2,\ i_3 : T_3,\ i_4 : T_4,\ i_5 : T_5,\ i_6 : T_6,\ i_7 : T_7,\ i_8 : T_8\ .$$

(2.37)

The terms T_i are defined by $T_1 = 2(a + b + c)/3 - 3r/4$, $T_2 = 2(b + c)/3 + \left[\pi bca^2/2 - a^4/12 - (b + c)f(a, r)\right]/r^3$,

$$T_3 = \frac{2c}{3} + \frac{3r}{4} + \frac{\pi abc(a + b)/2 - (a^4 + b^4)/12 - (b + c)f(a, r) - (a + c)f(b, r)}{r^3},$$

$$T_4 = \frac{2c}{3} + \frac{\pi abc(a+b)/2 - a^2 b^2/2 - c(f(a,r) + f(b,r) - f(s_c,r))}{r^3} - \frac{abc\left(b\tan^{-1}\left[\frac{\sqrt{r^2 - s_c^2}}{a}\right] + a\tan^{-1}\left[\frac{\sqrt{r^2 - s_c^2}}{b}\right]\right)}{r^3},$$

$$T_5 = \frac{3r}{2} + \frac{\pi abc(a+b+c)/2 - (a^4 + b^4 + c^4)/12 - (b+c)f(a,r) - (a+c)f(b,r) - (a+b)f(c,r)}{r^3},$$

$$T_6 = \frac{3r}{4} + \frac{\pi abc(a+b+c)/2 - c^4/12 - (f(a,r) + f(b,r) - f(s_c,r))c - (a+b)f(c,r) - a^2 b^2/2}{r^3} - \frac{abc}{r^3}\left(b\tan^{-1}\left[\frac{\sqrt{r^2 - s_c^2}}{a}\right] + a\tan^{-1}\left[\frac{\sqrt{r^2 - s_c^2}}{b}\right]\right),$$

$$T_7 = \frac{a^4/12 + \pi ab^2 c/2 - a^2(b^2 + c^2)/2 + c(f(s_c,r) - f(b,r)) + b(f(s_b,r) - f(c,r))}{r^3}$$
$$+ \frac{abc}{r^3}\left[c\tan^{-1}\left[\frac{a}{\sqrt{r^2 - s_b^2}}\right] - b\tan^{-1}\left[\frac{\sqrt{r^2 - s_c^2}}{a}\right] + a\left(\tan^{-1}\left[\frac{c}{\sqrt{r^2 - s_b^2}}\right] - \tan^{-1}\left[\frac{\sqrt{r^2 - s_c^2}}{b}\right]\right)\right],$$

$$T_8 = \frac{abc}{r^3}\left[b\left(\tan^{-1}\left[\sqrt{r^2 - s_a^2}, c\right] - \tan^{-1}\left[a, \sqrt{r^2 - s_c^2}\right]\right)\right.$$
$$+ c\left(\tan^{-1}\left[\sqrt{r^2 - s_b^2}, a\right] - \tan^{-1}\left[b, \sqrt{r^2 - s_a^2}\right]\right)$$
$$\left. + a\left(\tan^{-1}\left[\sqrt{r^2 - s_b^2}, c\right] - \tan^{-1}\left[b, \sqrt{r^2 - s_c^2}\right]\right)\right]$$
$$+ \frac{(a^4 + b^4 + c^4)/12 - (a^2 b^2 + c^2 b^2 + a^2 c^2)/2 + af(s_a,r) + bf(s_b,r) + cf(s_c,r)}{r^3} - \frac{3r}{4}.$$

The *Mathematica* function $\arctan[u/v] \equiv \tan^{-1}[v, u]$ is applied in T_8 (see [234]). Based on Eq. (2.37), typical cases are illustrated in Fig. 2.5.

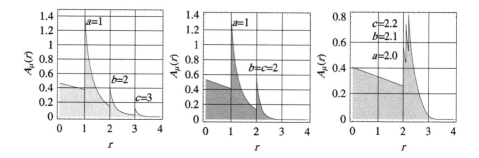

FIGURE 2.5

CLDDs $A(r) \equiv A_\mu(r)$ of parallelepipeds significantly reflect the edge lengths of the body by finite jumps at $r = a$, $r = b$ and $r = c$.

The jump sizes [at a : $4bc/(aS)$, at b : $4ac/(bS)$ and at c : $4ab/(cS)$] are additive [50].

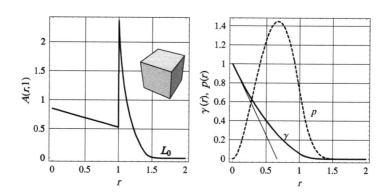

FIGURE 2.6

Comparison of the distribution densities $A(r,a) \equiv A_\mu(r,a)$ and $p(r,a) = 4\pi r^2 \cdot \gamma(r,a)/a^3$ [right: dashed line] for the unit cube $a = 1$, $L_0 = \sqrt{3} \cdot a$. Both the distribution densities are connected with the scattering pattern I. The moments of $A_\mu(r,a)$ and $p(r,a)$ are interrelated [2, 50, 3, 43]. The CF $\gamma(r,a)$ and its tangent in the origin are involved (see right side). The relation $\gamma'(0,a) = -S/(4V) = -3/(2a)$ holds.

2.3.1 The unit cube

For the cube $a = b = c$ holds. Equation (2.37) simplifies. The formulation of the unit cube CLDD, i.e., $a = 1$ [see Eq. (2.32)] requires three r-intervals, as given in $i_1 = \{0 \le r < 1\}$, $i_2 = \{1 \le r \le \sqrt{2}\}$ and $i_3 = \{\sqrt{2} \le r \le \sqrt{3}\}$, in which $f(r) = A_\mu(r, a = 1)$ differs from 0. The terms belonging to these intervals are

$$f(r) = \begin{cases} i_1 : (8 - 3r)/(3\pi) \\ i_2 : \dfrac{6\pi + 6r^4 - 1 - 8\sqrt{r^2 - 1}\left(1 + 2r^2\right)}{3\pi r^3} \\ i_3 : \dfrac{8\left(1 + r^2\right)\sqrt{r^2 - 2} - 5 - 3r^4}{3\pi r^3} + \dfrac{4\arctan\left[\frac{1}{\sqrt{r^2 - 2}}\right] - 4\arctan\left[\sqrt{r^2 - 2}\right]}{\pi r^3}. \end{cases} \qquad (2.38)$$

The remaining spike marks a [see Eq. (2.38) and Fig. 2.6]. It is not possible to detect the largest particle dimension L_0 from such a simple plot because chord lengths near $r = L_0-$ are seldom and $A_\mu(r, \sqrt{3}a) \to 0$.

The determination of the largest particle diameter can be a complicated matter (see Damaschun, Müller et al., 1980 [177, 180]; Gille, 2000 [66, 67]). For the first three $(n = 1, 2, 3)$ derivatives, $A_\mu^{(n)}(r, a\sqrt{3}) = 0$ results. Higher derivatives have been investigated for general particle shapes [67] and in the cuboid case [73, 95].

2.4 Right circular cylinder

This case has been investigated by many authors. The cylinder model has been frequently used for describing porous materials (nearly cylindrical pores) [78]. Circular cylinders with diameter d, height h, $L = \sqrt{h^2 + d^2}$ and mean chord length $l_1 = \bar{l}_1 = dh/(d/2 + h)$ are considered. The CLDD involves a pole at $r = d$ and a finite jump at $r = h$. There is a nonlinear behavior in the first r-interval.

In order to fix simple CLDD formulas, it is useful to distinguish the following two basic cases: The flat cylinder case 1, $0 < d < h < \infty$, and the long cylinder case 2, $0 \le h < d < \infty$. By use of the substitutions

$$T(t, r, d, h) = \frac{4[2d^2 \cos^2(t)\sin(t) + r\cos^4(t) \cdot (h - 3r\sin(t))]}{d^2 h \pi \sqrt{d^2 - r^2 \cos^2(t)}}, \qquad (2.39)$$

$$R(r, d, h) = \frac{2h}{\pi r^3}\left(\arccos\left[\sqrt{r^2 - h^2}\big/d\right] - \frac{\sqrt{r^2 - h^2}\sqrt{d^2 + h^2 - r^2}}{d^2}\right), \qquad (2.40)$$

the CLDD can be expressed by parametric integrals.

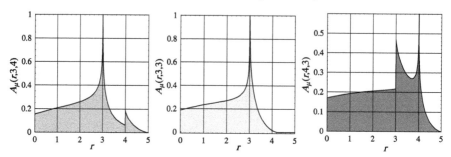

FIGURE 2.7
Three well selected CLDDs of circular cylinders: The finite jump at $r = h$ and the pole at $r = d$ are typical. For $h \to \infty$, $A_\mu(0) \to 0$ results. In this case, there is no finite jump (see [74, 75, 79]).

For case 1, the intervals $i_1 = \{0 \le r < d\}$, $i_2 = \{d < r < h\}$ and $i_3 = \{h < r \le L\}$ must be distinguished, which results in

$$A_{\mu 1}(r, d, h)) = \begin{cases} i_1 : l_1 \cdot \int_0^{\frac{\pi}{2}} T(t, r, d, h)dt \\ i_2 : l_1 \cdot \int_{\arccos(d/r)}^{\frac{\pi}{2}} T(t, r, d, h)dt \\ i_3 : l_1 \cdot \int_{\arccos(d/r)}^{\arcsin(h/r)} T(t, r, d, h)dt + l_1 \cdot R(r, d, h). \end{cases} \tag{2.41}$$

In case 2 (flat cylinders and lamellas), the intervals $i_1 = \{0 \le r < h\}$, $i_2 = \{h \le r < d\}$ and $i_3 = \{d < r \le L\}$ are used and

$$A_{\mu 2}(r, d, h)) = \begin{cases} i_1 : l_1 \cdot \int_0^{\frac{\pi}{2}} T(t, r, d, h)dt \\ i_2 : l_1 \cdot \int_0^{\arcsin(h/r)} T(t, r, d, h)dt + l_1 \cdot R(r, d, h) \\ i_3 : l_1 \cdot \int_{\arccos(d/r)}^{\arcsin(h/r)} T(t, r, d, h)dt + l_1 \cdot R(r, d, h) \end{cases} \tag{2.42}$$

is obtained. In the first r-interval, $0 < r < \min(d, h)$, $A_\mu(r, d, h)$ is a nonlinear function (see Fig. 2.7).

2.5   Ellipsoid and limiting cases

An analytic expression for the CLDD of ellipsoids was determined and published by H. Wu and P.W. Schmidt in 1968 [236, 237]. In addition, collaborative work in this area was done by D. Fanter and P.W. Schmidt. Unfortunately, the results of this cooperation were not published jointly. Fanter (1977) [34] developed computer programs which simulate random ellipsoid chords. The achievements of these authors are of current interest.

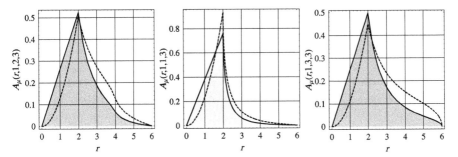

FIGURE 2.8
CLDDs of three ellipsoids with $L_0 = 6$. The semiaxes are: (left) $a = 1$, $b = 2$, $c = 3$; (center) $a = 1$, $b = 3$, $c = 3$; (right) $a = 1$, $b = 1$, $c = 3$.

The CLDD of a sphere of radius a, given by $A_\mu(r, a) = r/(2a^2)$ with $0 \leq r < 2a$, is a special case of the following: Let the ellipsoid be fixed by $x^2/a^2 + y^2/b^2 + z^2/c^2 = 1$, where $0 < a \leq b \leq c < \infty$. Thus, $L_0 = 2c$ results. A three-interval splitting, given by $i_1 = \{0 < r \leq 2a\}$, $i_2 = \{2a \leq r \leq 2b\}$ and $i_3 = \{2b \leq r \leq 2c\}$, where the CLDD differs from zero, is usual. Let $S = S(a, b, c)$ be the surface area of the ellipsoid

$$S(a, b, c) = 8ab \int_0^{\frac{\pi}{2}} \int_0^1 t \cdot \sqrt{1 + \frac{c^2 t^2}{1 - t^2} \left(\frac{\sin^2(\varphi)}{b^2} + \frac{\cos^2(\varphi)}{a^2} \right)} \, dt d\varphi.$$

See also the section entitled "Formula Gallery" in *The Mathematica book* by Stephen Wolfram [234, 95]. Operating with the abbreviations $u = b/a$ and $v = c/a$ gives

$$A_\mu(r, u, v) = \begin{cases} i_1 : \frac{4uvr}{S} \int_0^{\frac{\pi}{2}} \int_0^{\frac{\pi}{2}} P(x, y) dx dy \\ i_2 : \frac{4uvr}{S} \left[\int_0^{y_0(r)} \int_0^{x_0(r)} P(x, y) dx dy + \int_{y_0(r)}^{\frac{\pi}{2}} \int_0^{\frac{\pi}{2}} P(x, y) dx dy \right] \\ i_3 : \frac{4uvr}{S} \int_0^{\frac{\pi}{2}} \int_0^{x_0(r)} P(x, y) dx dy \end{cases}.$$

$$(2.43)$$

The parametric integrals given in Eq. (2.43) include abbreviations. The integrand $P(x, y)$ and the integration limits $x_0(r)$ and $y_0(r)$ are defined by

$$P(x, y, a, b, c) = \sin(x) \left(\sin^2(x) \cdot \left(\cos^2(y) + \frac{a^2 \sin^2(y)}{b^2} \right) + \frac{a^2 \cos^2(x)}{c^2} \right)^{3/2},$$

$$x_0(r, a, u, v, y) = \cos^{-1} \left[\sqrt{\frac{\cos^2(y) + \sin^2(y)/u^2 - 4a^2/r^2}{\cos^2(y) + \sin^2(y)/u^2 - 1/v^2}} \right] \quad and$$

$$y_0(r, a, u) = \sin^{-1} \left[\sqrt{\frac{1 - 4a^2/r^2}{1 - 1/u^2}} \right].$$

$$(2.44)$$

Three special cases are analyzed in Fig. 2.8 which are based on Eqs. (2.43) and (2.44). The μ-chords (the filled plots) and ν-chords (the dashed lines), in particular, are compared. Since the ellipsoid is a smooth body, both CLDD types disappear at $r = 0$. In the first interval, $0 \leq r < 2a$, $A_\mu(r)$ is a linear function and $A_\nu(r)$ is a parabola.

A further analysis of Eq. (2.43) leads to elaborate representations, which are much less concrete than the original relation in Eq. (2.43). Computer simulations allow CLDDs to be established in an effective way [34]. D. Fanter and the author had some useful discussions about this matter in 1980 and 1981. D. Fanter (Institute of Polymer Chemistry, Teltow-Seehof) was probably the first one to determine the ellipsoid CLDD in 1966 and 1967.

The CLDD of the parallel body to an ellipsoid

The ellipsoidal shape has now been modified. An intermediate case between the pure ellipsoid and a very big sphere is analyzed by way of an example. Generally, to an *initial set* (or body) X_0 the *parallel set* X_ρ can be constructed.

The *starting body* X_0 is smaller in volume and surface area than the final parallel body X_ρ (see Fig. 2.9). If the initial body possesses a mean caliper diameter b_0, the mean caliper diameter b_ρ of the parallel body equals $b_\rho = b_0 + \rho$. In the geometric illustration, the case $\rho = 0.5$ is considered, i.e., the sphere diameter d_s for the construction of the parallel body equals $d_s = 2\rho = 1$.

Let the set $B = (0, \rho)$ be an exact sphere of radius ρ with $0 \leq \rho < \infty$. Based on a given ellipsoid X_0, the *parallel ellipsoid* $X_\rho = X_0 \oplus B$ results (Fig. 2.9). Figure 2.10 compares the CLDDs $A_\mu(r)$ and $A_{\mu\rho}(r)$ of the sets X_0 and X_ρ. Both the CLDDs possess common properties and essential differences as well. For more about this matter, see [205, 210, 212] and Figs. 1.3, 1.4 and Fig. 3.8. The maximum particle diameter L_ρ of the parallel body X_ρ is $L_\rho = L_0 + 2\rho$.

Starting with the parallel body X_ρ, the construction of the body X_0 is not so simple. There exist restrictions for the back direction. Depending on the size and shape of the set X_ρ, not all radii ρ are possible.

In Fig. 2.10, $L_\rho = 2c + 2\rho = 6 + 6 = 12$ holds, i.e., the largest diameter of X_0 equals $L_0 = 6$ and that of X_ρ equals $L_\rho = 12$. While the function $A_\mu(r)$ disappears for $L_0 < r < \infty$, the function $A_{\mu\rho}(r)$ disappears for $L_\rho < r < \infty$. In the case X_0, the CLDD $A_\mu(r)$ is a linear function for $0 \leq r \leq 2$. In the case X_ρ, the CLDD $A_{\mu\rho}(r)$ is a linear function for $0 \leq r \leq 8$.

The first three moments of $A_{\mu\rho}(r)$ are: $M_0 = 1$, $M_1 = l_{\rho 1} = 5.863...$, $M_2 = 38.98...$, $M_3 = 278.7...$ Since the maximum particle diameters of the three sets compared differ significantly, very different scattering patterns result. The scattering pattern of a particle sensitively depends on its size and shape.

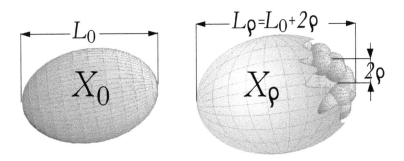

FIGURE 2.9
Construction of the *parallel set* of an ellipsoid with semiaxes $a = 1$, $b = 2$, $c = 3$. This set is *the envelope of all the outer borders of the spheres B* given as $X_\rho = X_0 \oplus B$ (see also Fig. 3.8). A small number of spheres $B = (0, \rho)$ is used for this illustration. Hence, the initial ellipsoid can still be clearly seen.

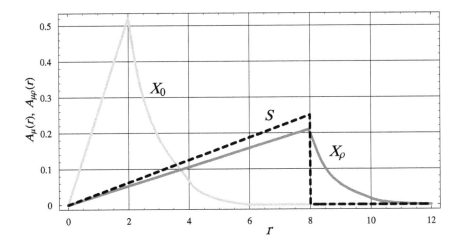

FIGURE 2.10
CLDD $A_\mu(r)$ of an ellipsoid, set X_0, with semiaxes $a = 1$, $b = 2$, $c = 3$ and CLDD $A_{\mu\rho}(r)$ of the parallel set X_ρ for $\rho = 3$ (solid lines).
Additionally, the CLDD of a sphere (S) of diameter $L_s = 8$ (dashed line) is included. The curves were calculated by use of a *Mathematica* program written by the author.

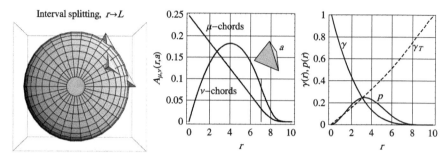

FIGURE 2.11

CLDD of a tetrahedron with edge length $a = 10$ for μ-chords and ν-chords. In the first r-interval ($0 \le r \le a/\sqrt{2}$, marked in the figure on the left), $A_\mu(r, a)$ is a linear function and $A_\nu(r, a)$ is a parabola.

The function γ_T is predestined for detecting L via $\gamma_T(L-) = 1$, which is impossible via $\gamma(r, a)$, $p(r, a)$ or $A_{\mu,\nu}(r, a)$.

The figure on the left illustrates one approach to detect the interval-splitting mechanism. A different way was chosen by Ciccariello (2005) [19], which, however, is not less complicated.

2.6   Regular tetrahedron (unit length case $a = 1$)

Much work has been devoted to investigating this matter. In the regular tetrahedron case, an averaging procedure over two independent direction angles φ and ϑ of the IUR chord direction $\vec{r}(\varphi, \vartheta)$ is required. There exist similarities between the case of a tetrahedron and a pyramid. Figure 2.14 illustrates the meaning of the direction angles for a pyramid.

Using a clever approach, for the tetrahedron an analytic CLDD expression was published by Ciccariello (2005) [19]. The case $a = 1$ is now explained without any restrictions [see Eq. (2.32)]. Three substitutions given as $d_1 = d_1(x) = \sqrt{2x^2 - 1}$, $d_2 = d_2(x) = \sqrt{3x^2 - 2}$, $d_3 = d_3(x) = \sqrt{4x^2 - 3}$ and five abbreviations $T_{1,2,\dots 5}$ given as

$$T_1(x) = \frac{12\sqrt{2}}{\pi}\left(1 + \frac{\pi - \cos^{-1}[1/3]}{2\sqrt{2}}\right) - \frac{3\left(6 + 5\sqrt{3}\pi\right)x}{2\sqrt{2}\pi},$$

$$T_2(x) = T_1(x) - 3\left(4 - 3\sqrt{2}x - \frac{\sqrt{2}}{4x^3}\right),$$

$$T_3(x) = T_2(x) - \left(12 - 9x\sqrt{3/2} - \frac{2\sqrt{2/3}}{x^3}\right),$$

$$T_4(x) =$$

$$\frac{12}{\pi}\left(\sqrt{2} - \tfrac{1}{2}\cos^{-1}[1/3] - \cos^{-1}\left[\frac{\sqrt{2/3}\cdot d_3}{d_1{}^2\cdot d_2}\right] - \cos^{-1}\left[\frac{2x^4+x^2-2}{d_1{}^2\cdot d_2{}^2}\right]\right) +$$

$$\frac{36\sqrt{2}x}{\pi}\left(\cos^{-1}\left[\frac{d_3-1}{2d_1}\right] - \frac{\pi}{2} - \frac{1}{8}\right) + \frac{\sqrt{3/2}x}{\pi}\left(2\cos^{-1}\left[\frac{3}{2x^2} - 1\right] - 9\cos^{-1}\left[\frac{d_3}{2x}\right] - \right.$$

$$16\cos^{-1}\left[\frac{\sqrt{3}(2x^2+d_3-3)}{4x^2}\right] - 9\cos^{-1}\left[\frac{9(d_3-1)x^2-9d_3+7}{4d_2{}^3}\right] + 67\pi/3\right) -$$

$$\frac{3}{\sqrt{2\pi}x}\left(5d_3 + 3\pi - 12\cos^{-1}\left[\frac{1-d_3}{2d_1}\right] + 3\cos^{-1}\left[\frac{4x^4-12x^2+7}{d_1{}^4}\right]\right) + \frac{9\sqrt{2}+16\sqrt{6}}{12x^3} -$$

$$\frac{2\sqrt{2/3}}{\pi x^3}\left(\frac{3\sqrt{3}}{8}\cos^{-1}\left[\frac{4x^4-12x^2+7}{d_1{}^4}\right] + 4\cos^{-1}\left[\frac{3\sqrt{3}(x^2-1)d_3}{2d_2{}^3}\right] - \right.$$

$$3\cos^{-1}\left[\frac{9(d_3-1)x^2-9d_3+7}{4d_2{}^3}\right]\right),$$

$$T_5(x) =$$

$$-\frac{3\sqrt{2}}{2\pi}\left(3 - \pi\left(12 - \sqrt{3}\right)\right)x - \frac{15\sqrt{2}d_3}{2\pi x} + \frac{6}{\pi}\left(\sqrt{8} - \cos^{-1}[1/3]\right) +$$

$$\frac{9}{2\pi}\left(\sqrt{6}\cos^{-1}\left[\frac{d_3(9-18x^6+46x^4-36x^2)}{2x^5d_2{}^3}\right] - 8\sqrt{2}\cos^{-1}\left[\frac{1-d_3}{2d_1}\right]\right)x -$$

$$\frac{12}{\pi}\left(\cos^{-1}\left[\frac{6x^4-x^2+(1+6x^4-9x^2)d_3-3}{\sqrt{3}d_1{}^3d_2\sqrt{1+6x^2}-d_3}\right] - \right.$$

$$\cos^{-1}\left[\frac{\sqrt{2}(15x^2+13x^4-36x^6+(2-9x^4+9x^2)d_3-2)}{d_1d_2{}^2(1+6x^2+d_3)\sqrt{1+6x^2}-d_3}\right]\right) +$$

$$\frac{\sqrt{6}}{3\pi x^3}\left(2\pi + 4\cos^{-1}\left[\frac{9x^2-7}{2d_2{}^3}\right] + \frac{3\sqrt{3}}{4}\cos^{-1}\left[\frac{12x^2-4x^4-7}{d_1{}^4}\right] - 6\cos^{-1}\left[\frac{3d_3-1}{4d_2}\right]\right)$$

are applied. After splitting the chord length x into five ranges of distances i_k, where the CLDD $f(x)$ differs from zero, given as $i_1 = \{0 < x < \sqrt{2}/2\}$, $i_2 = \{\sqrt{2}/2 < x < \sqrt{2/3}\}$, $i_3 = \{\sqrt{2/3} < x < \sqrt{3}/2\}$, $i_4 = \{\sqrt{3}/2 < x < \sqrt{7}/3\}$, $i_5 = \{\sqrt{7}/3 < x \leq 1\}$, it finally results in

$$f(x) = l_1 \cdot \{i_1 : T_1, i_2 : T_2, i_2 : T_3, i_4 : T_4, i_5 : T_5, i_6 : T_6, i_7 : T_7. \qquad (2.45)$$

See Figure 2.11. The first moment of $A_\mu(r, a)$, $\bar{l}_1 = 4V_0/S_0 = \sqrt{6}/9a$, is ascertained from Eq. (2.45) [see Eq. (2.32)]. Corresponding to the powers of the edge length a, the moments of A_μ: $M_0 = a^0$, $M_1 = 0.272166 \cdot a$, $M_2 = 0.111734 \cdot a^2$, $M_3 = 0.0553487 \cdot a^3$, $M_4 = 12V_0{}^2/(\pi S_0) \approx 0.0306294 \cdot a^4$ result.

In fact, this CLDD representation requires an elaborate interval splitting analysis, although the Platonic bodies (and regular polyhedra, as well) only involve *the* elementary shape. The reason for this interval splitting effort becomes evident by considering *only one* of those configurations resulting from the intersection *tetrahedron/sphere* (centered at a tip) (left of Fig. 2.11).

Up to now, a rigorous simplification of Eq. (2.45) has not yet been achieved. However, a simple two-interval approximation of the tetrahedron CF is known

(Gille, 2003) [91], which is

$$\gamma(r,a) = \begin{cases} 0 \le r \le \frac{a}{\sqrt{2}} : 1 - 3.67423 \cdot \frac{r}{a} + 4.52547 \cdot \frac{r^2}{a^2} - 1.86855 \cdot \frac{r^3}{a^3} \\ \frac{a}{\sqrt{2}} \le r \le a : -\frac{13093(r-a)^5}{270\sqrt{2}\pi a^5} + \frac{425983(r-a)^6}{540\sqrt{2}\pi a^6} + 0[r-a]^7 \end{cases}.$$

(2.46)

Equation (2.46) is based on the Taylor series of $\gamma(r,a)$ at $r = a-$ connected with the continuity of the first two derivatives at $r = a/\sqrt{2}$. In fact, Eq. (2.46) yields $A_\mu(r,a) = l_1 \cdot \gamma''(r, a = const.)$ with high precision. The maximum deviation of this approximation [compared with Eq. (2.46)] is smaller than 10^{-4}, which is sufficient for the interpretation of experiments. Polydisperse quasi-diluted arrangements of tetrahedrons can be analyzed based on Eq. (2.46).

Besides $A_\mu(r, a = 10)$, Fig. 2.11 involves the distance distribution density $p(r,a) = 4\pi r^2 \cdot \gamma(r,a)/V_0$ and the transformed CF, $\gamma_T(r) = 2/\pi \cdot [1 - [\gamma(r,a)]^{1/3}]$, $0 \le r \le a$. Furthermore, this figure involves the CLDD for ν-chords $A_\nu(r,a) = r/l_1 \cdot A_\mu(r,a)$. While $A_\mu(r,a)$ is strictly monotonically decreasing, $A_\nu(r,a)$ possesses a maximum. In SAS, ν-chords are relatively seldom applied. The textbook by Guinier and Fournét [143] is an exceptional case.

2.7 Hemisphere and hemisphere shell

For a hemisphere of radius R, SAS CF and CLDD were analyzed in 2000 and 2005 [66, 103]. There exist two types of chords which are cap chords [lines which do not intersect the basic plane of the hemisphere, $A_{\mu cc}(r,R)$] and basic chords [lines which intersect the basic circle, $A_{\mu bc}(r)$]. These CLDDs fix the CLDD of the hemisphere via $A_\mu(r,R) = [2 \cdot A_{\mu bc}(r,R) + A_{\mu cc}(r,R)]/3$. The cap chord distribution density was investigated in detail in 2007 [106]. This function is written for the *whole interval* $0 \le r < 2R$ as follows:

$$A_{\mu cc}(r,R) = A_{cc}(r,R) = 2r \cdot \arctan\left[\sqrt{4R^2/r^2 - 1}\right]/(\pi R^2).$$

The formulation of the CLDD of the (whole) hemisphere $A_\mu(r,1)$ requires interval splitting to be done twice, $i_1 = \{0 \le r \le 1\}$ and $i_2 = \{1 \le r \le 2\}$,

$$A_\mu(r,1) = \begin{cases} i_1 : \frac{4\pi r^4 + r(2+3r^2)\sqrt{4-r^2} - 8\arcsin[r/2]}{6\pi r^3} \\ i_2 : \frac{r(2+3r^2)\sqrt{4-r^2} + 8\arccos[r/2]}{6\pi r^3} \end{cases}.$$

(2.47)

Outside of these intervals $A_\mu \equiv 0$. There exists a significant spike at the abscissa $r = R$ (Fig. 2.12). The initial value at $r = 0+$ is $A_\mu(0+,R) = 8/(9\pi R)$. From Eq. (2.47), a mean chord length $\bar{l}_1 = 4V/S = 8R/9$ results. Higher moments of $A_\mu(r,R)$ are $M_2 = R^2$, $M_3 = 1.278 \cdot R^3$ and $M_4 = 12V^2/(\pi S) = 16R^4/9$.

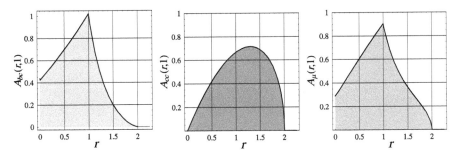

FIGURE 2.12
The CLDD of a hemisphere $A_\mu(r, R)$ (here $R = 1$; see right) can be traced back to the basic chord case $A_{bc}(r, R)$ and the cap chord case $A_{cc}(r, R)$.

2.7.1   Mean CLDD and size distribution of hemisphere shells

The *averaged* CLDD $A_m(r)$ of a quasi-diluted collection of hemispherical half shells of varying diameters $D = 2R$ with distribution density $f(D)$ (see Fig. 2.13), in terms of the function $A_{cc} = A_{\mu cc}$, is described by

$$A_m(r) = \frac{A_m'(0)}{4} \int_{D=r}^{\infty} D^2 \cdot A_{cc}(r, D) \cdot f(D) dD. \qquad (2.48)$$

The inversion of Eq. (2.48) results in

$$f(D) = \frac{1}{A_m'(0)} \int_{r=D}^{L} \frac{8A_m(r) - 5rA_m'(r) + r^2 A_m''(r)}{r^5} \cdot \frac{3D^3 - 2Dr^2}{\sqrt{r^2 - D^2}} dr. \qquad (2.49)$$

This is an explicit solution, i.e., Eq. (2.49) is formulated in terms of $A_m(r)$.

Explicit representations $f(x)$ of such particle sizing problems (random particle size parameter x, quasi-diluted ensemble) can also be easily traced back to other experimental functions – especially to the averaged CF $\gamma_m(r, x)$ and the averaged distance distribution density $p_m(r, x)$ [103, 104].

2.8   The Large Giza Pyramid as a homogeneous body

In this section, a simulation method will be explained which is well suited for determining the functions $\gamma(r)$ and $A_\mu(r)$ in a numerical way. The parameters a and h denote the edge length and the height of a square right pyramid, respectively. Existing holes inside the pyramid, which are of current interest in archeology, are neglected.

To investigate such a pyramid by a scattering experiment, a monochromatic

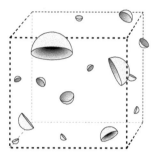

FIGURE 2.13
A quasi-diluted ensemble of spherical half shells with random diameter D. The averaged CLDD, $A_m(r)$, involves all the information to determine the size distribution density $f(D)$. The largest diameter of the largest shell, L, is an important parameter for data evaluation [see Chapter 1 and Eq. (2.49)].

radiation of a wavelength $\lambda \approx 1\ m = 10^9$ nm should be applied. However, multiplying the real size $a \approx 230$ m, $h \approx 147$ m by a factor 10^{-9} and taking the size parameters in nanometers instead of meters, such an experiment will succeed by use of neutrons or X-rays.

2.8.1 Approach for determining $\gamma(r)$ and $A_\mu(r)$

Similar to the cone (see Fig. 2.1), the CF of a square right pyramid can be developed considering specific overlapping volumes given as $V_c(r, \vartheta, \varphi, a, h)$ (see Fig. 2.14).

Based on a *Mathematica* program, V_c can be obtained by a numerical way. The (isotropized) CF is then obtained by averaging the volume V_c over two direction angles φ and ϑ with

$$\gamma(r, a, h) = 1 \Big/ \Big(\frac{1}{12}a^2 h\pi\Big) \cdot \int_0^{\pi/2} \int_0^{\pi/4} V_c(r, \vartheta, \varphi, a, h) \cdot \cos(\vartheta) d\varphi d\vartheta, \quad (2.50)$$

where $V_c(r, \vartheta, \varphi, a, h)$ is defined by the program below. A table $[r_k, \gamma(r_k, a, h)]$ results through step-by-step with arbitrary precision. After a suitable interpolation (e.g., application of cubic spline functions) of this data, the CLDD follows by numerical differentiation.

```
Vc[r_,theta_,phi_,a_,h_] :=(
(* section of the home pyramid at height z=r*Sin[theta];
   the coordinates {A(ax,ay),B(bx,by),C(cx,cy),D(dx,dy)}
   fix the corner points of the square  *)
ax=a/2 r Sin[theta]/h;      ay=a/2 r Sin[theta]/h;
bx=a - a/2 r Sin[theta]/h;  by=a/2 r Sin[theta]/h;
cx=a - a/2 r Sin[theta]/h;  cy=a - a/2 r Sin[theta]/h;
dx=   a/2 r Sin[theta]/h;   dy=a - a/2 r Sin[theta]/h;
```

```
dz=         r Sin[theta];

(* the coordinates of the ghost - basis - square:
   {At(atx,aty),Bt(btx,bty),Ct(ctx,cty),Dt(dtx,dty)} *)
atx=r Cos[theta]Cos[phi]; aty=r Cos[theta]Sin[phi];
atz=r Sin[theta];
btx=a + atx; bty=0 + aty; btz=r Sin[theta];
ctx=a + atx; cty=a + aty; ctz=r Sin[theta];
dtx=0 + atx; dty=a + aty; dtz=r Sin[theta];

(* the overlapping lengths a0 (x - direction)
                   and b0 (y -direction) *)
a0=Which[0 < ax < bx < atx < btx,
         0, 0 <= ax < atx < bx < btx, bx - atx,
         0 <= atx < ax < bx < btx, bx - ax, True, 0];
b0=Which[0 < by < cy < aty < dty, 0,
         0 <= by < aty < cy < dty, cy - aty,
         0 <= aty < by < cy < dty, cy - by, True, 0];

(* fixing the overlapping volume
   in terms of the actual parameters {r,theta,phi,a,h} *)
  h/(6*a)*(3*a0^2*b0 - a0^3) );
```

This procedure avoids complicated interval splitting for r as depending on all other variables. Selected special results for a numerical test are $V_c[0.01, \pi/20, \pi/20, 1, 1] = 0.326926$ and $\gamma(0.5, 1, 1) = 0.196$.

Such approaches are useful for many similar cases. They allow specific CLDDs to be determined for fixed size parameters. Each parameter set requires a new, separate simulation. A great deal of CPU time must be devoted to finding out relatively rough results, which yield the corresponding scattering pattern. In order to summarize the knowledge in this field, the author is preparing a lecture for students entitled *"The chord length distribution of the Giza Pyramid and other geometric figures; simulations and exact formulas."*

2.8.2 Analytic results for small chords r

In the first r-interval, i1:$\{0 \leq r \leq r_1\}$, the all-embracing Eq. (2.50) is simple to handle. Here, a compact analytic expression of the CLDD of the pyramid can be derived via $A_\mu(r) = \gamma''(r)/|\gamma'(0)|$. Since the integration limits in Eq. (2.50) do not depend on r, $A_\mu(r)$ is a linear function, which is expressed as $A_\mu(r, a, h) = A_\mu(0, a, h) + A'_\mu(0+, a, h) \cdot r$ (see Fig. 2.15). A calculation yields the terms $A_\mu(0, a, h)$ and $A'_\mu(0+, a, h)$. The latter is always non-positive. Detailed expressions for $A_\mu(0, a, h)$ and $A'_\mu(0+, a, h)$ [which fixes the Kirste-Porod contribution, $+2A'_\mu(0+)/q^2$, of the approximation $P_{1\infty}(q)$ of the scattering pattern of the pyramid, where q is the amount of the scattering vector]

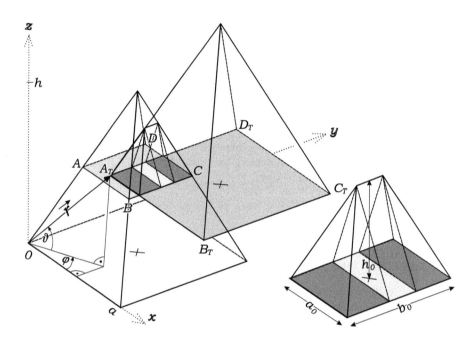

FIGURE 2.14

Model configuration of home pyramid (edge length a and height h) and its r-translate.

The base is $\{A_T, B_T, C_T, D_T\}$. A rectangular x, y, z coordinate system is used (compare with Fig. 2.1).

Home pyramid and r-translate are limited by five faces each. At a height of $h = r \cdot \sin(\vartheta)$, the r-translate intersects the home pyramid. The points $\{A, B, C, D\}$ and $\{A_T, B_T, C_T, D_T\}$ are in the same plane. The typical shape of the overlapping volume consists of two half pyramids separated by an intermediate "triangular stripe" of breadth $b_0 - a_0$. The common positive volume V_c of both the overlapping figures can be described by the lengths a_0, b_0 and h_0. It is $V_c = h/(6a) \cdot \left(3a_0{}^2 \cdot b_0 - a_0{}^3\right)$. Depending on the other variables inserted, the spectrum of possible cases is immense and cannot be described by a simple analytic formula. The fundamental configurations $a_0 = b_0$ and $a_0 < b_0$ must be distinguished.

are

$$A_\mu(0,a,h) = \frac{2a^2}{3\pi hh^2} \cdot \left(\sqrt{1+\frac{4h^2}{a^2}} - 1\right) \cdot$$

$$\left(\frac{2h}{a} - \frac{\sqrt{2}\left(\frac{h}{a} + \frac{(\pi-4)h^3}{a^3} - \frac{4h^5}{a^5}\right)}{\left(1+\frac{2h^2}{a^2}\right)^{3/2}} + \pi + \cot^{-1}\left(\frac{2h}{a}\right) - \tan^{-1}\left(\sqrt{1+\frac{a^2}{2h^2}}\right)\right)$$

and

$$A'_\mu(0+,a,h) = \frac{\frac{4}{3}\cdot h}{1+\sqrt{1+\frac{4h^2}{a^2}}} \cdot \left(\frac{9+\frac{54h^2}{a^2}-\frac{24h^4}{a^4}}{32h^3\left(1+\frac{2h^2}{a^2}\right)^2} + \frac{9-\frac{112h^2}{a^2}-\frac{48h^4}{a^4}}{8\pi a^2 h\left(1+\frac{2h^2}{a^2}\right)^2} - \frac{3\left(1+\frac{4h^2}{a^2}\right)}{2h^3\left(1+\frac{2h^2}{a^2}\right)^2} + \right.$$

$$\left. \frac{3\sqrt{2}\tan^{-1}\left[\frac{\sqrt{2}h}{a}\right]}{\pi a^3} - \frac{9\left(1+\frac{4h^2}{a^2}+\frac{8h^4}{a^4}\right)\tan^{-1}\left[\sqrt{1+\frac{4h^2}{a^2}}\right]}{8\pi h^3\sqrt{1+\frac{4h^2}{a^2}}}\right).$$

For $0 \le r < r_1$, Eq. (2.50) reduces to three overlapping terms. Thus, $\gamma(r,a,h) = \gamma_B(r,a,h) + \gamma_{VAT}(r,a,h) + \gamma_{Vt}(r,a,h)$ results.

Compared with cone and tetrahedron, the last term corresponds to the *simple total overlapping case*. Here, the overlapping volume is $V_t(r,\vartheta,a,h) = h/(3a) \cdot [a - a \cdot r \cdot \sin(\vartheta)/h]^3$. Altogether,

$$\gamma(r,a,h) = \frac{1}{\frac{1}{12}a^2h\pi} \cdot \int_0^{\vartheta_e(a,h)} \int_0^{\varphi_{ATB}(a,h,\vartheta)} V_B(r,\vartheta,\varphi,a,h)\cos(\vartheta)d\varphi d\vartheta +$$

$$\frac{1}{\frac{1}{12}a^2h\pi} \cdot \int_0^{\vartheta_e(a,h)} \int_{\varphi_{ATB}(a,h,\vartheta)}^{\frac{\pi}{4}} V_{AT}(r,\vartheta,\varphi,a,h)\cos(\vartheta)d\varphi d\vartheta +$$

$$\frac{1}{\frac{1}{12}a^2h\pi} \cdot \int_{\vartheta_e(a,h)}^{\frac{\pi}{2}} \int_0^{\frac{\pi}{4}} V_t(r,\vartheta,a,h)\cos(\vartheta)d\varphi d\vartheta$$

$$(2.51)$$

follows. The integration limits of Eq. (2.51) are $\varphi_{ATB}(a,h,\vartheta) = \arcsin[a \cdot \tan(\vartheta)/(2h)]$ and $\vartheta_e(a,h) = \arctan\left[h \cdot \sqrt{2}/a\right]$. The overlapping volumes V_B and V_{AT} are defined by Eqs. (2.52) and (2.53) below.

$$V_B(r,\vartheta,\varphi,a,h) = h \cdot (3a_0^2 b_0 - a_0^3)/(6a) \quad with$$
$$a_0 = a - \frac{a\cdot r\sin(\vartheta)}{2h} - r\cos(\vartheta)\cos(\varphi), \qquad (2.52)$$
$$b_0 = a - \frac{a\cdot r\sin(\vartheta)}{h}$$

$$V_{AT}(r,\vartheta,\varphi,a,h) = h \cdot (3a_0^2 b_0 - a_0^3)/(6a) \quad with$$
$$a_0 = a - \frac{a\cdot r\sin(\vartheta)}{2h} - r\cos(\vartheta)\cos(\varphi), \qquad (2.53)$$
$$b_0 = a - \frac{a\cdot r\sin(\vartheta)}{2h} - r\cos(\vartheta)\sin(\varphi).$$

For the limiting cases $h \to \infty$ (with $a = const$) and $a \to \infty$ (with $h = const$), $\lim_{h\to\infty} A_\mu(0,a,h) = 8/(3a\pi)$ and $\lim_{a\to\infty} A_\mu(0,a,h) = 4/(3h)$ result. This "coincides" with the square rod case.

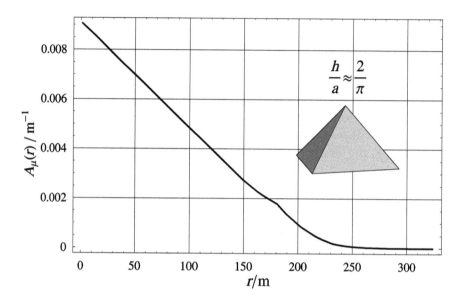

FIGURE 2.15

Chord length distribution density $A_\mu(r)$ of the *Large Giza Pyramid*, i.e., $a \approx 230$ m and $h \approx 147$ m. The original building was probably constructed with the intention to fulfill the relation $h/a = 2/\pi$ – 4500 years ago! See also www.world-mysteries.com/mpl_2.htm#Geometry/. The largest particle diameter is $L = \sqrt{2} \cdot a \approx 325$ m. The actual CLDD is a linear function in the interval $0 \le r \le r_1 = h$. The mean chord length is $\bar{l}_1 \approx 74.7$ m. The moments $\bar{l}_{2,3,4}$ of the CLDD equal $\bar{l}_2 \approx 8520$ m^2, $\bar{l}_3 \approx 1.17 \times 10^6$ m^3 and $\bar{l}_4 \approx 1.85 \times 10^8$ m^4.

In the actual case given as $0 \le h \le a/\sqrt{2}$, any analytic representation requires the analysis of the situation $0 < h < h_s < s < h_e < e < a < d = L$. Thus, seven r-intervals must be well-distinguished. In greater detail, in the actual case the order relation $0 \le h \le 1/\sqrt{\frac{1}{a^2} + \frac{1}{4h^2}} \le \sqrt{h^2 + \frac{a^2}{4}} \le 1/\sqrt{\frac{1}{a^2} + \frac{1}{4h^2 + a^2}} \le \sqrt{h^2 + \frac{a^2}{2}} \le a \le \sqrt{2} \cdot a$ is fulfilled. In other words, the shape of the *Large Giza Pyramid* belongs to the special pyramid shape type(1). This type is characterized by $0 \le h \le a/\sqrt{2}$. Furthermore, there exist five other basic shape types, which are designed as type(2)...type(6). Probably, the interval splitting analysis of all these cases will require years.

For the large Giza pyramid (see Fig. 2.15), $A_\mu(0) = 0.00905$ m^{-1}, $A'_\mu(0+) = -4.22 \times 10^{-5}$ m^{-2} results. These numbers completely describe the behavior of the CLDD in the first r-interval. There exists a certain similarity with the tetrahedron case. Thus, it can be concluded that the scattering patterns of a tetrahedron and pyramid can be very similar. The scattering pattern of a geometric figure is in direct relationship to its CLDD [see Fig. 1.7, Eq. (2.29)]. The inverse transformation, i.e., the step from $I(h)$ to $A(r)$, can be represented by Eq. (2.57). In order to demonstrate this in detail, the case of a lens will be elaborated on. In this case, the starting point is the analytic expression of the scattering intensity $I(h)$ of this body. A model case and experiments will be analyzed in Section 2.10.

2.9 Rhombic prism Y based on the plane rhombus X

Recently, a team of Armenian mathematicians derived the CLDD $B(x) = B(x, a, \alpha)$ of a rhombus X of side length a and angle α (see (Harutyunyan & Ohanyan, 2011) [149]). The density function $B(x)$ characterizes all rhombi possible (see first part of Fig. 2.16) and fixes the CLDD $A(r) = A(r, a, \alpha, H)$ of the spatial (right) rhombic prism Y of height H in $0 < H < \infty$. The functions $A(r)$ result in terms of $B(x)$ via an integral transformation where two r-intervals must be distinguished: In the first case, $0 \le r \le H < \infty$, the transformation is written

$$A(r) = \frac{\beta'(0)}{\gamma'(0)} \cdot \left[\int_0^r \left(\frac{x^3}{r^3} \frac{1}{\sqrt{r^2 - x^2}} - \frac{2r^3 + x^3}{3Hr^3} \right) \cdot B(x)dx + \frac{2}{3H} \right]. \tag{2.54}$$

In the second case, where $0 \le H \le r < \infty$, the transformation is written

$$\begin{aligned} A(r) = \frac{\beta'(0)}{\gamma'(0)} \cdot \Bigg[& \int_{\sqrt{r^2 - H^2}}^r \frac{x^3}{r^3} \left(\frac{1}{\sqrt{r^2 - x^2}} - \frac{1}{3H} \right) \cdot B(x)dx \\ & + \frac{H^2}{r^3} \int_{\sqrt{r^2 - H^2}}^{L_0} \left(\frac{x}{H} - \frac{\sqrt{r^2 - H^2}(H^2 + 2r^2)}{3H^3} \right) \cdot B(x)dx + \frac{2}{3H} \int_r^{L_0} B(x)dx \Bigg]. \end{aligned} \tag{2.55}$$

Denoting the perimeter of the plane region X by u and the surface areas of the regions X and Y by S_2 and S_3, respectively, $\beta'(0)/\gamma'(0) = 4Hu/(\pi S_3)$ follows.

The maximum diameter L_0 of the convex plane figure X fixes the integration limits. The largest diameter L of the right prism Y, i.e., the largest existing chord length r_{max}, is written $r_{max} = L = \sqrt{L_0{}^2 + H^2}$. Hence, in the second case the relation $0 \le H \le r \le L = \sqrt{L_0{}^2 + H^2}$ holds. Furthermore, Eqs. (2.54) and (2.55) include the special case of an infinitely long prism, $H \to \infty$ (see Chapter 3). Chapter 3 was introduced because the limiting case $H \to \infty$ is of practical relevance (in the field of microwires and cylinder arrays) and leads to enormous simplifications of the analytic expressions.

With regard to Eqs. (2.54) and (2.55), as early as 2002, S. Ciccariello published the general connection between the SAS CF $\beta(x)$ of a plane set X and the corresponding CF of the spatial set Y, $\gamma(r)$ (see Ciccariello, 2002, 2009 [18, 21]). Based on this, but assuming a single convex plane region X, the author derived Eqs. (2.54) and (2.55), which have many applications. This way, all CLDDs known for convex sets X (triangle, parallelogram, trapezoid, etc.) can be transformed. CLDDs of three rhombic prisms are illustrated in Fig. 2.16. The behavior of these functions is similar to the cases already discussed in Sections 2.3.1 and 2.3. Furthermore, there is a certain similarity to the case of a lens (see Section 2.10).

In this regard, this CLDD described here is a combination of most of the various investigations: The results of the Armenian scientists who worked with the *Pleijel identity* (coming from the field of stochastic geometry) were combined with general formulas for the SAS CFs by Ciccariello. Altogether, this is of practical relevance in SAS.

2.10   Scattering pattern $I(h)$ and CLDD $A(r)$ of a lens

The scattering pattern $I(h)$ of a particle fixes the CLDD $A(r)$ of the particle. This is illustrated in the case of a lens in the following (see Fig. 2.17). From Kaya's results [see Eq. (1.7) and [160, 161]], the analytic expression for $I(h)$ in terms of H and R results in

$$I(h, H, R) = \quad I(Q, H, R) = \int_0^{\frac{\pi}{2}} [T_A(Q, H, R, \theta)]^2 \cdot \sin(\theta)d\theta, \ with$$

$$T_A(Q, H, R, \theta) = 4\pi R^3 \int_{H/R}^1 \frac{\cos[Q \cdot \cos(\theta) \cdot (Rt - H)] \cdot (1 - t^2) \cdot J_1\left(QR\sin(\theta) \cdot \sqrt{1 - t^2}\right)}{QR\sin(\theta) \cdot \sqrt{1 - t^2}} dt.$$

(2.56)

Figure 2.18 shows the scattering patterns for the two cases $(R = 5, H = 4)$ and $(R = 10, H = \sqrt{91})$ (see also the size/shape illustration in Fig. 2.17). Both the parameter combinations lead to the same maximum particle diameter, $L = 6$. The inverse transformation of Eq. (1.34) goes back to $A(r) = l_1 \cdot \gamma_0''(r)$. The connection

$$A(r) = \frac{\int_0^\infty \left[\frac{(2 - h^2 r^2) h I(h)}{r^3} \cdot \sin(hr) - \frac{2h^2 I(h)}{r^2} \cdot \cos(hr) \right] dh}{|\gamma_0'(0)| \cdot \int_0^\infty h^2 I(h) dh}$$

(2.57)

results. Thus, the CLDD is traced back to $I(h)$ (see Fig. 2.19). Truncation errors are reduced by use of the *Mathematica* function SequenceLimit[] [69, 99, 111, 113, 114, 234]. The application of Wynn's algorithm [234] requires a

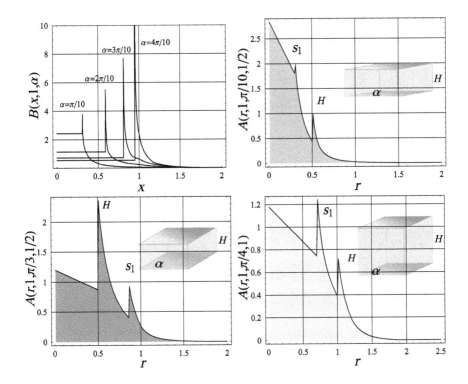

FIGURE 2.16

Investigation of the CLDDs of rhombi (see four curves on the left above) and right rhombic prisms (see three filled plots).

The CLDDs $B(x) = B(x, a, \alpha)$ of four rhombi with $a = 1$ and varying angle α, $[\alpha = k \cdot \pi/10, k = 1(1)4]$ possess a pole at $x = s_1 = a \sin(\alpha)$. There exists the one significant *parallel distance*, $s = s_1$, in the case of this plane figure X. On the other side, for the right rhombic prism of height H, there exist two significant distances, which are $s_1 = a \cdot \sin(\alpha)$ and $s_2 = H$ (two parallel interfaces). Based on this, the CLDDs $A(r) = A(r, a, \alpha, H)$ of these prisms possess two spikes (no poles, but finite jumps) at $r = s_{1,2}$. The roles of s_1 and s_2 can be interchanged. There exists the special case $s_1 = s_2$ (superimposition of both spikes for the same r). The case of the cube results if $\alpha = \pi/2$. Near the origin, $A(r)$ is a linear function, which is given as $A(r) = A(0) + r \cdot A'(0)$. The size of the spikes sensitively depends on α and H. For $H \to \infty$, the H spike more or less disappears. For $\alpha \to 0$, but with the conditions $0 < H = const.$ and $a = const.$, $A(r)$ concentrates to a very narrow region near the origin. Obviously, here chords perpendicular to the base (right section) of the cylinder, i.e., nearly parallel with the space diagonal, are relatively rare.

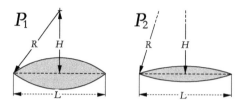

FIGURE 2.17

Illustration of the shape parameters H and R, $0 \leq H \leq R < \infty$ for a thick lens (P_1) and for a relatively thin lens (P_2). Both cases $P_{1,2}$ will be analyzed in the following.

The maximum particle diameter is $L = 2\sqrt{R^2 - H^2}$. The mean chord length of a lens l_1 is written $l_1 = l_1(R, H) = 4V_0/S_0 = 2 \cdot (R - H)(2R + H)/(3R)$. In summary, if $H \to R$, then $l_1 \to 0$. If $H \to 0$, the limiting case of a sphere results, $l_1 = 4R/3$. Here, the thickness of the lens body, $r_{max} = 2(R - H)$, equals the sphere diameter $d = 2R$.

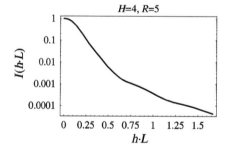

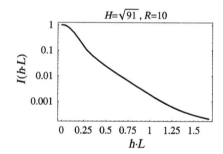

FIGURE 2.18

Normalized scattering pattern of two lenses, $I(0, H, R) = 1$.

In both cases, the largest particle diameter equals $L = 6$. The patterns involve information about the CLDD $A(r)$ of the body [see Eqs. (1.33) or (1.34) and their inverse transformations in Eq. (2.57)].

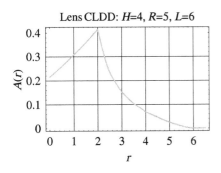
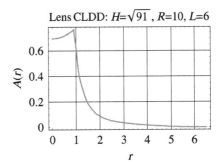

FIGURE 2.19

CLDDs of two lenses possessing the same parameters as in Figs. 2.17 and 2.18. The CLDD of a lens possesses a certain similarity to that of a hemisphere (Fig. 2.12). There is a maximum (a spike) at $r_{max} = 2(R - H)$. The order relation $0 \leq r_{max} < l_1$ holds true. In the whole interval $0 \leq r \leq L$, the function $A(r, R, H)$ has a curvature to the left. If the ratio H/R approaches 1 and $r_{max} < r < \infty$, the behavior is similar to that of a lamella of thickness $\tau = 2(R - H)$.

separation of sine and cosine terms [see Eq. (2.57)]. For this transformation, the resolution limit of the scattering pattern is an essential parameter. All the characteristic CLDD curves presented in this chapter were derived from an analytic expression which directly results from a *geometric chord length analysis*. Much farther reaching is the actual example, which makes clear that *the detailed behavior of a CLDD can be detected in terms of the pattern* $I(h)$. Certainly, this requires a high numerical effort. The CPU time for determining a complete $A(r)$ function in terms of $I(h)$ is seldom shorter than an hour.

Recognition of particle parameters: In all the cases known, significant geometric length parameters of the particle shape are closely interrelated with intrinsic details of the CLDD. Actually, from $A(r, R, H)$ the parameter $L = 2\sqrt{R^2 - H^2}$ can hardly be detected (see Fig. 2.19). However, operating with the first moment $l_1 = 1/|\gamma_0'(0)|$ and the spike position $r = r_{max}$, from

$$l_1 = \frac{2}{3}R \cdot \left(1 - \frac{H}{R}\right)\left(2 + \frac{H}{R}\right) \quad and \quad r_{max} = 2(R - H),$$

the shape parameters

$$H = \frac{r_{max}(3l_1 - 2r_{max})}{6(r_{max} - l_1)} \quad and \quad R = \frac{r_{max}^2}{6(r_{max} - l_1)}$$

definitely follow. Such approaches are not limited to the lens. The shape parameters of isotropic ensembles of many other single bodies can be detected analogously (see [118, 119]).

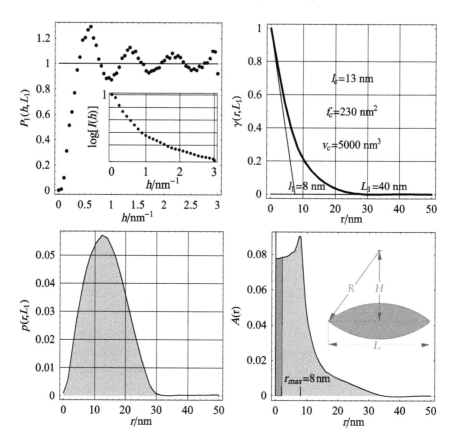

FIGURE 2.20

Investigation of lens-like precipitates in an Al-Zn(6.8 at.-%) alloy via SAS.
Heat treatment of the sample 400°C/DQ to 120°C (oil)/120°C: 120 h.
The scattering pattern is shown as an insert of the normalized Porod plot
$P_1(h)$, where the Porod asymptote equals 1. The behavior of $\gamma(r, L_1)$ confirms
the order range inserted and yields the parameters l_1, l_c, f_c and $v_c = 5\,000$
nm^3. The latter is a useful final check for the set of parameters estimated. The
(normalized) distance distribution density $p(r) = 4\pi r^2 \gamma(r)/V_0$ and the CLDD
$A(r) = l_1 \cdot \gamma''(r)$ (see the second line) yield the moments $l_1 = 8$ nm, $l_2 = 100$
nm^2. As a consequence of the resolution limit of the experiment, π/h_{max}, the
extrapolation of the $A(r)$ plot to the origin $r \to 0$ (see the part of the curve
near the origin) is an estimation. Nevertheless, the behavior of $A(r)$ clearly
indicates the lens-like shape. The maximum position $r_{max} = 8$ nm results.
Putting these SAS parameters together, the lens parameters $R = 52$ nm and
$H = 48$ nm result. The relations $H = r_{max}(3l_1 - 2r_{max})/[6(r_{max} - l_1)]$, $R =$
$r_{max}^2/[6(r_{max} - l_1)]$ are applied (see the $A(r)$ subfigure for the meaning of the
lens parameters).

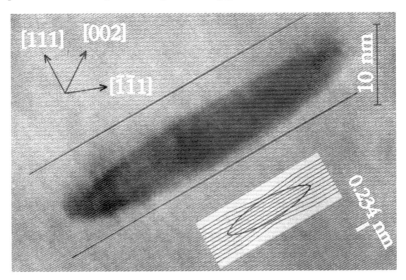

FIGURE 2.21

High resolution micrograph (HREM) of orientation $[1,\bar{1},0]$ representing a lens-like precipitate of an Al-Zn(6.8 at.-%) alloy (see Ramlau, 1985, part II, page 128, figure 6-20) [196]).

The shape of such precipitations can be approximated by lenses or ellipsoids [by a lens in terms of R and H or by an ellipsoid $x^2/a^2 + y^2/a^2 + z^2/c^2 = 1$]. From the HREM image analysis, $a \approx 20$ nm and $c \approx 4$ nm follow. Hence, this approximation yields a precipitation volume of $V \approx 6700$ nm^3.

Without any assumption about the particle shape, from the function $\gamma(r)$ a characteristic volume $v_c \approx 5\,000$ nm^3 results. This v_c coincides with the volume that results by inserting the lens shape parameters $R \approx 52$ nm and $H \approx 48$ nm (see Fig. 2.20). The lens shape approximation is the better one.

Analysis of α'_m, α'_R precipitates in Al-Zn alloys

The CLDD analysis described was applied for the shape size characterization of metastable phases in Al-Zn alloys (Figs. 2.20, 2.21) (see Gille, 1983; Ramlau 1985 [48, 196]). These authors investigated the sequence of precipitation (GP zones $\to \alpha'_R \to \alpha'_m$) in an Al-Zn(6.8 at.-%) alloy. The shape of precipitates of the phase α'_R that have a rhombohedral distorted fcc-lattice can be approximated by the lens shape (Fig. 2.17). To investigate these precipitates, scattering experiments for a maximum particle diameter of $L = L_1 = 40$ nm were performed using a Kratky plant. The only (numerical) parameter inserted is L_1 (see Chapter 1). The patterns $I(h)$ were recorded up to $h_{max} = 3$ nm^{-1} [48]. From the structure functions $\gamma(r, L_1)$, $p(r, L_1)$ and $A(r, L_1)$ (see Fig. 2.20), a sequence of structure parameters was estimated.

The parameter $r_{max} = 8$ nm together with $L_1 = 40$ nm yield the lens parameters $R = 52$ nm, $H = 48$ nm. Hence, the volume of the typical precipitate is $V = 2\pi/3(R-h)^2(2R+H) \approx 5\,000$ nm^3. This result is confirmed by the behavior of the SAS CF (see the parameters l_c, f_c, v_c). The first moment of $A(r)$ is $l_1 = 8$ nm, which also follows from $l_1 = -1/\gamma'(0)$ for L_1. One year later, these parameters were reproduced from HREM images (Ramlau & Löffler, 1984; Ramlau, 1985) [195, 196].

In summary, these examples demonstrate the applicability of the CLD concept in materials science. Furthermore, the chord length distributions discussed have importance in other fields [174, 211, 212, 66, 175, 110, 113]. The last reference opens up application possibilities in archeology.

Care must be taken with the fact that *there exist geometric figures with the same chord length probability density function.* See examples given in (Gille, 2009) [115]. Indeed, no equivalent CLDD shape seems to exist for the lens shape. These kinds of comparative investigations are more complicated than the analysis of the CLDD of a specific body.

In concluding Chapter 2, it must be mentioned that by no means are all the known cases of CLDD included here. The main intention of this chapter was to give the reader a "feeling of experience" for the behavior of CLDDs of basic geometric bodies. In most cases, interval splittings are present and cannot be avoided. Infinitely long figures of constant right section (Chapter 3 includes some applications) were studied in detail in the literature [74, 81, 88, 217].

In addition, the equivalence between chord length distribution and scattering pattern is illustrated in the very last part of this chapter. G. Porod, the great SAS expert, described this as follows (Porod, 1982) [194]:

"Thus, the chord length distribution is equally well suited to represent a particle with respect to its diffraction pattern ..."

The applications connected with chord length distributions are universal: As already briefly explained (see part A of Fig. 1.9 in Section 1.3), particle phase 1 is embedded in another phase 2. All phases, i.e., phase 2 (and probably subsequent phases 3, 4, etc.) involve a specific chord length distribution.

In a typical two-phase sample, there are at least two chord length distributions. By use of the discussed function $A_\mu(r)$, only the particle phase can be described. The following chapters include several extensions of this concept. Two different distribution functions of random chord lengths l_i or m_i, their moments, the whole set of SAS parameters and structure functions and the scattering pattern, are interrelated. Connections with the volume fractions of the phases are made accessible to the reader.

3

Chord length distributions of infinitely long cylinders

Infinitely long particles, where $L \to \infty$, do not exist in realistic samples. Nevertheless, the model considered in this chapter is a useful practical approach. A typical length $L = L_1$ must be adapted to the diameter of the region of the right section. Simple expressions of small-angle scattering structure functions are obtained which possess the essential properties of realistic samples. An isotropic uniform random (IUR) chord length analysis of long stretched particles (see [74]) is of interest for a large spectrum of materials, e.g., for controlled porous glasses [78, 92].

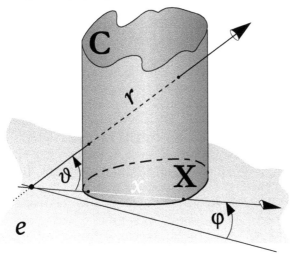

An infinitely long cylinder **C** with a constant oval convex right section **X**.
The *orthogonal right section*, i.e., the base region in $\mathbb{R}^2$ of the cylinder **C**, is denoted by **X**. It is easy to recognize that the chord length distribution density (CLDD) $B(x)$ of **X** and the CLDD $A(r) = A_\mu(r)$ of **C** are interrelated. On the one hand (transformation 1), the entire whole CLDD of **C** in $\mathbb{R}^3$ is defined in terms of the CLDD of the region **X**. In other words, the two-dimensional CLDD of the base defines the CLDD of the three-dimensional cylinder. On the other hand, the inverse transform (transformation 2) exists and is known. Both transformations are useful for analyzing the scattering pattern of relatively long cylinders and of tightly packed cylinder arrays.

3.1 The infinitely long cylinder case

By analogy with the case of flat particles, which was fully investigated by Glatter [134, 129, 136, 178] (limiting case of a layer), much is known about rods of constant right section (RS) $\mathbf{X}$. In this regard, the following three remarks are useful before analyzing the rod case in detail.

1. Layer case and rod-like particle case
A layer is described by its thickness H, i.e., by one parameter. The whole sequence of structure functions is fixed as depending on H, e.g., the CLDD $A(r, H)$ (first moment $l_1 = 2H$) and the CF $\gamma_0(r, H) = 1 - r/(2H)$ if $0 \leq r < H$ and $\gamma_0(r, H) = H/(2r)$ if $H < r < \infty$ (see Section 10.2.5). In contrast, the description of rod-like particles requires a larger parameter set. It must involve certain parameters which define the cylinder's RS $\mathbf{X}$, e.g., L_0, surface area $S_{RS} = S$, perimeter u, mean caliper diameter $\bar{d}$ [48], mean curvatures [59], and so on. For the following (besides certain special cases), let L_0, S and u be the specific parameters selected.

2. Rods with oval, convex right section $\mathbf{X} = \mathbf{X}(L_0, S, u)$
In 1987, the Lithuanian mathematician Geciauskas [44, 45, 46] investigated the plane convex oval domain $\mathbf{X}$ (free of vertices and line segments) of maximum diameter L_0 for relatively small x, where $0 \leq x \leq x_H$, and published a well-developed approximation formula for the distribution function $F(x) = F(x, u, S)$ of the distance x between two random points, which are uniformly and independently distributed in $\mathbf{X}$. For example, in the case of an ellipse, x_H is twice the minor semiaxis. For a circle, x_H is the diameter.

3. Correlation function and CLDD of X
In this regard and independent of point 2, the first and the third moment of $B_0(x)$ are $x_1 = \pi S/u$ and $x_3 = 3S^2/u$, respectively. Hence, the interrelation between the first CLDD moments of $\mathbf{X}$ and $\mathbf{C}$ is $x_1/l_1 = \pi/4$. The CF $\beta_0(x)$ and the CLDD $B_0(x)$ of the domain $\mathbf{X}$ are connected with its distance distribution $F_0(x)$, where $F_0'(x) = p_0(x) = 2\pi x \beta_0(x)/S$. Furthermore, $B_0(x) = x_1 \cdot \beta_0''(x)$. Analytic expressions of these functions in $0 \leq x < x_1$ (see [45, p. 122/428]) result from $F_0(x)$ as shown by

$$
\begin{aligned}
\beta_0(x) &= \tfrac{2}{\pi} \arccos\left(\tfrac{\pi}{4} \cdot \tfrac{x}{x_1}\right) - \tfrac{x}{x_1} \cdot \frac{\sqrt{16 - \pi^2 \cdot x^2/x_1{}^2}}{8} \\
B_0(x) &= \frac{\pi^2 x}{4x_1{}^2 \sqrt{16 - \pi^2 x^2/x_1{}^2}}.
\end{aligned}
\tag{3.1}
$$

For $u = \pi \cdot d$ and $S = \pi \cdot d^2/4$, the structure functions of a circle of diameter d, [e.g., $B(x) = x/\sqrt{d^4 - d^2 x^2}$, $x_1 = \pi d/4$] result. Equations (3.1) represent fine approximations for the ellipse in the first x-interval (Gates, 1987) [43].

The parameters and functions in points 1–3 are connected with the $\mathbb{R}^3$ structure functions of the rods $\gamma_0(r)$ and $A(r) = A_\mu(r)$ [95, 76]. This has been verified in several cases [72, 81, 88, 217]. In the following subsections, triangular, rectangular, pentagonal and hexagonal rods are explained in detail. At the end of the chapter, applications of Eqs. (3.1) are studied. This is useful for the analysis of scattering patterns of oval rods **C** with RS **X** in practice.

3.2 Transformation 1: From the right section of a cylinder to a spatial cylinder

Second order characteristics of **X** define those of **C**. First, the theoretical aspect is discussed, which is followed by examples. The analysis includes the determination of the spatial CLDD $A(r) = A_\mu(r)$ starting from $B(x)$. There exists a large spectrum of well-known functions $B(x)$ "waiting to be transformed" (see the Stoyans' textbook [211, pp. 135–140] and many of its references).

Theory of transformation 1: Let $A(r)$ be the CLDD (for IUR chords) of an infinitely long cylinder of convex, fixed RS. Denoting the first moment of $A(r)$ by l_1 gives the relation with the cylinder CF $\gamma_0(r)$ as $A(r) = l_1 \cdot \gamma_0''(r)$. Let $B(x)$ be the CLDD of the RS region (planar bounded convex domain of surface area S_{RS} and perimeter u) of the cylinder. Denoting the maximum RS chord length by L_0 interrelates the functions A and B (see Gille, 2003; Sukiasian & Gille, 2007) [81, 217], as shown below:

$$A(r) = \begin{cases} \frac{4}{\pi} \cdot \int_0^r \frac{x^3}{r^3 \cdot \sqrt{r^2 - x^2}} \cdot B(x)dx, & \text{if } 0 < r \leq L_0 \\ \frac{4}{\pi} \cdot \int_0^{L_0} \frac{x^3}{r^3 \cdot \sqrt{r^2 - x^2}} \cdot B(x)dx, & \text{if } L_0 \leq r < \infty \end{cases} \quad (3.2)$$

In the limiting case of long cylinders of height H, where $(H \to \infty)$, and volume V, where $(V \to \infty)$, the result is $l_1 = 4S_{RS}/u$. The function $A(r)$ defines the dimensionless, normalized scattering intensity $I(h)$ of the three-dimensional particle as

$$I(h) = \frac{4\pi}{V \cdot l_1} \int_0^\infty A(r) \cdot \frac{2 - 2\cos(hr) - hr\sin(hr)}{h^4} dr. \quad (3.3)$$

With this normalization of the scattering pattern $I(h)$, Equation (3.3) fulfills the invariant relation of SAS, where $inv = V \int_0^\infty h^2 I(h)dh = 2\pi^2$ (see also Section 1.2). For infinitely long cylinders, Eq. (3.3) possesses the property $I(0) \to \infty$ (Porod, 1949; Porod, 1982) [194].*

It is possible to analyze the asymptotic behavior of $I(h)$ for relatively large h to detect some geometric features of the particle surface (Wu & Schmidt,

---

*Thus, a radius of gyration $R_g = R_g(L_1)$ [143] is not defined by the initial part of the corresponding scattering pattern $I(h)$.

1973, 1974; Ciccariello, 1985, 1991; Sobry et al., 1991, 1994; Ciccariello & Sobry, 2002) [207, 14, 15, 18]. By breaking down Eq. (3.3) into two parts and normalizing the Porod contribution to unity, the (normalized) Porod plot is obtained plotting the function

$$P_1(h) \equiv \frac{h^4 \cdot I(h) \cdot V \cdot l_1}{8\pi} = 1 - \frac{1}{2}\int_0^\infty A(r) \cdot [2\cos(hr) + hr\sin(hr)]dr \quad (3.4)$$

versus h. The function $P_1(h)$ possesses the property $P_1(h) \to 1$ if $h \to \infty$ (for some examples of infinitely long particles, see [74]) and specifies the limit $A(0+)$, where $A(0+) = 2/(3\pi)\int_0^\infty[1 - P_1(h)]dh$. Equation (3.4) is better suited to analyze the scattering behavior of long cylinders compared to a simple $I(h)$ plot. There exist clear asymptotic descriptions of $P_1(h)$ for $h \to \infty$, where $P_{1\infty}(h) \approx P_1(h)$, which involve non-oscillating and oscillating terms, depending on the particle's interface constellation. For example, with Eq. (3.4), the Kirste-Porod contribution of $I(h)$, where primary $\sim 1/h^6$, results in $2 \cdot \gamma_0'''(0+)/(\gamma_0'(0+) \cdot h^2)$. Well-founded terms of $P_{1\infty}(h)$ appear for plane particle interfaces (Ciccariello, 1991; Ciccariello & Sobry 1995; Gille, 1999; Ciccariello, 2002) [14, 17, 60, 18]. In summary, parallel particle interfaces at a distance d_0 lead to the undamped oscillations $\sim \sin(h \cdot d_0)$ or $\sim \cos(h \cdot d_0)$. The latter is the case with hexagonal rods (3 pairs of parallel interfaces, separated by a dihedral angle $\alpha = 2\pi/3$). In contrast, parallel edges with distance d_1 result in the damped oscillating terms $\sim \cos(h \cdot d_1)/h^1$ or $\sim \sin(h \cdot d_1)/h^1$ (see the pentagonal rod case). These interrelations will be verified for the CLDD $B_n(x,a)$ of regular n-polygons with side length a for $n = 5, 6$ and $n = 3, 4$.

3.2.1 Pentagonal and hexagonal rods

Recently, Aharonyan and Ohanyan [1] succeeded in establishing an analytic expression for the CLDD of a regular pentagon. In 2007, Harutyunyan [148] calculated the explicit CLDD expression of a regular hexagon. Both formulas can be transformed into infinitely long regular homogeneous pentagonal and hexagonal cylinders. The CLDDs of the pentagon or hexagon are denoted by $B_5(x,a)$ and $B_6(x,a)$, respectively. In order to fix these functions, interval splittings for x in terms of a are required (see Chapter 2).[†] The CLDDs of the infinitely long cylinders $A_5(r,a)$ and $A_6(r,a)$ (see Fig. 3.1) result from transformation 1, as shown in Eq. (3.2). Both CLDDs start in the origin with the constant values $A_5(0+,a) = \frac{4}{3a\pi}[1 - \frac{2\pi(\sqrt{5}-1)}{5\cdot\sqrt{2(5+\sqrt{5})}}]$ and

$A_6(0+,a) = \frac{4}{3a\pi}[1 - \sqrt{3}\pi/9]$, where $\gamma_0'''(0+) = 0$.

[†]Due to similarities, the standard $a = 1$ is sufficient and $B(x,a) := B(x/a,1)/a$, $A(r,a) := A(r/a,1)/a$ holds. Nevertheless, typical length parameters are fixed in terms of a.

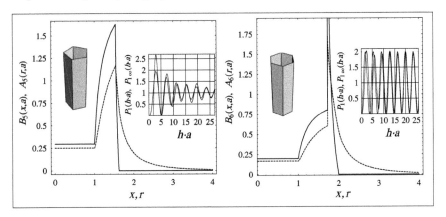

FIGURE 3.1
Left: The CLDDs $B_5(x,a)$ (regular pentagon, solid line) and $A_5(r,a)$ (infinitely long pentagonal cylinder, dashed line) for $a = 1$. The function $P_1(h \cdot a)$ (solid line) and its approximation $P_{1\infty}(h \cdot a)$ (dashed line) of the pentagonal cylinder are compared in the insert.
Right: The CLDDs $B_6(x,a)$ (regular hexagon, solid line) and $A_6(r,a)$ (infinitely long hexagonal cylinder, dashed line) for $a = 1$. The functions $P_1(h \cdot a)$ (solid line) and $P_{1\infty}(h \cdot a)$ (dashed line) of the hexagonal cylinder are compared in the insert.

Pentagonal rod: The function $A_5(r,a)$ involves the first moment $\overline{l_5} = 4S_{RS}/(5a) = \sqrt{1 + 2/\sqrt{5}}a$ and is continuous. The damped leading asymptotic term is written $P_{1\infty}(h \cdot a) = 1 - 5 \cdot \sin(h \cdot d_5(a))/h$ (see insert of Fig. 3.1). Here, $d_5(a)$ denotes the shortest distance between an edge and the opposite interface.

Hexagonal rod: The function $A_6(r,a)$ possesses a discontinuity at $r = \tan(\pi/3) \cdot a/2 = \sqrt{3} \cdot a$. The size of the finite jump at this abscissa equals $2\sqrt{3}/(3a)$. Furthermore, $A_6(\sqrt{3}+) \approx 1.76151\dots/a$. The first moment $\overline{l_6} = \sqrt{3} \cdot a$ results. The distance $d_6(a)$ fixes the undamped leading asymptotic term $P_{1\infty}(h \cdot a) = 1 - \cos(h \cdot d_6(a))$, which approximates $P_1(h \cdot a)$ (see insert of Fig. 3.1). These CLDDs and scattering patterns correspond to the line of thought first expressed by Porod and Mittelbach [193, 194]. In the sequence of cases where $n = 3, 4, 5, 6$, the CLDDs B_n and A_n approach the limiting cases of a circle and circular rod. The behavior near the origin $A_{5,6}(0+)$ fits with the following results obtained by Ciccariello and Benedetti (1982) [11]: For pentagonal and hexagonal rods, the $\gamma_0''(0+)$ contribution of one edge between two facets of the dihedral angle α results from $\gamma_0''(0+)/n = A(0+)/(nl_1) = 1/(6\pi S_{RS}) \cdot [1 - (\pi - \alpha)/\tan(\pi - \alpha)]$. In fact, for $n = 5$ and 6 infinitely long edges, the expected constant terms of $A_{5,6}(r,a)$ in the first r-intervals result. The Kirste-Porod term disappears for both kinds of rods. The differences

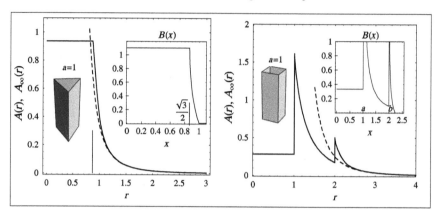

FIGURE 3.2

Left: The CLDDs $B(x)$ (triangle) and $A(r)$ (infinitely long triangular rod) for $a = 1$. The dashed line represents the behavior of the asymptotic approximation $A_\infty(r) \simeq 3a^3/(4\pi r^4) + 5a^5/(24\pi r^6) + \dots$ of the CLDD.

Right: The CLDDs of a rectangle and a rectangular rod, $B(x)$ and $A(r)$ for $a = 1, b = 2$. The dashed line represents the behavior of the asymptotic approximation $A_\infty(r) \simeq 6a^2b^2/[(a + b)\pi r^4] + 5a^2b^2(a^2 + b^2)/[3(a + b)\pi r^6] + \dots$ of the CLDD.

between A_5 and A_6 lead to distinct differences in the scattering behavior. For pentagonal rods, $P_1(h \cdot a)$ is a damped oscillating function (similar to the case of a triangular rod). However, by analogy with the rectangular rod case, the term $P_1(h \cdot a)$ of hexagonal rods involves undamped oscillations. Conflating the results for $n = 3, 4, 5$ and 6, the parameters L_0 and a can be estimated and n can be fixed by operating with the functions B, A and P_1.

3.2.2 Triangle/triangular rod and rectangle/rectangular rod

The CLDDs of these figures are illustrated in Fig. 3.2. A sequence of CLDDs of rods with several other RSs was considered by Gille [72, 74, 105].

3.2.3 Ellipse/elliptic rod and the elliptic needle

Application of transformation 1 to the CLDD $B(x, a, b)$ of an ellipse, semi-minor axis a and semi-major axis b, where $a \leq b$, yields the CLDD of the elliptical rod $A(r, a, b)$ (Fig. 3.3). An analytic expression for $B(x, a, b)$ is Eq. (3.5). A detailed analytic representation of $A(r, a, b)$ is suppressed here since the integral representation given is much more compact. This is typical of many CLDD representations. The same holds true for the integral type in Eq. (3.5) [95] (see other CLDD representations given by Piefke, 1979 [189]

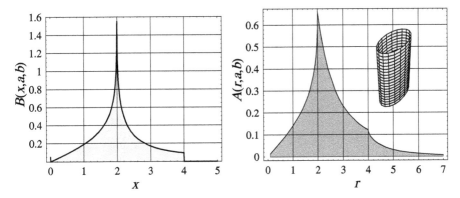

FIGURE 3.3

From an ellipse to an elliptical rod via transformation 1.
Left: CLDDs $B(x,a,b)$ of an ellipse, where $L_0 = 2b$ (see left) and $A(r,a,b)$ of an infinitely long elliptic cylinder (right) with semi-minor axis $a = 1$ and semi-major axis $b = 2$. In the case of an ellipse, there exists a pole at $x = 2a$ and a finite jump at $r = 2b-$. The first moment of $B(x,a,b)$ is $x_1 = \frac{a\pi^2}{4 \cdot E[1-a^2/b^2]}$, where $E = E[m]$ is the complete elliptic integral (see [234]). In contrast, the function $A(r,a,b)$ is continuous, but involves singularities at $r = 2a$ (spike) and $r = 2b$ (tip). This is typical of the transition from plane figures to the corresponding infinitely long rods [48, 50, 73, 74].

and Gates, 1987 [43]). An integral representation of $B(r,a,b)$, is written

$$B(x,a,b) = \begin{cases} 0 \le x < 2a : \ \frac{32}{\pi^2} abx_1 \cdot x \cdot \int_{2a}^{2b} \frac{1}{t^3 \cdot \sqrt{t^2-4a^2}\sqrt{t^2-x^2}\sqrt{4b^2-t^2}} dt \\ 2a < x < 2b : \ \frac{32}{\pi^2} abx_1 \cdot x \cdot \int_{x}^{2b} \frac{1}{t^3 \cdot \sqrt{t^2-4a^2}\sqrt{t^2-x^2}\sqrt{4b^2-t^2}} dt \\ 2b < x < \infty : 0 \end{cases}$$

(3.5)

This CLDD decreases for $x = 2b$ to $B(2b-,a,b) = \frac{\pi a^2}{4b^3 \sqrt{b^2-a^2} E\left[1-a^2/b^2\right]}$

(see [48, p. 112]). Here, $E(m) = \int_0^{\pi/2} \sqrt{1 - m\sin(t)^2} dt$ denotes the complete elliptic integral of the second kind $E(m)$ by use of the definitions used in *Mathematica*.[‡] This case, which was carefully investigated by Fanter, 1967 [34], does not meet the actual assumption of a constant RS but is mentioned nonetheless This case does not meet the actual assumption of a constant RS. The elliptic needle is a special case of the ellipsoid with fixed semiaxes a and b for $c \to \infty$. For analytic expressions and a plot, see [74, pp. 323–324], figure 5 and equations (13) and (14) of that paper.

[‡] For practical reasons, Wolfram Research introduced special (modified) denotations for elliptic integrals and elliptic functions. In this content, the perimeter $u(a,b)$ of an ellipse $0 < a \le b$ is $u(a,b) = 4b \cdot E[1 - a^2/b^2]$, but not $u = 4b \cdot E[\sqrt{1 - a^2/b^2}]$.

3.2.4 Semicircular rod of radius R

In the origin, this CLDD starts with $A(0+, R) = 8/[3\pi \cdot (\pi R + 2R)]$. A finite spike at $r = R$ and a tip at $r = 2R$ indicate the radius and diameter. More details can be found in [74, p. 327] [see figure 8 and equations (22) to (24) of that paper]. If $b = 2a$, there is a certain similarity to the elliptic rod case.

3.2.5 Wedge cases and triangular/rectangular rods

Analyses of triangular/rectangular rods can be found in [72] and [74, pp. 322–323]. The latter paper also analyzes the CLDD $B(x, b)$ of a plane stripe.

Similarly, a detailed investigation of wedge cases (60° and 90° wedge) can be found in [105]. In 2009, a comparison of these CLDDs with the triangular and square rod case was performed. The outcome was the following: The CLDDs of the corresponding body pairs agree [115]. That paper investigates six pairs of elementary geometric figures possessing the same CLDD. These results are of importance for automatic shape recognition apparatuses. A CLDD is a *fingerprint* of the particle shape, *but not more*.

3.2.6 Infinitely long hollow cylinder

Analytic CLDD expressions of selected non-convex geometric figures [66, 75] were investigated. In this regard, at least two types of CLDDs must be distinguished; they are the *One-Chord Distribution* (OCD) and the *Multi-Chord Distribution* (MCD). For example, in [75] analytic representations and useful plots of the hollow cylinder case are used (see also Fig. 3.4).

More intrinsic details, including analytic expressions for the asymptotic behavior of the scattering intensity, are given in [74, pp. 325–326, figure 7, equations (17) to (21) and equations (A7) to (A8)].

OCD and MCD: What does "chord length" mean in the case of non-convex and hollow particles?

In 1982, Glatter and Kratky [129] described the "serious difficulties" regarding the concept of chord length distribution as follows:

> "Thus, the chord length distribution is equally well suited to represent a particle with respect to its diffraction pattern. However, there are serious difficulties, in much more complicated cases like hollow or composite particles."

The terms MCD and OCD were probably introduced in 2000 [66]. There exist other principles for describing chords in non-convex figures (see Stoyan & Stoyan, 1992 [211, p. 140] or Ambartzumian, Stoyan & Mecke, 1995 [3]). In the case of particles containing one or more hollow parts, it is a good idea to introduce these two different types of chord length distribution for IUR chords as follows:

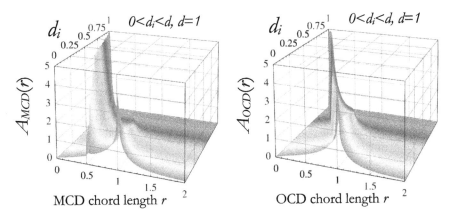

FIGURE 3.4
CLDDs of infinitely long circular hollow cylinders with constant outer diameter d and varying inner diameter d_i (Gille, 2001) [75].
The MCD case (see left) corresponds to the experimental equipment of SAS and is directly related to the scattering pattern of these cylinders. In this case, each chord length r_i is taken by itself. The sums of the chord lengths are not considered at all. Thus, the whole MCD CLDD is positioned at relatively small r. The first moment results from $l_{1MCD} = 4V/S = d - d_i$ (see Mazzolo et al., 2003 [175]). In the OCD case (right), all the lengths along the chord direction are summed up. Compared with the MCD case, the whole CLD is shifted to the right to larger r. Nevertheless, both the CLDDs are normalized by the same first principles. See [66] for the hollow sphere case.

The MCD type, which is the natural case for SAS experiments, is generated if each chord length segment is taken by itself.

In the other cases, it may be useful to consider the *sum of all chord segments through the particle* individually on one straight line as *the random variable*. This is the OCD type. There exist physical apparatuses where the sum $l = l_1 + l_2 + l_3 + \ldots + l_n$ is recorded and the summands $l_1, l_2, l_3, \ldots$ are not important. For this case, the mean chord length l_{OCD} results from the modified Cauchy theorem, $l_{1OCD} = 4(V_c + V_2)/S_w$. Here, S_w denotes the whole (inner plus outer) surface, V_2 is double the volume of the non-convex region and V_c is the volume of the convex region of the particle.

Naturally, $l_1 = l_{1MCD} = 4V/S < l_{1OCD}$. A sphere $d = 4$ with central void $d_i = 2$ gives $l_1 = 1.87$ and $l_{1OCD} = 2.35$. In [66] plots and analytic representations of this case are given.

In [115, 175], several several non-trivial examples of IUR chords are explained. Here, the difference between OCD and MCD chords becomes clear. Actually, Fig. 3.4 illustrates both chord types for the infinitely long hollow cylinder $0 < d_i \leq d < \infty$.

3.3 Recognition analysis of rods with oval right section from the SAS correlation function

After the sequence of model examples given in the last parts of this chapter, the structure functions β_0 and $B(x)$ describing the domain $\mathbf{X}$ will be reconsidered in order to describe infinitely long rods by use of a tricky approximation [44, 118]. Thus, approximations of the scattering patterns of rod-like particles result without assuming a certain, fixed shape of the right section beforehand. This approach consists in applying transformation 1 to the oval domain $\mathbf{X}$ via Eq. (3.1) (see the figures on the following pages). By analogy with the cases considered, $\beta_0(x)$ and $B(x)$ yield

$$\gamma_0(r, l_1) = 1 - \frac{r}{l_1} \cdot {}_2F_1\left(-\frac{1}{2}, \frac{1}{2}; 2; \frac{r^2}{l_1^2}\right) \approx 1 - \frac{r}{l_1} + \frac{r^3}{8l_1^3} + O[r]^5; \ 0 \le r < l_1,$$

$$A(r, l_1) = \frac{3r}{4l_1^2} \cdot {}_2F_1\left(\frac{1}{2}, \frac{5}{2}; 3; \frac{r^2}{l_1^2}\right) \approx \frac{3r}{4l_1^2} + \frac{5r^3}{16l_1^4} + O[r]^5; \ 0 \le r < l_1.$$

$$(3.6)$$

Without further assumptions, the first moment l_1 of the CLDD of these rods can be fixed by plotting the functions $\gamma_0(r)$, $r \cdot \gamma_0(r)$ or $A(r)$ (see Fig. 3.5). This is verified for $d = 1$ in Fig. 3.5 (see also Fig. 3.6).

3.3.1 Behavior of the cylinder CF for $r \to \infty$

Figure 3.6 illustrates a useful property which holds true regardless of the RS shape, e.g., for single (possibly hollow) cylinders, parallel infinitely long cylinders [76] and tightly packed cylinder arrays. The CF of rods with RS $\mathbf{X}$ has some fundamental characteristics which allow l_1, x_1 and the derived parameters to be recognized. The behavior of $A(r)$ for large r can be approximated by an asymptotic formula $A_\infty(r, x_i)$ in terms of the moments x_i of $B(x)$ [88, p. 187, figures 3 and 4] as shown by

$$A_\infty(r, x_i) = \frac{4}{\pi}\left(\frac{x_3}{r^4} + \frac{x_5}{2r^6} + \frac{3x_7}{8r^8} + \cdots + \cdots\right),$$

$$\gamma_\infty(r) = \frac{1}{x_1} \cdot \left(\frac{x_3}{6r^2} + \frac{x_5}{32r^4} + \frac{x_7}{112r^6} + \cdots + \cdots\right), \qquad (3.7)$$

$$\gamma_\infty(r) = \frac{S_{RS}}{2\pi r^2} + \frac{u}{\pi S_{RS}} \cdot \left(\frac{x_5}{32r^4} + \frac{x_7}{112r^6} + \cdots + \cdots\right).$$

Based on Eqs. (3.7), the right section surface area S_{RS} can be fixed from a plot $4\pi r^2 \gamma_0(r)$ vs. r via $S_{RS} = 1/2 \cdot \lim_{r \to \infty} 4\pi r^2 \gamma_0(r)$. Figure 3.6 illustrates the geometric deliberations. The parameters c, m_1, u, S_{RS}, x_1 and x_3 can be fixed for an $L = L_1$ from $\gamma(r)$ by use of $\gamma'(0) = -1/[l_1 \cdot (1-c)]$ [besides l_1 from

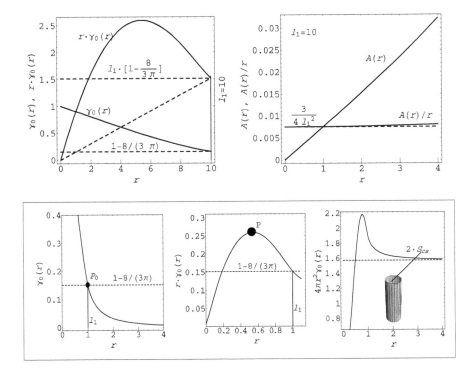

FIGURE 3.5
Approaches for fixing l_1 of long cylinders with constant oval right section X.
Upper: The behavior of the functions $\gamma_0(r)$, $r\gamma_0(r)$ and $A(r)$ of the rods
fixes l_1 (here $l_1 = 10$). While $\gamma_0(r)$ is plotted in the entire definition region,
$0 \leq r \leq l_1$, $A(r)$ is investigated closer to the origin, where $A(0+) = 3r/(4l_1^2)$.
If $0 \leq r < L_1/2$, an approximation of $A(r) = const.$ is obtained.
Lower: Analysis of the functions $r \cdot \gamma_0(r)$ and $4\pi r^2\gamma_0(r)$ for a circular rod of
diameter $d = 1$.
Left and center: The mean chord length l_1 of the three-dimensional cylinder's
right section X can be detected via $l_1 \cdot \gamma_0(l_1) = l_1 \cdot (1 - 8/(3\pi))$. The coordinates of point P are approximately $(\approx l_1/2, \approx l_1/4)$.
Right: Twice the surface area of X results from the plot $4\pi r^2\gamma_0(r)$ for relatively large r. This property holds true for any straight rod, hollow cylinders
or N parallel rods, where $2N \cdot S_{RS}$ is displayed. The first property is limited
to those X which fulfill the assumptions made by Geciauskas [45]. In the
general case, $\gamma'(0) = -1/[l_1(1 - c)]$; in the quasi-diluted case for $0 < r < r_0$,
$\gamma(r) = [\gamma_0(r) - c]/(1 - c)$ holds true. These relations fix l_1 and c. Furthermore, $l_1 = 4S_{RS}/u$, $x_1 = \pi S_{RS}^2/u$ and $x_3 = 3S_{RS}/u$ (see Chapter 9) hold
true for selected moments of the CLDDs $A(r)$ and $B(x)$. Thus, a sequence of
rod parameters can be obtained without assuming a (very) special shape of
the RS region beforehand (Fig. 3.6).

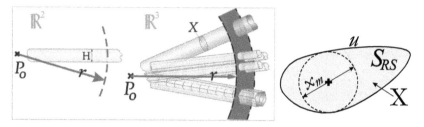

FIGURE 3.6

Asymptotic behavior of the CFs $\beta_{0\infty}(r)$ and $\gamma_{\infty}(r)$ for long stretched regions with const. RS in $\mathbb{R}^2$ and $\mathbb{R}^3$. A typical oval RS is shown on the right.
Left: A spherical shell of large radius r intersects several rods of fixed RS **X**. Point $\mathbf{P}_0$, the center of the shell, is placed somewhere close to the rod axis. The analogous situation in $\mathbb{R}^2$ is explained for a plane stripe.
Regardless of the shape of the RS of an infinitely long cylinder (or a cylinder array) for sufficiently large r, the surface area of the intersection of the shell of radius r with N cylinders equals $2N \cdot S_{cs}$ [see Eq. (3.8)].
Right: Typical parameters of an oval domain **X**. These well-chosen parameters give an approximate description of the RS shape of a rod. For $x_m = 1$ (unit length), the actual shape parameters $u = 5.3$ and $S_{RS} = 1.8$ result. Based on these parameters, $x_1 = 1.07$ follows. Finally, the mean chord length of the rod of RS **X** is $l_1 = 1.36$.

$r \cdot \gamma_0(r)$] (see upper and lower parts of Fig. 3.5).[§] The important step here is the analysis of $r \cdot \gamma_0(r)$. In general, independent of the cylinder packing, the basic behavior of the CF is clearly defined for relatively small and relatively large r as shown by

$$\gamma_0(r, l_1) = \begin{cases} 0 \leq r \leq l_1 : & 1 - \frac{r}{l_1} + \frac{r^3}{8l_1^3} + O[r]^5 \\ l_1 \leq r < r_e : & different\ behavior \\ r_e \ll r < \infty : & \frac{2S_{cs}}{4\pi r^2} + \frac{1}{x_1} \cdot \left(\frac{x_5}{32r^4} + \cdots + \cdots\right). \end{cases} \qquad (3.8)$$

The preceding sections considered the transformation $\mathbf{X} \Rightarrow \mathbb{R}^3$ for perfect RS CFs and for an approximation introduced by the mathematician Geciauskas (see [44, 45, 46]).

In the following, the inverse transformation from $\mathbb{R}^3 \Rightarrow \mathbf{X}$ will be explained. In fact, the scattering pattern of the cylinder defines that of its RS.

[§]The moment l_1 of the function $A(r) = l_1 \cdot \gamma_0''(r)$ is the highest existing moment. The limit of the sample CF $\gamma(r)$ results in $-1/\gamma'(0) = l_p = l_1(1-c) = m_1 c$.

3.4 Transformation 2: From spatial cylinder C to the base X of the cylinder

The structure functions of **C** define those of **X**. Transformation 2 fixes the properties of **X** in terms of an experimental scattering pattern of **C**. This is essential for practical applications. The inverse transformation of Eq. (3.2) is

$$B(x) = \int_0^x \left[\frac{3rx^2 - r^3}{4x^2\sqrt{x^2 - r^2}} - \frac{3}{4}\arcsin\left(\frac{r}{x}\right)\right] A'(r)dr - \frac{3\pi}{8}\int_x^\infty A'(r)dr. \quad (3.9)$$

For more details and other synonymous formulas, see (Gille, 2003) [88]. Equation (3.9) is useful for reconstructing the functions $B(x)$ and $\beta_0(x)$ as well as the parameters L_0, x_1, S_{RS} and u based on the isotropic scattering intensity $I_R(h)$ of homogeneous long cylinders directly from SAS data. A typical example of such a data evaluation is given in [88, pp. 128–129].

The function $\beta(x)$ from $\gamma(r)$ from the SAS data of a porous glass

A micropowder of porous VYCOR glass PVG (code 7930) was investigated by use of SAS, see (Gille, Enke & Janowski, 2002) [86]. An idealized isotropic ensemble of nearly monodisperse cylindrical particles is assumed. According to the Babinet theorem, a cylindrical pore can be substituted by a cylindrical particle. Application of the normalized Porod plot $P_1(h)$, $P_1(h) = h^4 I(h) \cdot V l_1 (1 - c)/(8\pi) = h^4 I(h) \cdot V/(8\pi|\gamma'(0)|)$ is useful. The characteristic volume term $V = v_c$ can be traced back to $V = 2\pi^2/\int_0^L h^2 I(h)dh$, where $I(0) = 1$. This results in the normalized P_1 plot $P_1(h) = \pi l_1(1 - c) \cdot h^4 I(h)/[4\int_0^L h^2 I(h)dh]$. Actually, the pore diameter is about 1/6 of the mean length of the pores. There is 30% porosity and particle-to-particle interferences dominate. The CLDD of the pores was separated from the CLDD of the walls via a chord length analysis and taking into account additional information from other experimental methods. Thus, the remaining idealization of the $\gamma''(r)$ and $A(r)$ data sets is a certain polydispersity of the nearly cylindrical pores (varying pore diameter). This leads to a relatively large transition region $x \to L$ of the functions $\beta''(x) \to 0$ and $B(x) \to 0$. Nevertheless, the largest RS diameter can be nearly fixed.

For separating contributions of pores and walls, $\gamma''(r)$ was obtained for 2 nm $\leq r \leq 40$ nm from $I(h)$. The lower resolution limit of the isotropic SAS (relative measurement performed with a Kratky plant and CuK_α radiation) was $r = r_{min} \approx 2/$nm. The CLDDs of walls and pores are intermixed with $I(h)$ and $\gamma''(r)$. Based on the methods *mercury intrusion* and *nitrogen adsorption*, the compact function $\gamma''(r) \sim A(r)$ was separated into a pore term $\gamma_R''(r)$ and a wall term $\gamma_W''(r)$. The term $\gamma_R''(r)$ was approximated using an exponential approach with $\gamma_R''(r) = r \cdot 0.00211 \exp(-0.00853 \cdot [0.989 + r]^2)$ (length unit

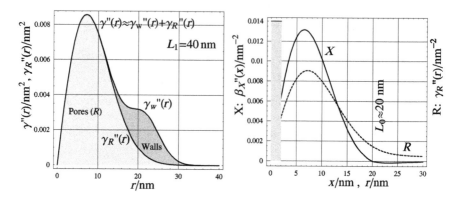

FIGURE 3.7

Approximation of pores by rods in VYCOR glass and analysis of the right section of this pore model.

The VYCOR glass is approximated by a random two-phase system with cylindrical pores (experimental resolution limit $r_{min} \approx 2$ nm).

Left: The experimental function $\gamma''(r)$ is a superimposition of a pore part and a wall part $\gamma'' \approx \gamma_R'' + \gamma_w''$, where $r < L_1 = 40$ nm. Based on the mean chord length of the wall chords $m_1 \approx 20$ nm and that of the pore(cylinder) chords, $l_1 \approx 10$ nm, a linear simulation model can be constructed (Gille, 2001) [82]) and (Enke et al., 2002) [32].

Right: Operating with the pore part (R) [with the function $\gamma_R''(r)$, see dashed line], the right section **X** of the pores (rods) is described by the second derivative of the correlation function $\beta_X''(x)$. This function (see solid line) results from transformation 2. The mean chord length and the maximum diameter of the right section, respectively, are $x_1 \approx 7.8$ nm and $L_0 \approx 20$ nm. Altogether, the experimental function $\gamma_R''(r)$ (dashed line) describes relatively long cylindrical pores. The function $\beta_X''(x)$ (solid line) obtained from $\gamma_R''(r)$ characterizes the mean right section **X** of these pores.

used is nm) (see dashed line in Fig. 3.7). This describes the VYCOR-pore. With Eq. (3.9), the right section function $\beta_X''(x) \sim B(x)$ of the pores results. The largest RS-diameter of the largest pores is $L_0 \approx 20$ nm; see solid line Fig. 3.7. After normalizing the functions $\gamma_R''(r)$ and $\beta_X''(x)$, the first moments $l_1 = 9.2$ nm $= 4S_{RS}/u$ and $x_1 = 7.5$ nm $= \pi S_{RS}/u$ result. The ratio of these mean chord lengths is $1.23 \approx 4/\pi$. This is a piece of circumstantial evidence for the existence of cylindrical and relatively long mesopores. Furthermore, L_0 is much smaller than L_1.

Using the same approach, the more specific structure functions of the RS can be determined from a large class of scattering experiments.

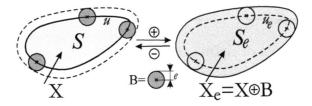

FIGURE 3.8

Dilation (left) and erosion (right) of a plane oval region by a structure element **B**, where **B**={circle of radius e}.

The denotation $\rho = e$ is usual in stochastic geometry and used in the text. This includes the dilation of a set by a sphere of radius ρ (see the ellipsoid case discussed in Chapters 1 and 2). Based on Steiner's formula (see Matheron, 1975 [174, p. 85] and Stoyan & Stoyan, 1992 [211, pp. 126]), the surface area S_ρ and the perimeter u_ρ of the dilated region $\mathbf{X}_\rho$ are defined in terms of surface area S and perimeter u. The set $\mathbf{X}_\rho = \mathbf{X} \oplus \mathbf{B}$ involves the properties $S_\rho = S + \frac{u}{2\pi} \cdot 2\pi\rho + \pi\rho^2$ and $u_\rho = u + 2\pi\rho$. This is not an approximation.

3.5 Specific particle parameters in terms of chord length moments: The case of dilated cylinders

For a plane oval region, the CLDD moments x_1 and x_3 can be traced back to the perimeter u and surface area S, where $x_1 = \pi S/u$, $x_3 = 3S^2/u$. These result in the terms $u(l_1, l_3) = \pi^2 x_3/(3x_1^2)$ and $S(l_1, l_3) = \pi x_3/(3x_1)$. In analogy with this, for a particle of volume V and surface area S, where $l_1 = 4V/S$ and $l_3 = 12V^2/(\pi S)$, $S = 4\pi l_3/(3l_1^2)$ and $V = \pi l_3/(3l_1)$ follow.

This idea can be specified for rods with oval RS. In the following, the transition from the RS $\mathbf{X}$ to the *dilated* RS $\mathbf{X}_\rho$ is investigated (see Fig. 3.8). Based on the fundamental parallel body constellation illustrated in Fig. 3.8, the CLDD moments $x_{1\rho}$ and $x_{3\rho}$ result in

$$x_{1\rho} = \frac{\pi\left(S + u \cdot \rho + \pi\rho^2\right)}{u + 2\pi\rho} \quad and \quad x_{3\rho} = \frac{3\left(S + u \cdot \rho + \pi\rho^2\right)^2}{u + 2\pi\rho}. \tag{3.10}$$

Clearly, $x_1 < x_{1\rho}$ and $x_3 < x_{3\rho}$. These interrelations describe the connection between u, S and the moments $x_{1\rho}$, $x_{3\rho}$ and ρ. After eliminating the parameters S, u and ρ, the result is the compact relation

$$\frac{12 \cdot x_1{}^3}{\pi^2 \cdot x_3} - \frac{12 x_1{}^4 \cdot x_{3\rho}}{\pi^2 \cdot x_{1\rho} \cdot x_3{}^2} + \frac{x_1{}^4 \cdot x_{3\rho}{}^2}{x_{1\rho}{}^4 \cdot x_3{}^2} = 1. \tag{3.11}$$

This interrelates the four moments x_1, $x_{1\rho}$, x_3 and $x_{3\rho}$ in the parallel body case. One of these moments can be obtained from Eq. (3.11) if the other three are known. For $\rho \to 0$, Eq. (3.11) reduces to the triviality $1 \equiv 1$.

3.6   Cylinders of arbitrary height H with oval RS

In this chapter, the main focus has been cylindrical particles of right section RS. The considerations discussed up to this point have been limited to infinitely long or at least very long cylinders. In this context, the term rod was introduced. In addition, it was determined that an analysis of right cylinders of finite height H, where $0 \leq H < \infty$, requires much greater effort. An exercise demonstrating this characteristic is given in Section 10.2.5. Explicit formulas for cylinder parameters from the SAS CF are summarized below (see also Gille & Kraus, 2010) [119].

The analysis of geometric cylinder parameters in terms of the CF $\gamma_0(r) = \gamma_0(r, H)$ is a long-standing problem. This becomes clear when comparing the general transformation formulas from the right section CF β_0 to the corresponding cylinder CF γ_0, where $\{\beta_0(x), H\} \to \gamma_0(r, H)$ (Ciccariello, 2009) [21] with the simpler transformation $\{\beta_0(x), \infty\} \to \gamma_0(r, \infty)$. A set of formulas (see Table 3.1) was established which fix the significant independent cylinder parameters in terms of $\gamma_0(r)$. Starting with the single cylinder case, the considerations will be extended to quasi-diluted cylinder ensembles.

The initial step of such a project is to fix H. This can be traced back to an equation of second degree, as given by

$$\gamma_0''(0) = -\frac{8}{3\pi H}\left(\gamma_0{}'(0) + \frac{1}{2H}\right) = \frac{8}{3\pi H}\left(\frac{1}{l_1} - \frac{1}{2H}\right), \qquad (3.12)$$

in which H is connected with the experimental terms $\gamma_0'(0) = -1/l_1$ and $\gamma_0''(0)$. The numerical analysis of the latter term has been investigated in depth in connection with the limit $\Phi(0)$ of the puzzle-fitting function $\Phi(r) = \gamma_0''(r)/[\gamma_0'(r)]^2$ in Chapter 7. From Eq. (3.12), both solutions for $H_{1,2}$, where $0 < H_1 \leq H_2 < \infty$, follow, as given by

$$H_{1,2} = -\frac{1}{T \cdot \gamma_0'(0)}\left(1 \mp \sqrt{1 - T}\right), \; \text{with } T = \frac{3\pi\gamma_0''(0)}{4[\gamma_0'(0)]^2} = \frac{3\pi}{4}\Phi(0), \qquad (3.13)$$

which may prove to be of practical importance. From Eq. (3.13), in the case $T \to 1$, the result is always the limiting solution $H = l_1$. Nevertheless, plate-like and rod-like cylinders must be distinguished. The minus sign leads to $H = H_1$ of a plate-like cylinder, where $l_1/2 < H_1 < l_1$. The plus leads to $H = H_2$ of rod-like cylinders, where $l_1 < H_2 < \infty$. An analysis of the term $w = H/l_1$ is illustrated in Fig. 3.9. Each case is considered independently based on Eq. (3.13).

Plate-like cylinders: The limiting case $H = H_1 \to 0$ (lamella) leads to $\gamma_0'(0) \approx -1/(2H) = -1/l_1$, $\gamma_0''(0) \to 0$ and $T \to 0$. Thus, Eq. (3.13) reduces in terms of T to an indeterminate expression $[0/0]$ in $\lim_{T \to 0} H(T) =$

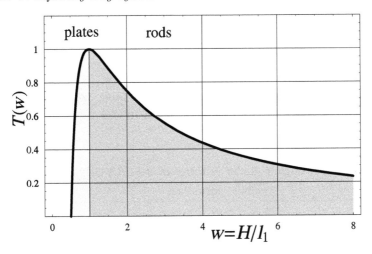

FIGURE 3.9
Analysis of the function $T = T(w) = T(H/l_1)$, i.e., of the inverse function of
$w = H/l_1 = (1 \mp \sqrt{1-T})/T$, $0 \leq T \leq 1$.
In both cases (i.e., $\mp$), $H = l_1$ follows for $T \to 1$. For each $T < 1$, two different
$w = H/l_1$ result. The positive sign leads to a rod-like cylinder. The negative
sign leads to a plate-like cylinder. In the limiting case $T \to 0$, the result is
$H/l_1 \to 0$ or $H/l_1 \to \infty$.

$l_1 \cdot \lim\limits_{T \to 0} \frac{1-\sqrt{1-T}}{T} = \frac{l_1}{2}$. Consequently, $H_1 \to 1/(2|\gamma_0'(0)|)$ and $l_1/2 \leq H_1 \leq l_1$.

Rod-like cylinders: In the limiting cases $H = H_2 \to \infty$, which means
$\gamma_0(r) \to 1 - r/H_2$, where $0 \leq r \leq H_2$, the mean chord length l_1 approximately
agrees with H, where $l_1 \approx H$ (see the special case of a circular cylinder). There
is a dominating spike at $r \approx l_1$ in the CLDD $A(r)$ of rod-like cylinders. The
greater H_2 is, the more the CLDD collapses to $A(r, H_2) \to \delta(r - H_2)$. This
fact can be verified from Eq. (3.13): As is to be expected, based on $\gamma''(0) \to 0$
and $T \to 0$, the result is $H(T) = l_1 \cdot \lim\limits_{T \to 0, H \to \infty} \frac{1+\sqrt{1-T}}{T} = \infty$. The only
difference from the plate-like case investigated before is the positive sign in
front of the root term (see Fig. 3.9).
Compact results for the infinitely diluted case and quasi-diluted cylinder
arrangements of the cylinder parameters are summarized in Table 3.1. Hence,
based on the terms $\gamma_0'(0)$ and $\gamma_0''(0+)$, the problem of parameter estimation
of right oval cylinders from the SAS correlation function can be handled in a
surprisingly easy manner.

The results illustrated in Tables 3.1 and 3.2 hold true for any right cylinder
of oval RS (see the illustration for two right circular cylinders in Fig. 3.10).

In the quasi-diluted case, the functions $\gamma(r)$ and $\gamma_0(r)$ are interrelated by

$$\gamma(r) = \frac{\gamma_0(r) - c}{1 - c}, \quad \gamma'(r) = \frac{\gamma_0'(r)}{1 - c}, \quad \gamma''(r) = \frac{\gamma_0''(r)}{1 - c}. \qquad (3.14)$$

Equation (3.14) is one of the fundamentals of Table 3.2. For more details about quasi-diluted particle ensembles, see Chapter 4.

TABLE 3.1
Parameters of a single cylinder in terms of the SAS CF $\gamma_0(r)$.

parameter	symbol	formula
1 height	H	$H_{1,2} = -\frac{1}{T \cdot \gamma_0'(0)} \left(1 \mp \sqrt{1-T}\right),\ T = \frac{3\pi}{4}\Phi(0)$
2 volume	V_0	$V_0 = \int_0^{L_0} 4\pi r^2 \cdot \gamma_0(r)dr$
3 total surface area	S_0	$S_0 = -4V_0 \cdot \gamma_0'(0)$
4 RS surface area	S_{RS}	$S_{RS_{1,2}} = \frac{V_0}{H_{1,2}}$ (*two solutions*)
5 RS perimeter	u	$u = \frac{S_0 - 2S_{RS_{1,2}}}{H_{1,2}}$ (*one solution*)
6 RS chord length	x_1	$x_1 = \frac{\pi \cdot S_{RS_{1,2}}}{u}$ (*two solutions*)

TABLE 3.2
Recognition of cylinder parameters from the SAS correlation function in the isotropic, quasi-diluted cylinder case of volume fraction c. All parameters are traced back to the experimental function $\gamma(r)$ (see the Section "Quasi-diluted particle arrangements" in Chapter 4).

parameter	symbol	formula
1 characteristic volume	v_c	$v_c = \int_0^{\infty,L} 4\pi r^2 \gamma(r)dr$
2 mean chord length	l_1	$l_1 = \int_0^{L_0} \frac{r\gamma''(r)}{|\gamma'(0)|}dr == \frac{1 - L_0 \cdot \gamma'(L_0)}{|\gamma'(0)|}$
3 volume fraction	c	$1 - c = \frac{1}{l_1 \cdot |\gamma'(0)|} = \frac{1}{\int_0^{L_0} r \cdot \gamma''(r)dr}$
4 cylinder volume	V_0	$V_0 = (1-c) \cdot v_c = \frac{\int_0^\infty 4\pi r^2 \cdot \gamma(r)dr}{\int_0^{L_0} r \cdot \gamma''(r)dr}$
5 single cylinder CF	γ_0	$\gamma_0(r) = (1-c)\gamma(r) + c,\ \gamma_0' = (1-c)\gamma'$
6 height	H	$H_{1,2} = -\frac{1}{T \cdot \gamma_0'(0)} \left(1 \mp \sqrt{1-T}\right),\ T = \frac{3\pi\Phi(0)}{4}$
7 RS surface area	$S_{RS_{1,2}}$	$S_{RS_{1,2}} = \frac{V_0}{H_{1,2}}$ (*two solutions*)
8 total surface area	S_0	$S_0 = \frac{4V_0}{l_1} = 4v_c \cdot (1-c)^2 \cdot |\gamma'(0)| = \frac{4v_c \cdot l_p}{l_1^2}$
9 RS perimeter	u	$u = \frac{S_0 - 2S_{RS_{1,2}}}{H_{1,2}} = \frac{3\pi V_0 \gamma_0''(0)}{2}$ (*one solution*)

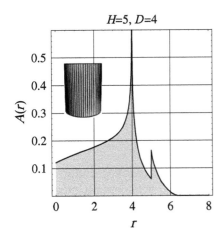

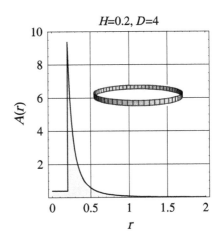

FIGURE 3.10

The function $A(r, H, D)$ of a circular cylinder ($H = 5$, $D = 4$) and a circular lamella ($H = 0.2$, $D = 4$) (see also Fig. 3.9).

Three r-intervals must be distinguished in order to obtain a compact analytical expression (see Chapter 2). The CLDD is not a linear function in the first interval of representation, where $0 \leq r \leq Min(H, D)$. There is always a pole at $r = D$ (see Ciccariello, 1990) [13].

Rod-like cylinder 1: Height $H = 5$ and diameter $D = 4$, where $\gamma_0''(0) = 0.0424$. Furthermore, the parameters $S_{RS} = 4\pi$, $V_0 = 20\pi$, $l_1 = 20/7$ and $\gamma_0'(0) = -7/20$ result. Inserting these parameters into Eq. (3.13) gives $H = 5$, as it must be. Actually, $A(0) = 0.121$ holds. For the spike on the left-hand side $\gamma_0''(r) \sim A(r \to D, H, D) \to \infty$ holds true, but the spike on the right-hand side is limited. There exists a finite jump at $r = H$ (Gille, 1983) [48, pp. 108–113].

Lamellar cylinder 2: For a lamella with $H = 0.2$ and diameter $D = 4$, the result is $\gamma_0''(0) = 1.06$. The parameters $S_{RS} = 4\pi$, $V_0 = 2.51$, $l_1 = 0.3636$ and $\gamma_0'(0) = -2.75$ are obtained. In this case, $A(0) = 0.386$. The curve is very similar to that of a lamella. However, in contrast to the lamella, there is actually a spike at $r = D = 4$. Even if the interval of representation includes the pole position $r = D = 4$, the spike (not far from a Dirac delta function) cannot be recognized in this kind of plot.

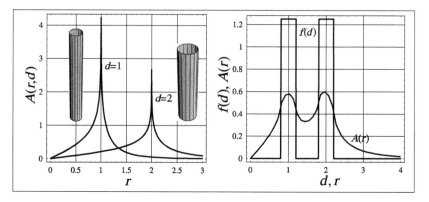

FIGURE 3.11
Ensembles of long cylinders with different diameters and the resulting CLDDs.
Left: CLDDs $A(r, d)$ of two infinitely long circular cylinders with fixed diameters $d = 1$ and $d = 2$. The first moment of $A(r, d)$ for one single cylinder of diameter d equals $2 \cdot d$.
Right: Assuming a simple bimodal diameter distribution density $f(d)$ of the cylinder diameter d, a modified function $A(r) = \overline{A}(r)$ results (see dashed line). This function can also be used to estimate the mean diameters $\overline{d} = 1, 2$ of both cylinders. Operating with scattering data, such an approach is very effective. The CLDD is fixed without the assumption of a special particle shape, but $f(d)$ is not.

3.7   CLDDs of particle ensembles with size distribution

Figure 3.11 illustrates a special case of the connection between polydispersity and CLDD for an ensemble of circular rods of diameters d_i, possessing a diameter distribution density $f(d)$. Similarly, rods (trigonal, pentagonal, etc.) of varying thicknesses can be studied. A model function, representing a size distribution density $f(a)$ of a random edge length a, could be introduced. An approach for detecting $f(a)$ is the following: Based on an averaged scattering intensity $\overline{I}(h, f(a))$, an averaged CLDD $\overline{A}(r, f(a))$ results. Hence, a mean CLDD of the RS $\overline{B}(x, f(a))$ results [86, 89]. Finally, $f(a)$ is defined in terms of $\overline{B}$. If $f(a)$ is nearly constant in a relatively large region, all the oscillations in P_1 are smoothed out. This makes it more complicated to give a direct interpretation of scattering pattern details. Generally, the interpretation of smooth scattering curves requires more assumptions about their origin.

The CLDDs of tightly packed arrangements of parallel rods (also referred to as cylinder arrays) and their scattering patterns are considered in Chapter 4. In order to simplify the approach, circular rods are assumed.

4

Particle-to-particle interference – a useful tool

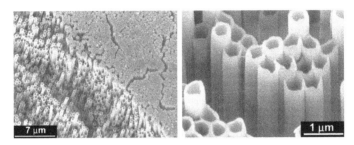

Analysis of carbon tubes in the μm region. With geometric models, averaged structure parameters of such tightly packed cylinder samples can be investigated via scattering experiments (see the model figure below).

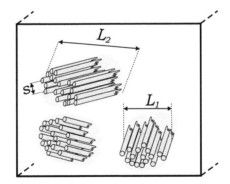

For interpreting the scattering properties of cylinder packages, it is absolutely necessary to consider the phenomenon of particle-to-particle interference. The length L_1 (which is much smaller than the "overall length" of the cylinders) denotes the largest diameter of a package perpendicular to the cylinder axes. The largest package diameter is L_2, where $L_1 \ll L_2$. In addition to the information about the cylinder diameters, the scattering pattern of such particle collections involves information about the spatial packing of the model particles. One component of the latter is reflected by the *pair correlation function* $g(s)$, which describes the distances between the parallel cylinders in a package. For anisotropic cases, the term *cylinder array* is used (Ruland & Smarsly, 2005) [200].

4.1 Particle packing is characterized by the pair correlation function $g(r)$

Particle-to-particle interferences allow an investigation of the packing law of the particles in $\mathbb{R}^3$, $\mathbb{R}^2$ and $\mathbb{R}^1$. Considering relatively small h values, each scattering pattern $I(\mathbf{h})$ involves information about the distances from one particle to the other. In this regard, the distances between the centers of the particles are an important special case. There exist many other distances between points that belong to different particles. It is not possible to separate all this distance information exactly based on one scattering experiment. Particle-to-particle interferences exist for all degrees of dilution $0 < c < 1$. However, tightly packed particle ensembles reflect these interferences starting at relatively small r values.

The function $g(r)^*$ has already been briefly considered in Section 1.2.2 and is of major importance for Chapter 4. The textbook by (Illian et al., 2008) [154] contains a long chapter on the pair correlation function, which helps to understand the information content of that function for describing the spatial arrangement of points. Furthermore, it gives explicit formulas for pair correlations of many models (see also Hermann, 1991 [150, chapter: WAXS experiments] and Stoyan & Stoyan, 1995 [212, pp. 275–276]).

The function $g(r)$ describes essential details of the packing arrangement, but is far from defining it. Most of $g(r)$ approaches known are approximations. The famous *Percus-Yevick model for hard spheres* is one of them (see Torquato, 2000, 2005) [225, 226].

4.1.1 Explanation of the function $g(r)$ and Ripley's K function

The following presents a heuristic explanation of $g(r)$. This function describes the distances between points. It is an important second-order characteristic of homogeneous and isotropic *point processes*. In the context of SAS, it is used to describe the centers of particles belonging to an isotropic particle ensemble.

Let λ be the intensity of the process, i.e., the mean number of points per volume. The probability of having a point of the point process in the infinitesimally small sphere $s(X)$ of volume dV_x centered at point X is λdV_x. The pair correlation function appears in the case of two such spheres as follows: Consider a second point Y at distance r from X and the probability that there is a point in both $s(X)$ and the small sphere $s(Y)$ of volume dV_y and center in Y. This probability is denoted by $Pr_2(r)$.

---

*The symbol $g(r)$ belongs to the established notation systems in physics and statistics. However, the same symbol is also used for other functions in an SAS context, like Porod's function $g(r) = l_p \gamma''(r)$ (see [7]).

If the point distribution is completely random, according to the multiplication theorem of probability theory, it holds that $Pr_2(r) = \lambda dV_x \cdot \lambda dV_y$. In the general case, $Pr_2(r)$ can be written formally for all r, where $0 < r < L$, as

$$Pr_2(r) = g(r) \cdot \lambda dV_x \cdot \lambda dV_y, \qquad (4.1)$$

where $g(r)$ plays the role of a factor of proportionality, which depends on the distance r. In general, the term $\lambda dV_x \cdot \lambda dV_y$ in Eq. (4.1) has to be multiplied by a correction factor to yield $Pr_2(r)$, which depends on r. This factor is exactly $g(r)$, the pair correlation function, which has the property $\lim_{r \to \infty} g(r) = 1$. It is always nonnegative, but can take arbitrary positive values. The function does not characterize a point process uniquely, i.e., there are examples of different point processes with the same $g(r)$.

Figure 1.21 illustrates the pair correlation of the centers of the atoms of an AlLa30at% alloy, where $0 \leq r \leq 3$ nm. The experiments and data evaluation were performed at the SAS laboratory of the University of Halle-Wittenberg by inserting the order range $L = 3$ nm. The big $g(r)$ values correspond to frequently existing distances r. In contrast, small $g(r)$ values mean that these distances r are relatively rare. A normalization $g(4nm) = 1$ was applied in Fig. 1.21.

Another function called the Ripley's K function, which is also defined for homogeneous and isotropic point processes, is closely related to $g(r)$, as given by

$$K(r) = \int_0^r 4\pi r^2 g(r) dr, \; g(r) = \frac{K'(r)}{4\pi r^2}, \; dK(r) = 4\pi r^2 g(r) dr; \; 0 \leq r \leq L.$$
$$(4.2)$$

The function $K(r)$ has a cumulative nature and plays an important role in point process statistics. It describes the mean number of points of the point process which are contained in the spheres of radius r centered in the points of the field. The functions $K(r)$ and $g(r)$ have been discussed in a few elementary cases [211, 213, 231], which have mostly been limited to Poisson fields and derived cases. The famous Percus-Yevick pair correlation function is of great importance in materials science and in practice; however, it is only an approximation, which breaks down for higher volume fractions c. The analytic $g(r)$ expression of this approach is complicated. The literature concerning these functions is extensive and difficult to understand for the beginner. In most cases, $g(r)$ functions are given implicitly. Such connections are illustrated at the end of the chapter (see Fig. 4.14).

The functions $g(r)$ or $K(r)$ are frequently applied in structure research. These functions are uniquely defined for a point process. For realistic particle ensembles, the functions result from *wide angle scattering* (WAS) experiments, see Eqs. (4.55)–(4.58). However, the inversion of this connection is wrong: Except for some special cases, $g(r)$ and $K(r)$ do not exactly define the law of particle packing.

4.1.2 Different working functions and denotations in different fields

In materials science and digital image processing several structure functions have been defined which are very similar to the basic set covariance $C(r)$. It is remarkable that $C(h)$ is frequently used by many authors in the field of stochastic geometry, i.e., here *distance* or *length* is frequently denoted by h.

However, not all mathematicians follow this convention. In the textbooks by *Stoyan*, the variable r consequently denotes the *distance*. This simplifies the application of the results and avoids queries about the denotation. In order to avoid Greek letters—but underestimating the resulting difficulties—some authors (for example G. Damaschun and team) denoted the sample CF by $C(r)$, i.e., $\gamma(r) \equiv C(r)$. Then the CLDD $A_0(r)$ of a convex single particle is written as $A_0(r) = \bar{l} \cdot C_0''(r)$.

Similarly, the use of the variable λ leads to contradictions. In physics, λ is reserved for the *wavelength*, but in the field of stochastic geometry, λ always denotes the *intensity of a point process*.

In physics, the symbol g is used for a sequence of functions. In this chapter, it is mainly reserved for the par correlation.

In all chapters of this book, the SAS CF of a sample is uniformly specified by $\gamma(r)$. This function is of current interest for the analysis of SAS scattering patterns. In this context, the convolution square of the density fluctuations $\gamma(r)$ (Debye & Bueche, 1949) [27] is an isotropic set covariance, which is transformed to unity in the origin, $\gamma(0) = 1$. The traditional denotation for a single particle is $\gamma_0(r)$.

For a given volume fraction c of the *non-connected particle phase* 1, $\gamma(r)$ and $C(r)$ are dependent functions, where $C(r) = c(1 - c) \cdot \gamma(r) + c^2$ (Serra, 1982) [205, pp. 271–280]. It holds that $\gamma(L) = 0$; however, $C(0) = c$ and $C(L) = c^2$ (see Fig. 4.25 and Fig. 4.8). The set covariance does not result directly from $I(h)$; however, $\gamma(r)$ does—based on *relative measurement* of the scattering pattern. In this case, c is an unknown parameter, i.e., the CF can be detected without taking c into consideration at all. Using suitable models as a basis, c can be investigated in separate, additional procedures. Such procedures are the main part of Chapter 8.

After starting with the intrinsic properties of the *quasi-diluted case* (Section 4.2) and inspecting the special case of *two (possibly touching) particles* (Section 4.3), an analysis of the function $\gamma = \gamma(r, g(r), L)$ is performed in Section 4.4. Sections 4.5 and 4.6 include extensions and applications for spheres and cylinders. In addition, the relation to the scattering pattern $I(h, g(r), L)$ is also covered. Approaches by Frisch and Stillinger, 1963 [39], who investigated the essential terms of the Taylor series of the CF in the origin, are discussed. Section 4.7 interrelates SAS and WAS (wide-angle scattering).

4.2 Quasi-diluted and non-touching particles

In many very realistic cases, *the smallest distance* r_0 between two hard particles is greater than *the largest diameter* L_0 of the largest particle. This special case of what is referred to as a quasi-diluted particle ensemble (QDPE) of particle volume fraction c is important in many fields.[†] Evidently, a QDPE excludes contact between any two objects.

Ensembles of tightly packed but *non-touching particles* represent another special case, which is also discussed here.

The basic idea of a QDPE is illustrated in Fig. 4.1 by the CF $\gamma(r, s, d)$ of two spheres of diameter d, the centers of which are separated by a distance s, where $d < s < \infty$. The function $\gamma(r)$ can be represented as the sum $\gamma(r, s, d) = \gamma_0(r, d) + \gamma_{AB}(r, s, d)$, where $0 \le r < \infty$. The first term, $\gamma_0(r, d)$, denotes the single sphere CF. The second term describes the particle-to-particle distances as depending on s and d as

$$\gamma_{AB}(r, s, d) = \begin{cases} s - d \le r \le d + s, & \frac{d^5 - 5(r-s)^2 d^3 + \left[5d^2 - (r-s)^2\right] \cdot |r-s|^3}{20 r s d^3} \\ -\infty < r < s - d, & 0 \\ d + s < r < \infty, & 0 \end{cases} \qquad (4.3)$$

The parameter s significantly influences the CF and the scattering pattern. The cases $s < d$ are not possible. If $d < s < 2d$, the CF is a superimposition of the random distances inside the particle (sphere) *and* the random distances from one particle (sphere) to another particle (sphere).

However, there is the huge s-interval $i_{QDPE} = [2d < s < \infty]$. Here, the function $\gamma_0(r, d)$ of the single sphere (particle) is completely separate. This special QDPE is characterized by $c = 0$. Thus, $\gamma_0(r, d) \equiv \gamma(r, s, d)$ results inside i_{QDPE}.

Extrapolating this special case to an isotropic QDPE of volume fraction c, the more general connection $\gamma_0(r) = \gamma(r) \cdot (1 - c) + c$ fixes $\gamma_0(r)$ in terms of the sample CF $\gamma(r)$ and c (Gille, 2003) [80]. On the right of Fig. 4.1, a plane particle ensemble illustrates the "quasi-dilution property." Here, the area fraction c is about 30%. The case of infinite dilution $c \to 0$ is a special case of QDPEs.

1. Non-touching particles and the QDPE case: A more general case than the QDPE is that of hard homogeneous non-touching particles embedded in a homogeneous matrix. Here, the only thing known about the minimum distance r_0 is the relation $0 < r_0 \ll L_0$, i.e., $0 < L_0 < r_0$ is not fulfilled. Let $\gamma'(0)$ be the first derivative of the sample CF, V the total sample volume and

---

[†]The motion of the planets is such a system.

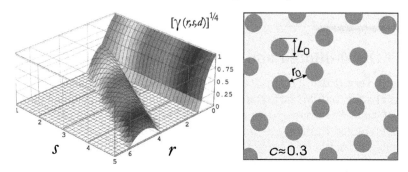

FIGURE 4.1
What is a QDPE? The parameters L_0, r_0 and c are explained on the right in the plane.
Left: The function $\gamma(r,s,d \to 1)$ of a pair of unit spheres. A transformation $[\gamma(r,s,d)]^{0.25}$ is applied, which improves the overview. The corresponding scattering pattern (for details see Section 4.5), $I_2(h,s,d) = 9/2[\sin(hd/2) - h(d/2)\cos(hd/2)]^2/[2h^6(d/2)^6]\cdot[1+\sin(hs)/(hs)]$, $0 \le h < \infty$, where $I(0) = 1$, has been known for a long time (see Guinier & Fournét, 1950 [143, pp. 141–142]).

S the total surface area of the sample. A well-known more general connection (Guinier & Fournét [143, pp. 141–142]) is,

$$\gamma'(0) = -\frac{S}{4Vc(1-c)},\tag{4.4}$$

which can be easily verified. Let N be the total number of particles with volume V_0 and surface area S_0 forming the non-connected particle phase of the sample. This gives $c = N \cdot V_0/V$ where $V = N \cdot V_0/c$. Furthermore, $S = N \cdot S_0$. After inserting S and V into Eq. (4.4) in terms of N, c, V_0 and S_0 the result is

$$\gamma'(0) = -\frac{N \cdot S_0}{4\frac{N \cdot V_0}{c}c \cdot (1-c)} = -\frac{S_0}{4V_0 \cdot (1-c)} = -\frac{1}{l_1 \cdot (1-c)}.\tag{4.5}$$

For a single particle, $\gamma'_0(0) = -1/l_1$ (Gille et al., 2005) [102]. Furthermore,

$$\frac{1-c}{V_0} = \frac{S_0}{4V_0^2|\gamma'(0)|} = \frac{1}{l_1V_0|\gamma'(0)|}.\tag{4.6}$$

Equation (4.6) is useful for determining the volume fraction c of non-touching particles of volume V_0. Obviously, Eqs. (4.4) to (4.6) also hold true in the QDPE case. These connections are applied for determining a set of independent geometric parameters of cylinders with oval RS (see Chapter 3).

The characteristic features of QDPEs are summarized below. The assumptions are the following: hard, convex homogeneous particles that do not touch, i.e., their surfaces are always separated by a distance r_0, where $0 < L_0 < r_0$.

2. Characteristic volume v_c, particle volume V_0 and volume fraction c: If the largest particle diameter is denoted by L_0,

$$\gamma(r) = \frac{\gamma_0(r) - c}{1 - c}, \quad \gamma'(r) = \frac{\gamma_0'(r)}{1 - c}, \quad \gamma''(r) = \frac{\gamma_0''(r)}{1 - c}, \quad 0 \le r < L_0 \qquad (4.7)$$

holds [143]. As $V_0 = \int_0^{L_0} 4\pi r^2 \gamma_0(r) dr$, a formal integration of $\gamma(r)$ in Eqs. (4.7) with upper limit L_0 includes *the single particle* and yields

$$\int_0^{L_0} 4\pi r^2 \gamma(r) dr = \frac{1}{1 - c}\left(V_0 - \frac{4}{3}\pi L_0{}^3 \cdot c\right). \qquad (4.8)$$

However, the term $\int_0^{L_0} 4\pi r^2 \gamma(r) dr$ does not represent v_c of the particle ensemble. The analysis of $v_c = v_c(L)$ requires information about the whole sample CF for the interval $0 < r < L_0 < L$.

Nevertheless, there is a relation for the total number of scattering particles N of diameter L_0. For a QDPE, the ratio

$$\frac{V_t}{N} = \frac{V_0}{c} = \frac{V_0}{1 - 1/[l_1 \cdot |\gamma'(0)|]} \qquad (4.9)$$

involves valuable information about N (see also Gille, 2002 [87, Appendix B, p. 475]). In the limiting case $c \to 0$, $V_t/N \to \infty$ results, but the limiting case $c \to 1$ is not realistic for a QDPE.

In certain special cases, Eqs. (4.8) to (4.9) allow an estimation of L_0. This length is directly connected with the first zero of the sample CF $\gamma(r)$.

3. Analysis of the first zero of $\gamma(r)$ in terms of L_0, c, l_1 and m_1: The largest particle diameter L_0 is characterized by $\gamma_0(L_0) = 0$. Starting in the origin $r = 0$, the abscissa $r = L_0$ is the first local minimum of $\gamma(r)$. Consequently, with Rosiwal's linear integration principle [197],

$$\frac{m_1}{l_1} = \frac{1 - c}{c}, \quad \gamma(L_0) = -\frac{c}{1 - c} = -\frac{l_1}{m_1} \qquad (4.10)$$

follows. Equation (4.10) implicitly fixes L_0 and the mean chord length m_1 of the connected phase of the QDPE in terms of c and l_1. Explicit equations for these parameters involve the moments M_1 and M_4 of the function $g(r) = \gamma''(r)/|\gamma'(0)|;^\ddagger$ see also the interrelations between higher moments of $g(r)$ and c in Chapter 9.

---

‡For these connections, the part $-2\delta(0)$ of $g(r)$ is unimportant.

4. First and fourth moment M_1 and M_4 of the function $g(r)$:
For a convex single particle possessing a CLDD $A(r)$ (see Chapter 2) and a maximum diameter L_0,

$$l_1 = \int_0^{L_0} r \cdot A(r)dr = \int_0^{L_0} r \cdot l_1 \gamma_0''(r)dr \qquad (4.11)$$

holds. With $l_1 = l_p/(1-c) = 1/[(1-c) \cdot |\gamma'(0)|]$ and $\gamma_0''(r) = (1-c) \cdot \gamma''(r)$, Eq. (4.11) is written

$$l_1 = \int_0^{L_0} \frac{r \cdot (1-c) \cdot \gamma''(r)}{|\gamma'(0)| \cdot (1-c)}dr = \int_0^{L_0} \frac{r \cdot \gamma''(r)}{|\gamma'(0)|}dr = \frac{1}{|\gamma'(0)|(1-c)}. \qquad (4.12)$$

Equation (4.12) traces the single particle parameter l_1 back to the sample CF $\gamma(r)$ and to the length L_0. This point leads to explicit equations for the parameters $1-c$, V_0 and S_0 of the single particle, where

$$1-c = \frac{1}{|\gamma'(0)| \cdot l_1}, \quad V_0 = (1-c) \cdot v_c, \quad S_0 = \frac{4V_0}{l_1} = \frac{4v_c(1-c)}{l_1}. \qquad (4.13)$$

The moment M_4 of $A(r)$ is not a parameter independent of this parameter set. Analogously to Eqs. (4.11) and (4.12)

$$l_4 = \int_0^{L_0} r^4 \cdot A(r)dr = \int_0^{L_0} r^4 \cdot l_1 \gamma_0''(r)dr = \int_0^{L_0} \frac{r^4 \cdot \gamma''(r)}{|\gamma'(0)|}dr \qquad (4.14)$$

results. Furthermore, the following connections are of importance (see Chapter 7):

$$M_4 = \frac{3}{\pi} \cdot l_p \cdot v_c = \frac{3l_p \cdot V_0}{\pi(1-c)} = \frac{3}{\pi} \frac{l_1(1-c)V_0}{(1-c)} = \frac{3}{\pi} l_1 V_0 = \frac{3}{\pi} l_1(1-c) \cdot v_c. \qquad (4.15)$$

The points enumerated above demonstrate that a sequence of parameters – rarely to be fixed for tightly packed ensembles of hard particles – can be explicitly detected by simple equations operating with $\gamma(r)$, i.e., in terms of the scattering pattern $I(h)$ of the QDPE.

Properties of two-particle ensembles: elementary examples

The structure functions of two close (possibly touching) single particles involve some basic information about ensembles of tightly packed particles possessing the same shape. There exist relatively small particle-to-particle chord lengths m_i between the surfaces of two nearly touching particles. Cases dealing with two *touching spheres*, two *touching cylinders*, the *hollow cylinder case* and the *hollow sphere case* are explained.

Structure functions for touching convex particles forming non-convex particles are derived for several cases (Mazzolo et al., 2003) [175] and (Gille et al.,
2005) [102]. For two touching spheres with diameters $0 < d_1$ and d_2, where
$0 < d_{1,2} < \infty$, the derivatives of the SAS CF in the origin are

$$\gamma'(0) = -\frac{3(d_1{}^2 + d_2{}^2)}{2(d_1{}^3 + d_2{}^3)}, \quad \gamma''(0) = \frac{2d_1 \cdot d_2}{(d_1 + d_2)^2(d_1^2 - d_1 d_2 + d_2^2)}. \quad (4.16)$$

Hence, $d_1 = d_2 = d$ gives $-\gamma'(0) = -\gamma_0'(0) = 3/(2d) = l_1$.§ The results for
a *hollow sphere* and a *hollow cylinder* are similar. A hollow sphere of outer
diameter d_a with a central hole of diameter d_i results in,

$$\gamma(r, d_i, d_a) = 1 - \frac{3(d_a{}^2 + d_i{}^2) \cdot r}{2(d_a{}^3 - d_i{}^3)} + \frac{r^3}{d_a{}^3 - d_i{}^3}, \quad (r \to 0, \ d_i < d_a). \quad (4.17)$$

The term $-\gamma'(0)$ agrees with the Cauchy theorem, as shown by

$$l_1 = \int_0^L l \cdot A(l)dl = \frac{2(d_a{}^3 - d_i{}^3)}{3(d_a{}^2 + d_i{}^2)} = 4V/S. \quad (4.18)$$

In [102], the profile of the function γ'' of the hollow sphere is analyzed. Certainly, in that non-convex case, $\gamma''(r)$ does not represent a CLDD $A(r)$ of one
random length variable r.

For an infinitely long circular hollow cylinder with $d_i < d_a$ [75]

$$\gamma(r, d_i, d_a) = 1 - \frac{r}{d_a - d_i} - \frac{r^3}{8d_a(d_a - d_i)d_i} + \frac{(d_a{}^2 - d_i d_a + d_i{}^2) \cdot r^5}{64d_a{}^3(d_a - d_i)d_i{}^3} - ... + ... \quad (4.19)$$

results. The CLDD $A(r, d_i, d_a)$ yields the mean chord length for IUR chords
$l_1 = d_a - d_i$. This can be confirmed by the Cauchy theorem, as shown by

$$l_1 = 4 \cdot V/S = 4 \cdot \left(\pi d_a{}^2/4 - \pi d_i{}^2/4\right)/(\pi d_a + \pi d_i) = d_a - d_i.$$

The typical CLDD spikes can be traced back to the parallel interfaces, [i.e.,
the height of the shell(s) $(d_a - d_i)/2$] inherent in these figures for relatively
small chord lengths.

Models of two-particle ensembles are of practical relevance [see Eqs. (4.18),
(4.19) and (4.17)]. Analytic expressions of the SAS CF, the CLDD and the
scattering patterns were derived for the cases referred to here. Figure 4.2
illustrates the behavior of the distance distribution density $p(r)$ of two spheres
of different radii. In some cases, such a two-particle ensemble is a simple but
useful model for tightly packed particles. Simply speaking, it is a first-order
approximation that considers *a certain next particle seen from a particle.*

Nevertheless, the volume fraction of such an approach still equals zero, i.e.,
the type of packing is not considered. This step is done next using the pair
correlation approach in detail.

---

§For two parallel, touching and infinitely long cylinders of equal diameter d, the connections
$-\gamma'(0) = -\gamma_0'(0) = 1/l_1 = 1/d$ hold true.

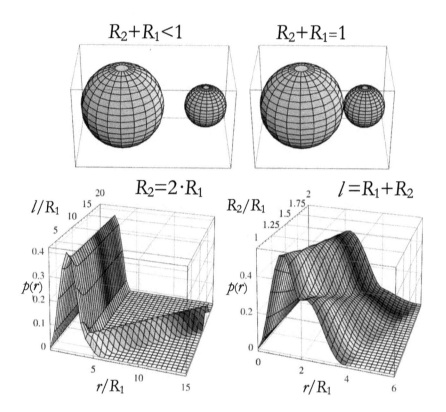

FIGURE 4.2

Two spheres, the centers of which are separated by a fixed distance l, where $R_1 < R_2$, in non-touching and touching configurations.

This geometry is reflected in the behavior of the SAS CF $\gamma(r) = \gamma(r, l, R_1, R_2)$ and in $p(r, l, R_1, R_2) = 4\pi r^2 \gamma(r)/(V_1 + V_2)$. On the left, $\gamma(r)$ is a simple average for $r \to 0$. The investigation of the case right for $r \to 0$ requires more effort.

The distance distribution densities $p(r, l, R_1, R_2)$, where $R_1 < R_2$, for non-touching (left) and for two touching (right) hard spheres.

The order range is $L = l + R_1 + R_2$. The non-overlapping condition is written $2R_2 \leq l - (R_1 + R_2)$. The cases $R_2 = 2 \cdot R_1$ (left) and $l = R_1 + R_2$ (right) are illustrated. For the latter case, the greater the ratio R_2/R_1 is, the smoother the $p(r)$ curves are. In cases where $R_1 \approx R_2$, there exist two maximums (see the curves in front on the right). In cases where $R_2/R_1 \approx 1.9$, $p(r)$ involves one maximum (see the curves on the right).

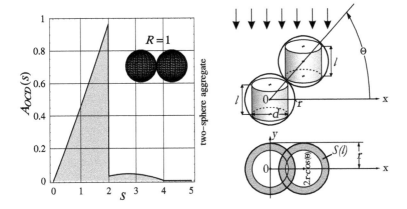

FIGURE 4.3

The case of a two-sphere aggregate using the example of an OCDD for IUR chords. Without restricting the generality, the sphere radius is $R = r = 1$; hence, $L = 4$ is the maximum diameter of the aggregate.

Left: The OCDD $A_{OCD}(s)$ is not a linear function; in $0 < s < 2$ either. The fact $A(0) = 0$ results from the very low probability that a chord will "touch" one of the spheres. Most of the IUR chords intersect only one of the spheres. The first moment of $A_{OCD}(s)$ is greater than $l_{1MCD} = 4V/S = 4 \cdot (2V_0)/(2S_0) = 4V_0/S_0 = 4R/3$. In summary, $l_{1OCD} = 1.44236 \cdot R$ results.

Right: The geometric situation for a derivation of $A_{OCD}(s)$. The OCD is traced back to the geometric probability that a random chord of length λ, where $0 \leq \lambda \leq L$, is smaller than a given chord length s where $0 \leq s \leq L = 4r$. The cylinders possess a height $H = l$. This fixes its diameter $d = d(l, r)$. For details, see (Gruy & Jacquier, 2008) [141].

The chord length distribution of a two-sphere aggregate (OCD case)

The *One-Chord Distribution Density* (OCDD) $A_{OCD}(s)$ of a two-sphere aggregate is illustrated in Fig. 4.3. For the geometric background of this investigation, see (Gruy & Jacquier, 2008 [141] and the right of Fig. 4.3). The *Multi-Chord Distribution* case (MCD case) considers each chord length separately. The sums of the chord lengths are analyzed in the OCD case (see Gille, 2000) [66]. This means (see also the Section *Infinitely long hollow cylinder* in Chapter 3) that if a random line intersects both spheres, *both line segments are added*. Hence, a relatively long sum chord length l results. Actually, $L = 4 \cdot sphere\ radius = 4 \cdot r$.

In general, by denoting the CF of the aggregate by γ, the aggregate can be described by different functions. The most popular are $l_{1MCD} \cdot \gamma''(x)$, $A(y) = A_{MCD}(y)$ and $A_{OCD}(z)$. These three functions involve length variables, which belong to different *structure elements* **B**. In the case of the autocorrelation function $\gamma(x)$, **B** is defined by *two points* separated by a distance x. In the

MCD case, **B** is a *line* of length y. The OCD case considers the same **B**, but here the sum $s = z = y_1 + y_2 + ...$ is analyzed. The actual consideration restricts to the one function $A_{OCD}(s)$.

Because of the rotational symmetry of the object, one orientation angle Θ describes the orientation of the aggregate related to a *line field* stretching from above to below, perpendicular to the x direction. In order to detect the probability $F(s)$ that a chord length λ is smaller than s, two cylinders are inscribed into the spheres of radius r. The projections of the cylinders into the x, y plane may overlap. The probability $Pr(\lambda < s)$ is proportional to the surface area $S(s)$ of the circular ring with inner diameter $d = d(s, r)$. The constant outer diameter is twice the sphere diameter $r = R$. Hence, the three-dimensional problem can be traced back to a study of two overlapping circular rings in terms of r, s and Θ. However, the averaging effort required with respect to the orientation angle Θ is enormous. Analytical results are hardly possible in this case. Nevertheless, the OCDD $A_{OCD}(s)$ is defined numerically (see the computer program below). The first digits of the OCDD moments are as follows: $l_{1OCD} = 1.44236 \cdot R$, $l_{2OCD} = 2.8847 \cdot R^2$ and $l_{3OCD} = 4.327 \cdot R^3$.

These results are of practical relevance in the fields of *Focused Beam Reflectance Measurement* (FBRM), *Turbidimetry* and *extinction measurements*.[¶] In the following *Mathematica* program, the sphere radius is r=R.

```
AOCD[s_,r_]:=Which[s==0,0,Inequality[0,LessEqual,s,Less,2*r],Dg[s,r]+Da[s,r]-
Dbn[s,r], Inequality[2*r,LessEqual,s,Less,4*r],Dal[s,r],True,0];

NN[r_]:= 1/(r^2*(Pi + 8/3)); T[l_,r_]:=l^2/(16*r^2);
t0[l_,r_]:= ArcSin[l/(4*r)]; x1[l_, r_]:= T[l, r]/Cos[t];
tp[l_,r_]:= ArcCos[(1+ Sqrt[1-l^2/(4*r^2)])/2];
tm[l_,r_]:= ArcCos[(1-Sqrt[1-l^2/(4*r^2)])/2];
Dbn[l_, r_]:=  -((1*Pi*(-4+4*Sqrt[1-(1/4)*(-1+Sqrt[1-l^2/(4*r^2)])^2]))/
   (4*(8/3+Pi)*r^2))-(2*NIntegrate[(-(1/2))*1*ArcCos[(Cos[t]-(l^2*Sec[t])/
   (16*r^2))/Sqrt[1-l^2/(4*r^2)]]*Cos[t],{t,ArcCos[(1/2)*(1+Sqrt[1-l^2/
   (4*r^2)])], ArcCos[(1/2)*(1-Sqrt[1-l^2/(4*r^2)])]}])/ ((8/3+Pi)*r^2);
Dg[l_, r_]  := 2*NN[r]*Pi*(1/2);
Da[l_,r_]:=NN[r]*NIntegrate[((2*r^2*x1[l,r])/1)*(Cos[t]/(x1[l,r]+Cos[t]))*
   (Sqrt[1-(x1[l,r]+Cos[t])^2]-((1-(Cos[t]+2*x1[l,r])*(x1[l,r]+Cos[t]))/
   (Sqrt[x1[l, r]]*Sqrt[x1[l,r]+Cos[t]]))*ArcSin[(Sqrt[x1[l,r]]*
   Sqrt[x1[l,r]+Cos[t]])/Sqrt[1-Cos[t]*(x1[l,r]+Cos[t])]])*Cos[t],
   {t,tp[l,r],tm[l,r]}] +
NN[r]*NIntegrate[((Pi/1)*r^2*Sqrt[x1[l,r]]*Cos[t]*(2*x1[l,r]^2+3*x1[l,r]*
   Cos[t]-1+Cos[t]^2)*Cos[t])/(x1[l,r]+Cos[t])^(3/2),{t,t0[l,r],tp[l,r]}] +
NN[r]*NIntegrate[((Pi/1)*r^2*Sqrt[x1[l,r]]*Cos[t]*(2*x1[l, r]^2+3*x1[l,r]*
   Cos[t]-1+Cos[t]^2)*Cos[t])/(x1[l,r]+Cos[t])^(3/2),{t,tm[l,r],Pi/2}];
Dal[l_,r_]:=(3*Pi*(-8*l^3*r+128*l*r^3-I*(3*l^4+32*l^2*r^2-256*r^4)*
   (Log[1+4*I*r]-Log[I*l + 4*r])))/(128*(8 + 3*Pi)*r^3*(l^2 + 16*r^2));

Plot[AOCD[s, 1], {s, 0, 5}, PlotRange->All,FrameLabel->{"s", "AOCD(s)"}];
```

[¶] If both spheres are partly light penetrable, the mean absorption coefficient for a suspension is connected with A_{OCD}. The optic properties of the suspensions of non-convex objects are modified, whereas those of convex particles are not.

4.3 Correlation function and scattering pattern of two infinitely long parallel cylinders

It is useful to analyze the scattering properties of two particles placed at a fixed distance s. This type of model involves at least two of the following parameters: One parameter describes the single particle (here the diameter d). The other length parameter describes a (typical) particle-to-particle distance s. In short, the pair correlation $g(r)$ (see Stoyan & Stoyan, 1992) [211, pp. 276–284] reduces to $g(r) = \delta(r - s_0)$.

In Section 4.2, the case of two (hard) spheres was analyzed [see Eq. (4.3)]. By analogy with this, the analytic expressions for two circular cylinders are known [76]. The Figs. 4.5 and 4.6 result from the program given below.

This program yields the SAS CF of two parallel infinitely long circular cylinders of const. diameter d, the axes of which are separated by a distance s. The calculation distinguishes between the two main configuration cases $0 < 2s < d < \infty$ and $0 < d < s < 2d$. The detailed geometric overlapping conditions (see Gille, 2001) [76, p. 183, figure 1] require the analysis of several sub-cases, which are involved in the following patterns: The utilizations [$d = 1, s = 10$ and $d = s = 1$], already included in Fig. 4.5, are as follows: Plot[{g1[r,1,10], g2[r,1,1]}, {r,0,15}]. The investigation shows that the "distance distribution" $p(r) \sim r^2 \gamma(r, d, s)$ is suitable for reflecting the geometric situation. The lengths s and d can be detected from this function (see the curves), especially for relatively big ratios s/d. The examples show that the CF is (sometimes) not very sensitive depending on the geometric constellation. Relatively small lengths s hardly result with high precision.

Based on $\gamma(r, d, s)$, the scattering pattern results via Fourier transformation from γ [see Section 1, Eq. (1.33)]. Figure 4.6 involves selected cases. The numerical realization is not trivial. Truncation errors can be reduced by use of the *Mathematica* function SequenceLimit[] (see Gille, 2007) [111] [||] and the examples given in the Appendix of the paper by (Gille, 2005) [99, pp. 59–61]. This approach is briefly explained in the following subsection.

Application of the *Mathematica* function *SequenceLimit* for reducing truncation errors of parametric integrals

Truncation errors are of great importance in the actual case of very long cylinders, where $L \to \infty$. The purpose of *SequenceLimit* is to take a finite sequence like $[f_1, f_2, f_3, f_4]$ and extrapolate it to "*the*" limit. Wynn's epsilon algorithm is used for this. This is perhaps the most abused *Mathematica*

[||]W. Gille, "Analyse endlicher Reihenterme, die Funktion *SequenceLimit* leistet Unglaubliches" (lecture, IX. Workshop Mathematica, Weierstrass Institute for Applied Analysis and Stochastics, Berlin, November 2007).

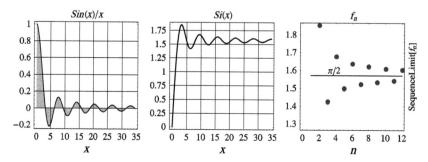

FIGURE 4.4

An analysis of the sequence f_n, where $n = 1(1)12$, allows an accurate approximation of the indefinite integral. Such an approach is useful for many practical cases. Clearly, the smaller the number of terms for f_n, the less precise the result. Example: If only 8 terms are inserted, i.e., $(0, 1.85194, 1.41815, 1.67476, 1.49216, 1.63396, 1.51803, 1.61609)$, the result is 1.57078. This is surprisingly close to the exact value of $1.5707963...$

function. In general, there is no way *that* a finite segment of an infinite sequence *can* determine the behavior of the sequence at infinity.

Let x_{max}, $x_{max} = (12-1)\pi$ for $n = 1(1)12$ be the upper integration limits of an indefinite integral (see Fig. 4.4).

$$f_\infty = \int_0^\infty \frac{\sin(x)}{x}\,dx, \quad f_n = \int_0^{x_{max}=(n-1)\pi} \frac{\sin(x)}{x}\,dx.$$

Actually, the exact result $f_\infty = \pi/2$ is known. The last point, f_{12}, is far from f_∞. Nevertheless, using 12 terms via SequenceLimit$[f_1, f_2, ..., f_{12}]$ results in an excellent approximation of the integral. The deviation from $\pi/2$ is smaller than 10^{-8}, as shown in the following two-liner:

```
trunc=Table[NIntegrate[Sin[x]/x,{x,0,(n-1)*Pi}], {n,1,12}];
SequenceLimit[trunc] - Pi/2 (* no additional options *)
Out = -2.43544 10^{-9}
```

This approach is tailor-made for long cylinders. The exact $P_1(h, d)$ functions are compared with the function $P_{1\infty}(h, d) = 1 - sin(hd)$, which is the simplest asymptotic approximation of the single infinitely long circular cylinder. For larger $h \cdot d$, the deviations between P_1 and $P_{1\infty}$ remain small regardless of the parameter s. The determination of the cylinder diameter is simpler than fixing an estimation of s [see also the corresponding functions $\gamma(r, d, s)$]. In general and near the origin (Taylor series), the normalized scattering pattern $I(h, d)$ of the infinitely long circular cylinder of height $H \to \infty$ is written

$$H \cdot I(h, d) = \frac{16\pi \cdot J_1(dh/2)^2}{d^2 h^3} = \frac{\pi}{h} - \frac{\pi d^2 \cdot h}{16} + \frac{5\pi d^4 \cdot h^3}{3072} - \frac{7\pi d^6 \cdot h^5}{294912} + 0[h]^6. \quad (4.20)$$

Since $I(h) \sim h^{-1}$, a simple normalization $I(0, d) = 1$ is not feasible. All infinitely long cylinders have this property.

```
(* Global case1: 0<2d<s *)
S[x_,d_]:=(d/4)*(d*Pi-2*d*ArcSin[x/d]-2*x*(Sqrt[d^2-x^2]/d));

gf1[r_, d_, s_]:=NIntegrate[2*S[r*Cos[\[Theta]],d]*Cos[\[Theta]]*(1/(Pi/2)),
{\[Theta],0,Pi/2},{\[Phi],0,Pi/2}]/(d^2*(Pi/2));

gf2[r_,d_,s_] :=NIntegrate[2*S[r*Cos[\[Theta]],d]*Cos[\[Theta]]*(1/(Pi/2)),
{\[Theta],ArcCos[d/r],Pi/2},{\[Phi],0, Pi/2}]/(d^2*(Pi/2));

gf3[r_,d_,s_]:=gf2[r,d,s]+
NIntegrate[1*S[Sqrt[s^2 + r^2*Cos[\[Theta]]^2 -2*r*s*Cos[\[Theta]]*
Cos[\[Phi]]],d]*Cos[\[Theta]]*(1/(Pi/2)),{\[Theta],0,ArcCos[(s-d)/r]},
{\[Phi],0,ArcCos[(r^2*Cos[\[Theta]]^2+s^2 - d^2)/(2*r*s*Cos[\[Theta]])]}]/
(d^2*(Pi/2));

gf4[r_,d_,s_]:=gf2[r,d,s] +
NIntegrate[1*S[Sqrt[s^2+r^2*Cos[\[Theta]]^2-2*r*s*Cos[\[Theta]]*
 Cos[\[Phi]]],d]*Cos[\[Theta]]*(1/(Pi/2)),
{\[Theta],ArcCos[(s+d)/r], ArcCos[(s-d)/r]},
{\[Phi],0,ArcCos[(r^2*Cos[\[Theta]]^2+s^2-d^2)/(2*r*s*Cos[\[Theta]])]}]/
(d^2*(Pi/2));
g1[r_,d_,s_]:=Which[Inequality[0,LessEqual,r,Less,d],gf1[r,d,s],
Inequality[d,LessEqual,r,Less,s-d],gf2[r,d,s],
Inequality[s-d,LessEqual,r,LessEqual,s+d],gf3[r,d,s],
Inequality[s+d,LessEqual,r,Less,Infinity],gf4[r,d,s]];

(* The global case 2: 0< d < s < 2d *)
gf12[r_,d_,s_]:=NIntegrate[2*S[r*Cos[\[Theta]],d]*Cos[\[Theta]]*(1/(Pi/2)),
{\[Theta],0,Pi/2},{\[Phi],0,Pi/2}]/(d^2*(Pi/2));

gf22[r_,d_,s_]:=NIntegrate[2*S[r*Cos[\[Theta]],d]*Cos[\[Theta]]*(1/(Pi/2)),
{\[Theta],0,Pi/2},{\[Phi],0,Pi/2}]/(d^2*(Pi/2)) +
NIntegrate[1*S[Sqrt[s^2+r^2*Cos[\[Theta]]^2-2*r*s*Cos[\[Theta]]*
  Cos[\[Phi]]],d]*Cos[\[Theta]]*(1/(Pi/2)),{\[Theta],0,ArcCos[(s-d)/r]},
{\[Phi],0,ArcCos[(r^2*Cos[\[Theta]]^2+s^2-d^2)/(2*r*s*Cos[\[Theta]])]}]/
(d^2*(Pi/2));

gf32[r_,d_,s_]:=NIntegrate[2*S[r*Cos[\[Theta]], d]*Cos[\[Theta]]*(1/(Pi/2)),
{\[Theta], ArcCos[d/r],Pi/2},{\[Phi],0,Pi/2}]/(d^2*(Pi/2)) +
NIntegrate[1*S[Sqrt[s^2+r^2*Cos[\[Theta]]^2-2*r*s*Cos[\[Theta]]*
  Cos[\[Phi]]],d]*Cos[\[Theta]]*(1/(Pi/2)),{\[Theta],0,ArcCos[(s-d)/r]},
{\[Phi],0,ArcCos[(r^2*Cos[\[Theta]]^2+s^2-d^2)/(2*r*s*Cos[\[Theta]])]}]/
(d^2*(Pi/2));

gf42[r_,d_,s_]:=NIntegrate[2*S[r*Cos[\[Theta]],d]*Cos[\[Theta]]*(1/(Pi/2)),
{\[Theta],ArcCos[d/r],Pi/2},{\[Phi],0,Pi/2}]/(d^2*(Pi/2)) +
NIntegrate[1*S[Sqrt[s^2+r^2*Cos[\[Theta]]^2-2*r*s*Cos[\[Theta]]*
    Cos[\[Phi]]],d]*Cos[\[Theta]]*(1/(Pi/2)),
 {\[Theta],ArcCos[(s+d)/r],ArcCos[(s-d)/r]},
 {\[Phi],0,ArcCos[(r^2*Cos[\[Theta]]^2+s^2-d^2)/(2*r*s*Cos[\[Theta]])]}]/
(d^2*(Pi/2));

g2[r_,d_,s_]:=Which[Inequality[0,LessEqual,r,Less,s-d],gf12[r,d,s],
Inequality[s-d,LessEqual,r,Less,d],gf22[r,d,s],
Inequality[d,LessEqual,r,Less,s+d],gf32[r,d,s],
Inequality[s+d,LessEqual,r,Less,Infinity],gf42[r, d, s]];
```

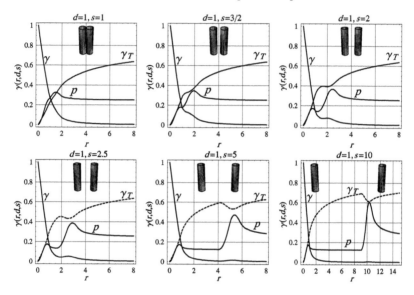

FIGURE 4.5
The functions $\gamma(r) = \gamma(r,d,s)$, $r^2\gamma(r,d,s)$ and $\gamma_T(r,d,s)$ for $d = const.$ and varying s (see Gille, 2001) [76, pp. 193–194].

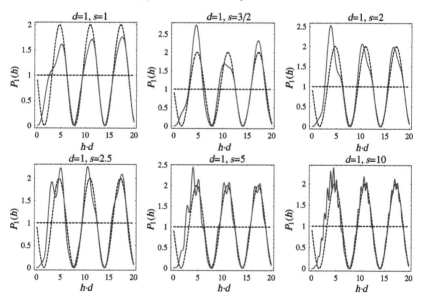

FIGURE 4.6
$P_1(hd)$ plots of two parallel, infinitely long cylinders of diameter d (full, highly oscillating curves). The Porod asymptote is the dashed horizontal line. Bigger values of s result in more oscillations. Dashed curve: $P_1(hd)$, single cylinder.

Applying the pair correlation function $g(s)$ to *two-particle ensemble models*

Subsections 4.2 and 4.2 are limited to the analysis of two particles, i.e., *two-particle ensemble models* were considered. To do this, a constant distance between the particle centers (from point O_1 to point O_2) was always inserted.

Now, the idea is to apply this approach to many *pairs of particles*. As a consequence of this modification, the particle-to-particle distance is no longer constant. Random distances s_i are introduced, stretching from one particle center to another one. These random distances s_i between a (very) large number of objects can be described by the functions $g(r)$ or $K(r)$ [see Eqs. (4.1) and (4.2) at the beginning of this chapter].

However, the meaning of $g(r)$ or $K(r)$ should not be overestimated. In fact, these functions are far from fixing the exact spatial position of the objects. Only the distances *between any two particle centers* are analyzed. Many theoretical g functions contradict the geometric realities.

The examples given above (see Section 4.2) involve the length variables s or l. The strategy is to drop the condition $s = const.$ or $l = const.$ This is a simple way to describe an ensemble of millions of particles. One consequence of this step is that all of the following considerations *hinge on the one function $g(s)$*. Analytic expressions for g (which is not a distribution function or distribution density function) are known in special cases only. There exist huge model calculations for $g(r)$ in statistical physics. Sometimes, approximations are mistakenly considered to be exact $g(r)$ functions. The *Percus-Yevick packing model* (see Torquato, 2002 [225]) is one of these seemingly perfect approaches.

The next sections/subsections are broken down as follows: Section 4.3 describes the relation between the CF $\gamma(r)$ in terms of $g(r)$ and c. In addition to some of the special cases discussed, the introduction of $g(r)$ in a certain interval $0 \leq r \leq r_{max}$ yields analytic expressions for the set covariance $C(r)$ of a system of equal spheres. Pair correlations consisting of a finite sum of Dirac δ functions describe special cases of sphere aggregates. Cylindrical particle cases are investigated in Section 4.4. All these considerations have been performed in $\mathbb{R}^3$ for a function $\gamma(r, L, g(r))$ or directly for a scattering intensity $I(h, L, g(r))$. Section 4.6 compares $g(r)$ in SAS with $g(r)$ in WAS (wide-angle scattering).

At the end of this chapter, Section 4.7 explains the connection between $g(r)$ and the CLDD $f(m)$. The surprising result is that $g(r)$ is not so far from $f(m)$. Both functions describe the intermediate space between the particles. There is a close connection to the volume fraction c, i.e., the smaller the particles at a fixed position are, the smaller c is and the more similar both of these different functions become.

4.4   Fundamental connection between $\gamma(r)$, c and $g(r)$

Let $g(r)$ be a well-known pair correlation function describing the distances between the centers of hard spheres of diameter d. Of course, no one will be able to write down a couple of functions at this point, because these types of functions are only known in selected special cases. A tessellation following from a Dead Leaves model (DLm) (see Serra, 1982, 1988 [205, 206] and Gille, 2005 [100]) results in the following $g(r, d)$ function

$$g(r,d) = \begin{cases} 0 \leq r < d: & 0 \\ d \leq r \leq 2d: & \dfrac{2}{2-\left[1-\frac{3}{2}\cdot\frac{r}{2d}+\frac{1}{2}\cdot(\frac{r}{2d})^3\right]} = \dfrac{32d^3}{(4d-r)(2d+r)^2} \cdot \\ 2d \leq r < \infty: & 1 \end{cases} \qquad (4.21)$$

For details, see (Stoyan & Schlather, 2000) [213].** It should be pointed out that Eq. (4.21), which involves the sphere diameter d, is not explicitly given in that paper. It is a conclusion based on the formulas developed by Stoyan and Schlather.

Equation (4.21) is useful because it is not an approximation. It makes it possible to create an analytic expression of the set covariance and SAS CF for arrangements of hard particles (see Gille, 2002) [80].

As early as 1950, Porod had developed the fundamental convolution connection between the CF of the particle ensemble $\gamma(r)$, the single particle CF $\gamma_0(r)$ and $g(r)$ for an inserted order range L. Porod's function of occupancy $Z(r)$ of the particle phase is the set covariance of the particles modified by the particle volume fraction c, where $C(r) = c \cdot Z(r)$. The connection with the CF is $\gamma(r)$, where $C(r) = c(1 - c) \cdot \gamma(r) + c^2$. Thus, the functions C, Z and γ involve the same information, which is slightly modified by c. Let the symbol $*$ denote a three-dimensional convolution integral extended over the whole space. The set covariance is then fixed in terms of γ_0, g, c and the (mean) hard particle volume V_0 by the fundamental connection

$$C(r) = c\left[\gamma_0(r) + \frac{c}{V_0}\cdot g(r)*\gamma_0(r)\right] \equiv c\gamma_0(r) + c^2 \cdot \frac{V_0 - [1 - g(r)]*\gamma_0(r)}{V_0}. \qquad (4.22)$$

Equation (4.22) involves two different terms which describe the random distances r_i between two points belonging to the particle phase. The term $\gamma_0(r)$ describes random distances *inside* the (same) single particle(s) of maximum diameter L_0. For spheres of constant diameter d, γ_0 exclusively depends on $d = L_0$, i.e., $\gamma_0 = \gamma_0(r, d)$. This term equals zero if $L_0 = d < r < \infty$. The convolution $1 * \gamma_0$ yields the particle volume $1 * \gamma_0 = V_0$.

**The following approach is based on considerations by the great physicist Günther Porod (Porod, 1951, 1952) [191, 192]. He introduced the so-called function of occupancy $Z(r)$ and the density autocorrelation function $\gamma(r)$ at the same time (see Chapter 1).

The *spatial convolution term* represents all the distances r_i between two independent points placed *in two different* single particles. The distribution law of these particle-to-particle distances r_i is traced back to a combination of $g(r) = g(r, d)$ with $\gamma_0(r) = \gamma_0(r, d)$. The convolution term disappears for $L < r$. The special pair correlation $g(r, d)$ given by Eq. (4.21) leads to $L = 3 \cdot L_0$.

A synonymous representation of Eq. (4.22) for spheres with $d = const.$ is

$$\gamma(r) = \gamma(r, d, c) = \frac{1}{1-c} \cdot \left[\gamma_0(r, d) - \frac{c}{V_0} [1 - g(r, d)] * \gamma_0(r, d) \right], \quad 0 \le r < L.$$
(4.23)

A detailed analysis of the convolution integral Eq. (4.23) leads to analytic representations. No vectors have to be considered for isotropic particle ensembles. With formal integration limits, Eq. (4.23) is written

$$\gamma(r) = \gamma(r, d, c) = \frac{1}{1-c} \cdot \left[\gamma_0(r, d) \right.$$
(4.24)

$$\left. - \frac{c}{V_0} \int_0^\infty \int_0^\infty \int_0^\infty [1 - g(x, y, z, d)] \cdot \gamma_0(x + x_0, y + y_0, z + z_0, d) dx dy dz \right].$$

The denotation $g(x, y, z, d)$ means $g(l, d)$ with $l = \sqrt{x^2 + y^2 + z^2}$. In detail, $1 * \gamma_0 = V_0$ gives

$$\int_0^\infty \int_0^\infty \int_0^\infty 1 \cdot \gamma_0(x + x_0, y + y_0, z + z_0, d) dx dy dz = V_0;$$
(4.25)

the integrand term $[1 - g(l, d)]$ simplifies, which results in the volume integral

$$\gamma(r) = \gamma(r, d, c) = \frac{1}{1-c} \cdot \left[\gamma_0(r, d) - c \right.$$
(4.26)

$$\left. + \frac{c}{V_0} \int_0^\infty \int_0^\infty \int_0^\infty g(x, y, z, d) \cdot \gamma_0(x + x_0, y + y_0, z + z_0, d) dx dy dz \right].$$

A new integration variable l describes all possible distances between the centers of two spheres A and B (see Fig. 4.7). Operating with $dV = dx dy dz = 4\pi l^2 dl$, the volume integral simplifies. Finally, by use of the probability $P_{AB}(r, l, d)$ (see the following pages), the following compact representation results from Eq. (4.26):

$$\gamma(r) = \gamma(r, d, c) = \frac{1}{1-c} \cdot \left[\gamma_0(r, d) - c + \frac{c}{V_0} \int_{r-d}^{r+d} P_{AB}(r, l, d) g(l, d) \cdot 4\pi l^2 dl \right].$$
(4.27)

With Eq. (4.27), the integration limits depend on r and $d = L_0$, which means that an immense interval splitting – as depending on l – must be considered. As a consequence of this interval splitting, each of the integral parts demands an intrinsic analysis [80, p. 19, Table 1]. Nevertheless, an analytic calculation of the integral in Eq. (4.27) (see Fig. 4.8) leads to unexpectedly simple expressions [80, p. 16]. A *Mathematica* program expression summarizes these formulas. Clearly, the scattering pattern derived must fulfill Porod's invariant [see Chapter 1, Eq. (1.36)], which it does.

Derivation of the geometric probability $P_{AB}(r, l, d)$

The term $P_{AB}(r, l, d)$ restricts to the r-interval $l - d \leq r \leq l + d$ (see Fig. 4.7). The calculation can be traced back to the geometrical covariogram $K(r, d) = V_0 \cdot \gamma_0(r, d)$ of a single sphere A of diameter d. By using this approach, the mean overlapping volume $V_{A'B}$ between sphere A' and B is analyzed in terms of the random angle ϑ with distribution density $\sin(\vartheta)$, where $2 \cdot P_{AB}(r, l, d) = \overline{V_{A'B}(r, l, d, \vartheta)}/V_0$. Applying the cosine theorem, this average results from

$$P_{AB}(r, l, d) = \frac{1}{2} \int_0^{\vartheta_2(r,l,d)} \left[1 - \frac{3}{2} \frac{t(r, l, \vartheta)}{d} + \frac{1}{2} \left(\frac{t(r, l, \vartheta)}{d} \right)^3 \right] \cdot \sin(\vartheta) d\vartheta,$$

$$t(r, l, \vartheta) = \sqrt{r^2 + l^2 - 2rl \cos(\vartheta)}. \qquad (4.28)$$

The upper integration limit ϑ_2 in Eq. (4.28) is defined by the condition that the spheres B and A' touch, i.e., their centers are separated by d. Hence, $\vartheta_2(r, l, d) = \arccos \left[(r^2 + l^2 - d^2)/(2rl) \right]$ results.

The evaluation of Eq. (4.28) requires two separate r-intervals, namely, $l - d \leq r$ and $r \leq l + d$. By use of the actual denotation, the compact result

$$P_{AB}(r, l, d) = \frac{d^5 - 5d^3(r - l)^2 + (5d^2 - (r - l)^2)|r - l|^3}{20ld^3r}, \ l - d \leq r \leq l + d,$$

$$\qquad (4.29)$$

is obtained. Equation (4.29) corresponds to Eq. (4.3). Analysis of $P_{AB}(r, l, d)$ (see [80, p. 14, Fig. 2]) shows the following: Smaller values of l result in higher specific peaks and a greater total probability of $\int_{l-d}^{l+d} P_{AB}(r, l, d) dr$.

The case $l \to \infty$ results in $\int_{l-d}^{l+d} P_{AB}(r, l, d) dr \to 0$. This is the limiting case of "two points" separated by a distance l. For a final check, Eq. (4.29) satisfies

$$V_0 = \int_{l-d}^{l+d} 4\pi r^2 \cdot P_{AB}(r, l, d) dr.$$

Similar spatial derivations for $P_{AB}(r, l, \ldots)$ should be possible for other particle shapes. The weakest and, at the same time, more complicated objective of such a project is handling the particle orientations. The angles of orientation depend on the particle shape(s). First steps in this direction restrict to numerical investigations based on *random mechanical particle packing* of ellipsoids, (Torquato, 2002) [225].

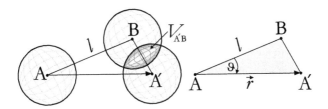

FIGURE 4.7
Geometric approach for the analysis of the function $P_{AB}(r) = P_{AB}(r, l, d)$ for two spheres of const. diameter d. The points A and B are the centers of the spheres, which are separated by l. The random angle ϑ possesses the distribution density $\sin(\vartheta)$ and describes the position of the r translated sphere A', where $AA' = r$. The overlapping volume depends on r, l, d and ϑ. The probability $P_{AB}(r, l, d)$ follows by averaging over the angle ϑ in the interval $0 \le \vartheta \le \pi/2$.

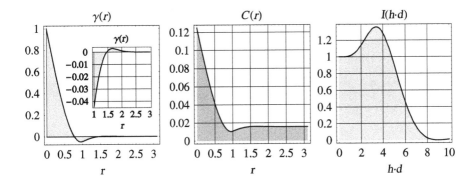

FIGURE 4.8
The functions $C(r)$, $\gamma(r)$ and the scattering pattern $I(h \cdot d)$, where $I(0) = 1$, derived for the ensemble of spheres [see the final Eq. (4.29) and the *Mathematica* program]. The special pair correlation inserted [Eq. (4.21)] results from a DLm. The positions of the sphere centers suffice for a strong short-order range. Thus, the pattern $I(h \cdot d)$ includes a clear maximum, which is approximately at $h \cdot d \approx 3.4$.
However, for larger r, the *disorder of the sphere centers* is much too big to allow an interpretation based on Bragg's equation. In contrast, the short-order range $r = d$ results from the non-overlapping condition of the hard particles.

The function $\gamma(r,d,c)$. Intrinsic properties of Eq. 4.27 are connected with the CLDDs of both phases (see the starting figure of Chapter 9 and Fig. 9.5). The initial slope of the CF is $\gamma'(0) = -12/(7d)$. By inserting the mean chord length outside the spheres $m_1 = 14d/3$ [the first moment of the CLDD $f(m)$] and the mean chord length inside the sphere(s) $l_1 = 2d/3$, the following relations are fulfilled:

$$|\gamma'(0)| = \frac{1}{l_p} = \frac{1}{l_1} + \frac{1}{m_1} \ \text{and} \ c = \frac{l_1}{l_1 + m_1}, \ 1 - c = \frac{m_1}{l_1 + m_1} \qquad (4.30)$$

(Rosiwal, 1898) [197]. Chapter 9 investigates the interrelations between the CLDD moments of the particle phase $\varphi(l)$ and outside the particles $f(m)$. In this connection, Eq. (4.30) is a special case. The following program represents the SAS CF $\gamma(r,d) = g[r,d]$ and the set covariance $C(r,d) = Co[r,d]$:

```
T1[r_, d_] :=
2/(45*d*r)*(r*(1080*d^4-1632*d^3*r + 160*d^2*r^2-259*d*r^3-4*r^4) -
4*(3*d + r)^2*(-3*d^3 + 32*d^2*r + 19*d*r^2 + r^3)*
Log[(3*d)/(3*d +r)]+4*(3*d-r)^3*(29*d^2-11*d*r+r^2)*Log[1-r/(3*d)]);


T2[r_, d_] :=
1/(5040*d^3*r)*(617107*d^7 - 1501255*d^6*r + 856982*d^5*r^2 -
237608*d^4*r^3 + 117047*d^3*r^4 + 1001*d^2*r^5 + 6*r^7 +
896*d^2*((5*d - r)^3*(5*d^2 - 5*d*r + r^2)*Log[-((3*d)/(-4*d + r))] +
(3*d - r)^3*(29*d^2 - 11*d*r + r^2)*Log[-((2*d)/(-4*d + r))] +
(d + r)^2*(131*d^3 + 108*d^2*r + 23*d*r^2 + r^3)*Log[(3*d)/(2*d + r)] -
(3*d + r)^2*(3*d^3 - 32*d^2*r - 19*d*r^2 - r^3)*Log[(4*d)/(2*d + r)]));


T3[r_, d_] :=
-(1/(5040*d^3*r))*(3*(116943*d^7 - 425523*d^6*r +
151438*d^5*r^2 - 67256*d^4*r^3 + 19747*d^3*r^4 + 35*d^2*r^5 +
2*r^7) + 896*d^2*((5*d - r)^3*(5*d^2 - 5*d*r + r^2)*
Log[(2*d)/(5*d - r)] + (d + r)^2*(131*d^3 + 108*d^2*r +
23*d*r^2 + r^3)*Log[(4*d)/(d + r)]));


c = 1/8;


g[r_, d_] :=
Which[Inequality[0, LessEqual, r, Less, d],
      1/(1-c)*(1-3/2*r/d+1/2*r^3/d^3-c+4*Pi*c/(1/6*Pi*d^3)*T1[r, d]),
      Inequality[d, LessEqual, r, Less, 2*d],
      1/(1-c)*(-c + 4*Pi*c/(1/6*Pi*d^3)*T2[r, d]),
      Inequality[2*d, LessEqual, r, Less, 3*d],
      1/(1-c)*(-c + 4*Pi*c/(1/6*Pi*d^3)*T3[r, d]),
      Inequality[3*d, LessEqual, r, Less, Infinity], 0];


Co[r_, d_] := c*((1 - c)*g[r, d] + c);


Plot[{Co[r, 1], g[r, 2]}, {r, 0, 6},
     PlotRange -> All, AxesLabel -> {"r", "g(r,1) and C(r,1)"}];
```

Up to this point, the equations do not include information about coordination numbers, particle numbers N or a particle number per volume. Sometimes, this is a disadvantage. The following subsections deal with synonymous representations of Eq. (4.27), which include specific particle numbers.

Smooth particles: $\gamma(r)$ for touching and non-touching particles

Smooth particles may or may not touch. This difference completely changes the behavior of $\gamma(r)$ near the origin. The CFs resulting from Eq. (4.27) include information about interparticle contacts in terms of the pair correlation and c. Let $z = g(d+)$ be the (mean) *coordination number of the first coordination sphere*, i.e., z is the average number of contacts that a given sphere has with its neighbors.

By assuming an ensemble of non-touching particles of volume V_0 and surface area S_0, inserting the corresponding pair correlation function $g(l)$ and operating with the integration limits $l = d$ to $l = r + d$,

$$\gamma(r) = 1 - \frac{S_0}{4V_0(1-c)} \cdot r + \left[\frac{c\pi \cdot g(d)}{4V_0(1-c)} + \frac{\gamma_0'''(0)}{6(1-c)}\right] \cdot r^3 + O\left(r^4\right) \quad (4.31)$$

for a specific c derived from Eq. (4.27). A quadratic term is missing in Eq. (4.31). Based on the total parameters $V_t = N \cdot V_0$ and $S_t = N \cdot S_0$ with $c = N \cdot V_0/V_t$, other representations for the linear term of the CF result. This can be written as

$$|\gamma'(0)| = \frac{1}{l_p} = \frac{1}{l_1(1-c)} = \frac{S_0}{4V_0 \cdot (1-c)} = \frac{N \cdot S_0}{4V_t c \cdot (1-c)} = \frac{S_t}{4V_t \cdot c(1-c)}. \quad (4.32)$$

The length l_1 denotes the mean chord length of the multi-chord distribution density of the particle [66]. Equation (4.32) coincides with the result in the quasi-diluted case for $r \to 0$ [see Eqs. (4.4) to (4.6)].

On the other hand, an ensemble of *touching* spheres (diameter d, volume V_0 and surface area S_0) possessing a *first-order coordination number* $z = g(d) = g(d+)/2 = z/2$ results in

$$\gamma(r) \approx 1 - \frac{S_0}{4V_0(1-c)} \cdot r + \frac{z}{24} \frac{\pi \cdot d}{V_0(1-c)} \cdot r^2 = 1 - \frac{3}{2d(1-c)} \cdot r + \frac{z}{4} \frac{1}{(1-c)} \cdot \frac{r^2}{d^2}. \quad (4.33)$$

This special formula for spheres of const. diameter d is in agreement with the investigations carried out by several other authors (Frisch & Stillinger, 1963 [39]; Torquato, 2005 [226, p. 38, (2.38) and references]). In fact, Eq. (4.33) involves a quadratic term, i.e., $0 < \gamma''(0)$. The special case of two single touching spheres (limiting case $c \to 0$ and $z \to 1$) is involved in Eq. (4.33). For this, the relation $\gamma''(0+) = 1/(2d^2)$ holds true (see the second derivative of Eq. (4.3) with $s = d$ and $r \to 0$).

Equation (4.33) demonstrates that the second derivative of the CF at the origin involves information about the mean coordination number and volume fraction. By combining the linear and quadratic term, a connection between the parameters d, c and z results: $g(d+) \sim \gamma''(0+)$. In detail,

$$g(d+) = z = 4d^2(1-c) \cdot \gamma''(0+) = 9l_1^2 \frac{m_1}{l_1 + m_1} \cdot \gamma''(0+) = 9l_1^2(1-c) \cdot \gamma''(0+).$$

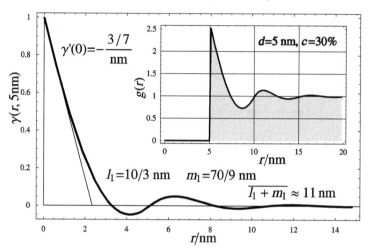

FIGURE 4.9

The function $\gamma(r)$ of an ensemble of hard spheres, where $d = 5$ nm and $c = 30\%$. The sphere packing is defined by the pair correlation function $g(r)$ (see insert of the Percus-Yevick model). The distance $r = d$ (see the finite jump) can be described by the coordination number(s) $z = g(d) \approx 1.295$ or $z = g(d+) \approx 2.59$. Based on the inserted parameters l_1, m_1 and c, it follows that $\gamma'(0) = \gamma_0'(0)/(1-c) = -3/(7 \text{ nm})$. The relation $\gamma''(0+) \approx 0.037/\text{nm}^2$ holds for the model. The coordination number $z = g(d+)$ is related to these model parameters via $z = 4d^2(1-c) \cdot \gamma''(0+) = 9l_1^2(1-c) \cdot \gamma''(0+)$. The estimation of the term $\gamma''(0+)$ has been explained by several authors in the literature [218, 219, 147, 110, 113].

This relation is of practical relevance. It fixes z in terms of the experimental parameters. This was investigated in detail for cases where $g(r)$ is known analytically, such as in the case of a Dead Leaves model (DLm) [80] and the Percus-Yevick approximation [188]. The latter model is applied in Fig. 4.9 for an ensemble of touching spheres of volume fraction c and coordination number z.

Analysis of non-touching fragment particles

The difference between Eq. (4.31) and Eq. (4.33) is of importance in connection with the *puzzle fitting function* $\Phi(r)$ (see Chapter 7). A clear geometric description of fragment particles, which could possibly belong to a puzzle originating from a tessellation, requires isolation of the particles. The scattering pattern of *touching* fragment particles does not contain enough information about the typical fragment shape(s). Even two touching fragments distort the function $\gamma''(r)$ near the origin.

Inclusion of the particle number N in the parameter set

The considerations analyzed up to now led to the SAS CF in terms of c, $g(s)$, $\gamma_0(r)$, as well as of the single particle $V_0 = v_{c0}$ in

$$\gamma(r) = \frac{1}{1-c} \cdot \left(\gamma_0(r) - c + \frac{c}{V_0} \cdot \int_0^\infty 4\pi s^2 g(s) \cdot P_{AB}(r,s)ds \right). \qquad (4.34)$$

The probability term $P_{AB}(r,s,d)$ in the integrand of Eq. (4.34) describes the probability of finding two points A and B, which are placed in two different single particles (spheres of maximum diameter $d = L_0$), and whose centers are separated by a (possibly random) distance s. The integration limits for two spheres can be specialized and $P_{AB} \equiv \gamma_{AB}$ is given explicitly by Eq. (4.3) [see also Eq. (4.27) and Fig. 4.7].[††]

For practice, especially for the cylinder packages analyzed in the next section, another representation of Eq. (4.34) is more effective. Let N, where $N = 1, 2, 3...$, be the particle number. This means that Eq. (4.34) is connected to an aggregate described by a specific order range L with a total volume $V_t = NV_0/c$, i.e., the following generalization describes a certain number of *interrelated particles* that belong to the sample. If $N = N(L)$ particles are involved in an agglomerate (or package) of total volume $V_t(L)$, further conclusions can be drawn for this order range L. First, there exists a certain average volume $V_a = V_a(L)$, which is available for a single particle belonging to the package. The introduction of this volume V_a is indispensable for establishing a synonymous representation of Eq. (4.34) which directly involves N. The limiting case *of one single particle*, i.e., $N = 1$, is allowed!

As an initial step, it is useful to distinguish between big and small particle numbers. Besides $c = N \cdot V_0/V_t$, the approximation $V_0/c \approx V_a/N$ is also valid for big N, such as $N > 10^4$ (in principle for $N \to \infty$). Hence, sometimes it is a clever step to substitute the term in front of the integral in Eq. (4.34) with $c/V_0 = N/V_a$. Clearly, the bigger N is, the better this holds.

However, if N is relatively small, i.e., for several particles, the (mean) particle volume V_0 cannot be neglected. Here, $V_0/c = V_a/N$ is definitely wrong, but

$$\frac{V_0}{c} = \frac{V_a - V_0}{N - 1}, \quad N = 1, 2, 3, ... \qquad (4.35)$$

holds true. In fact, $c \to 0$ follows for $N \to 1$. Thus, Eq. (4.34) is written with Eq. (4.35) for all N (including very small N down to $N = 1$) as

$$\gamma(r) = \frac{1}{1-c} \cdot \left(\gamma_0(r,d) - c + (N-1) \cdot \int_d^{r+d} \left[\frac{4\pi s^2 g(s)}{V_a - V_0} \right] \cdot P_{AB}(r,s,d)ds \right). \qquad (4.36)$$

The integrand term of Eq. (4.36) inside the brackets is of special interest. This term involves the dimension [1/m]. It represents a normalized distribution

---

[††]The case of cylinder packages is investigated in Section 4.4, where the cylinder specific function P_{AB} is also introduced.

density, which is the *pair correlation distribution density* $p(s) = p(s, L) = 4\pi s^2 g(s)/(V_a - V_0)$. The normalization $\int_0^L p(s)ds = 1$ holds.

In the plane case, the corresponding surface areas S_a and S_0 are inserted. Here, $p(s) = 2\pi s g(s)/(S_a - S_0)$ holds (see Section 4.4).

Hence, the steps from Eq. (4.34) to Eq. (4.36) lead to a separation of the factor $(N-1)$ in front of the integral. This is a useful normalization strategy. The special representation shown in Eq. (4.36) can be directly applied for $N = 1$, $N = 2$ and $N = 3...$ (see the following special constellations a), b), c) and d) below). The function $\gamma(r)$, which is derived in terms of the functions γ_0 and P_{AB} for IUR particle ensembles, is as follows:

a) For one single particle, in which $N = 1$ and possessing a limited volume $V_0 \ll V_t$, $c = 0$ and $\gamma(r) \equiv \gamma_0(r)$ result from Eq. (4.36). In this limiting case, the whole sample volume $V_t = V_t(L)$ is reserved for this one particle.

b) For two single particles, in which $N = 2$ and belonging to one package (for example two spheres of diameter d whose centers are separated by a fixed distance s_0, where $d \leq s = s_0$), the pair correlation distribution density equals $p(s) = \delta(s - s_0)$. In this special case, via the application of the shifting property of the δ function in $f(s_0) = \frac{\int_0^L \delta(s-s_0)\cdot f(s)ds}{\int_0^L \delta(s-s_0)ds}$, where $0 < s_0 < L$ and $c = 0$, Eq. (4.36) yields $\gamma_2(r, s_0, d) = \gamma_0(r) + P_{AB}(r, s_0, d)$.

c) For three single particles, in which $N = 3$ and whose centers form an equilateral triangle of side length s_0, Eq. (4.36) yields $\gamma_3(r, s_0, d) = \gamma_0(r) + 2 \cdot P_{AB}(r, s_0, d)$.

d) If four single particles, in which $N = 4$, are arranged in space with the pairs equidistant from each other, where $s = s_0$, the result is $\gamma_4(r, s_0, d) = \gamma_0(r) + 3 \cdot P_{AB}(r, s_0, d)$. This corresponds to a tetrahedral arrangement of the centers of the spheres. An analysis of this particle ensemble was done in Chapter 1, which is well-suited for discussing the data evaluation of an *all including scattering pattern* in terms of the order range parameter L (see Fig. 1.19). For geometric reasons, such a tetrahedral particle arrangement is not possible for long cylinders (see Section 4.4).

Equation (4.36) involves several parameters of practical relevance. Compared with the more general Eq. (4.34), this may have some advantages in terms of applicability. The following subsections deal with applications of Eq. (4.36) in selected special cases. First, a simple short-order model for hard spheres is presented. The approach of an isotropic two-phase ensemble used for this model is derived from Eq. (4.34). The relatively small volume fraction is denoted by c.

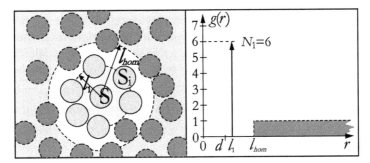

FIGURE 4.10

A *short-order range model* and the geometric parameters. The illustration of the spatial model applies circles instead of spheres and spherical shells.

Left: The (typical) sphere S of diameter d is surrounded by the N_1 spheres $S_1, S_2, \ldots, S_{N_1}$ of the same diameter. The length l_1 describes the typical distance from S to S_i. Actually, $N_1 = 6$ spheres belong to the *first coordination sphere*. However, a *second or even third coordination sphere* does not exist. No long order is assumed. This is illustrated by the set of many other spheres possessing a homogeneous spatial distribution. These particles are *placed at random positions* in a homogeneous region. This region starts at distances l from the sphere S, which possesses the property $l_{hom} < l < \infty$.

Right: The inserted pair correlation function $g(r)$ involves 4 parameters. Additionally, the volume fraction c belongs to the parameter set of the model. A δ function belongs to $g(r)$ at $r = l_1$. A nearly homogeneous particle distribution can be approximated by inserting $g(r) = const. = 1$ if $l_{hom} < r$. At the transition point, $g(l_{hom}) = 1/2$ holds.

4.3A Short-order range approach for hard spheres

A first *sphere of coordination* involves N_1 spheres. These requirements are modelled by the pair correlation function $g(r) = g(r, d, N_1, l_1, l_{hom})$, which includes 4 parameters. Figure 4.10 and Table 4.1 explain the model.

TABLE 4.1

Geometric meaning of the parameter set for the short-order range approach.

parameter	parameter	geometric meaning and limitations
1	d	sphere diameter; inside the whole order range L
2	l_1	size of the first coordination sphere
3	N_1	particles belonging to the *first coordination sphere*
4	l_{hom}	starting length for the homogeneous region of $g(r)$
5	c	volume fraction for the order range L; c is limited
6	L	order range of the approach $\gamma(L) = 0$

For geometric reasons, the parameters introduced in Table 4.1 are not independent. The exact formulation of limiting conditions is a complicated matter. The condition $0 < d < l_1 < l_1 + d$ is not sufficient at all. Naturally, the starting radius of the homogeneous region l_{hom} should be greater than d and l_1. Hence, a useful approach for the pair correlation is Eq. (4.37) below:

$$g(r) = g(r, d, N_1, l_1, l_{hom}) = N_1 \cdot \delta(r - l_1) + UnitStep[r - l_{hom}]. \qquad (4.37)$$

The SAS CF of the particle ensemble $\gamma(r)$ is defined by Eqs. (4.37) and (4.34). Therefore, it follows that

$$\gamma(r) = \begin{cases} 0 \le r \le l_{hom} + d : \frac{\gamma_0(r,d) - c + N_1 P_{AB}(r,l_1,d) + T_{hom}(r,l_{hom},d,c)}{1-c} \\ l_{hom} + d < r < \infty : 0 \end{cases} \qquad (4.38)$$

In Eq. (4.38), the term $(N - 1) \cdot P_{AB} = N_1 \cdot P_{AB}$ reflects all the particles belonging to the *first coordination sphere*.

Furthermore, the term $T_{hom}(r)$ reflects the homogeneous part of the pair correlation function [see Eq. (4.37)]. The following *Mathematica* pattern summarizes all of these terms [Eq. (4.38)]. This allows some room for playing around with the parameter set inserted. When doing this, it is important to include the scattering curve $I(h)$ in the calculation in order to detect the intrinsic limitations of the model. It is not difficult to find cases where the scattering pattern becomes negative. Such cases are neither considered nor plotted in Fig. 4.11. Clearly, cases like $N_1 = 20$ or $c = 0.7$ cannot be simulated.

```
(* Analysis of the CF gama[r,d,N1,l1,lhom,c] of the short order model *)
gamaf [r_, d_, n1_, l1_, lh_, c_] :=
    (g0[r, d] - c + n1*PAB[r, l1, d] + Th[r, d, lh, c])/(1 - c);
g0[r_, d_] := Which[0 <= r <= d, 1 - (3*r)/(2*d) + r^3/(2*d^3), True, 0];
PAB[r_, l1_, d_] := Which[l1 - d <= r <= l1,
    (d^2 - 5*(r - l1)^2 - ((5*d^2 - (r - l1)^2)*(r - l1)^3)/d^3)/(20*r*l1),
    l1 <= r <= l1 + d, (d^2 - 5*(r - l1)^2 +
    ((5*d^2 - (r - l1)^2)*(r - l1)^3)/d^3)/(20*r*l1), True, 0];
Th[r_, d_, lh_, c_] := Module[{x = r/d, y = lh/d},
    Which[x <= y - 1, 0, y - 1 <= x <= y,
    (c*(1 + x - y)^4*(9 + 2*x^3 + 8*x^2*(-1 + y) + 12*y*(1 + y)*(3 + y) -
    x*(1 + 40*y + 22*y^2)))/(70*x),
        y <= x <= y + 1,
    (1/(70*x))*(c*(9 + 35*x + 42*x^2 - 35*x^4 + 21*x^5 - 2*x^7 +
    42*(-1 + x)^3*(1 + x*(3 + x))*y^2 - 140*(-1 + x)^2*x*(2 + x)*y^3 +
    105*(1 - 3*x + 2*x^3)*y^4 - 84*(-1 + x^2)*y^5 + 70*x*y^6 -
    12*y^7)), True, c]];
```

The scattering patterns in Fig. 4.11, real and positive functions of practical relevance, result from these simulations. Figure 4.11 is far from illustrating every possible case. Hence, several tightly packed particle ensembles can be modeled on Eq. (4.34). However, all this hinges on the inserted pair correlation function. Following this line of reasoning, it is useful to make some remarks about the famous pair correlation approach by Perkus and Yevick, (1958) [188] (see next subsection).

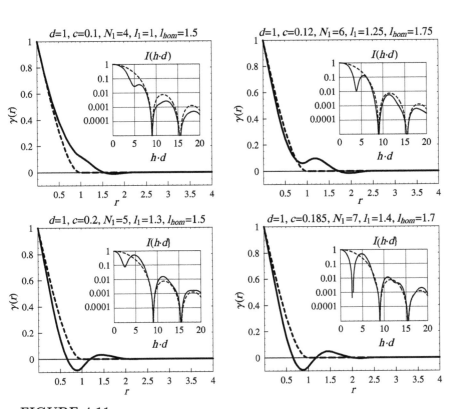

FIGURE 4.11

Application of the approach [Eq. (4.38)] by inserting different parameters in four cases.

The dashed curves show the single sphere case $d = 1$. The parameter set is reflected in the behavior of the functions $\gamma(r, d, c, N_1, l_1, l_{hom})$ and the scattering patterns $I(h \cdot d)$, $I(0) = 1$ (see the subfigures). The parameter cases represented fulfill Porod's invariant. However, it is not possible to insert arbitrary, independent parameters; i.e., there is a complicated interrelation between the parameters N_1, l_1, l_{hom}, d and c. It is interesting to study these interrelations via the *Mathematica* program given.

Obviously, a relatively large volume fraction c does not have to lead to a more extended negative part of γ.

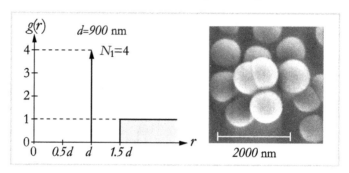

FIGURE 4.12

Cluster model of a typical sample of porous silica via the short-order range pair correlation introduced in this chapter (see Fig. 4.10). Based on such micrographs, the order range $L = 2\,000$ nm represents an optimum value for data evaluation of SAS experiments. Smaller L values decrease the packing information. Bigger L values do not essentially increase the information content about the packing of the spheres.

Modeling a nanoporous silica material for $L = 2\,000$ nm

Porous silica particles are universally applicable in materials science. Their morphology (see Figs. 4.12 and 4.13) can be analyzed with microscopes and by performing scattering experiments. When doing this, it is important to describe physical properties, such as particle size and specific surface area. For $L = 2\,000$ nm, the samples consist of nearly ideal spheres of unique size (Fig. 4.12). The spheres touch; however, a long-order range is not considered. The model applied does not involve a size distribution of the silica spheres.

The following parameters result from $I(h)$: $d = 900$ nm, $\gamma'(0) = -1/(600$ nm$)$, $l_p = 600$ nm, $c = 0.14$, $l_1 = 2d/3 = 600$ nm, $l_1 = l_p/(1 - c) = 700$ nm and $m_1 = l_p/c = 4\,000$ nm. Diameters of around $d = 1\,000$ nm exist in a sequence of micrographs.

The typical resolution limit of these experiments equals $2\,000$ nm (larger chords cannot be investigated). The specific surface area $[S/m]$ is about 0.8 m^2/g. This parameter can be traced back to l_p, c and the sample density ρ_L (the density for the specific L). It therefore holds that

$$[S/m] = 4|\gamma'(0)| \cdot c(1 - c)/\rho_L = 4c(1 - c)/(\rho_L \cdot l_p).$$

The cluster model described in Figs. 4.10, 4.12 and 4.13 can be applied for a large class of porous silica samples.

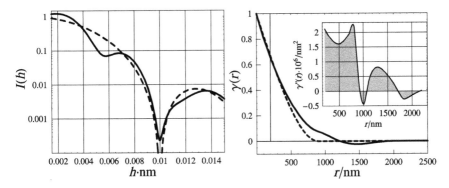

FIGURE 4.13

Typical scattering pattern $I(h)$ of porous silica samples in the interval $h_{min} = 0.0015$ nm$^{-1}< h < h_{max} = 0.015$ nm^{-1}.

In the interval 200 nm$< r < L = 2\,000$ nm, the functions $\gamma(r)$ (solid line) and $\gamma''(r)$ (filled plot) result. For this L, the particle volume fraction is $c(L) = 0.14$.

Furthermore, the experimental scattering pattern $I(h)$ and the function $\gamma(r)$ are compared with the case of a single sphere of diameter $d = 900$ nm (see dashed lines).

4.3B The function $g(r)$ from the Percus-Yevick approach

Based on the title of this section, a newcomer would most likely expect to find a direct analytic expression for $g(r)$, which is "similar" to the one given in Eq. (4.21). However, a direct analytic expression of $g(r)$ does not seem to exist. The pair correlation $g(r)$ results from two Fourier transformations (see the short summary below and Fig. 4.14).

Applying statistical mechanics, in 1958 the scientists Percus and Yevick [188] published a tricky convolution integral equation, which is related to the Ornstein-Zernike equation. The Percus-Yevick approximation (PYA) is commonly used in fluid theory. It gives an approximation for the pair correlation function of hard spheres.

An exact solution of the PYA integral equation for the hard sphere model exists (Wertheim, 1963, 1964)[232, 233]. This approach is mainly based on what is referred to as a *direct correlation function* $c_d(r)$, i.e., a function consisting of power terms in the variable r and the particle volume fraction c,

$$c_d(r) = c_d(r,c) = -\frac{(1+2c)^2}{(1-c)^4} + \frac{6\left(1+\frac{c}{2}\right)c}{(1-c)^4}\cdot\frac{r}{d} - \frac{c(1+c)^2}{2(1-c)^4}\cdot\frac{r^3}{d^3}. \qquad (4.39)$$

Equation (4.39) can be written in terms of the particle density (number of particles per volume), which is denoted by ρ. The interrelation between particle volume V_0, c and ρ is $c = V_0 \cdot \rho$, where $0 < V_0 < \infty$. [‡‡]

Based on the Fourier transformation theory for WAS, the direct CF is connected with the working function $C(Q)$ in reciprocal space by

$$c_d(r) = \frac{1}{2\pi^2\rho} \int_0^\infty C(Q)Q^2 \cdot \frac{\sin(Qr)}{Qr} dQ. \tag{4.40}$$

The inverse transformation is

$$C(Q) = 4\pi\rho \int_0^{L_0=d} c_d(r)r^2 \cdot \frac{\sin(Qr)}{Qr} dr. \tag{4.41}$$

The WAS pattern $I(Q)$ is connected with $C(Q)$ via $C(Q) = 1 - 1/I(Q)$. Hence, $I(Q) = 1/[1 - C(Q)]$.

Thus, $I(Q)$ is defined in terms of $c_d(r, c)$ and possesses the property $I(\infty) = 1$. Finally, the knowledge of $I(Q)$ via Eq. (4.41) is the starting point for representing $g(r) = g(r, d, c)$, which results from

$$g(r) = 1 + \frac{1}{2\pi^2\rho} \int_0^\infty [I(Q) - 1]Q^2 \cdot \frac{\sin(Qr)}{Qr} dQ. \tag{4.42}$$

Inserting $g(0) = 0$, Eq. (4.42) gives

$$\rho = \frac{1}{2\pi^2} \int_0^\infty [I(Q) - 1] \cdot Q^2 dQ = \frac{c}{V_0} = \frac{6c}{\pi d^3}. \tag{4.43}$$

Hence, the particle density is explicitly involved in the WAS pattern $I(Q)$.

By use of a *Mathematica* program, Eqs. (4.39) to (4.43) yield numerical results for the pair correlation $g(r, d, c)$ in terms of r, $d = const.$ and c, which is illustrated in Fig. 4.14 for varying volume fractions $c = 0.1$, $c = 0.2$, $c = 0.3$, $c = 0.35$ and $c = 0.365$. For higher volume fractions, the peaks in the WAS function $I(Q)$ [see Eq. (4.41)] increase (very) significantly. This is definitely wrong, i.e., the model breaks down for bigger c (see also Torquato, 2002) [225]. Small deviations can be detected when the PYA is compared with Eq. (4.21). For details consider the problem discussed in Section 10.3.2, where $c = 0.125$.

The PYA shows that the scientific results obtained in statistical physics can be readily applied across many fields, such as biology, crystallography, chemistry, physics and astrophysics. Even though the work by Percus and Yevick was published more than 50 years ago, it is still highly relevant today.

[‡‡]Actually, the amount of the scattering vector is denoted by Q instead of h in order to avoid confusion with the function $h(r)$, which is usually found in the literature for solving integral equations containing convolution terms.

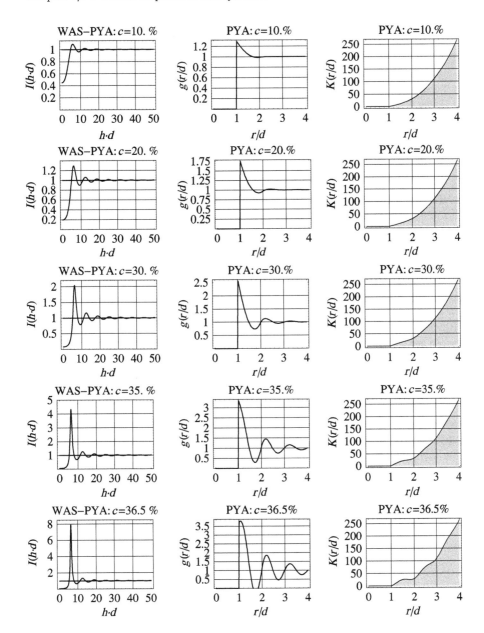

FIGURE 4.14

The PYA for selected volume fractions c, the WAS patterns and the derived functions g [see Eqs. (4.39) to (4.43)] and K [see Eq. (4.2)]. For $c \to 0$, g is a unit step function. The model is perfect for small c. However, see the contradiction $g(r/d) < 0$ in the last line case.

4.5 Cylinder arrays and packages of parallel infinitely long circular cylinders

Typical structure functions of packages of circular cylinders with parallel axes are considered. The paper by (Gille, 2004) [98, pp. 42–50] summarizes the *intrinsic properties of the SAS CF of packages of hard long parallel homogeneous circular cylinders.* In accordance with this, the function $g(s)$ describes the distribution law of the parallel cylinder axes, i.e., the three-dimensional problem is traced back to a two-dimensional one. The CF functions $\gamma(r, L, g(s))$ and scattering patterns $I(h, L, g(s))$ are discussed in (Gille, 2005) [99]. A short summary of those results follows.

The assumptions made about the cylinder ensemble are the following: The cylinders are arranged in more or less extended "packages" of largest diameter L and have a spatial IUR orientation. The initial figure of Chapter 4 illustrates a cylinder ensemble of four packages. Each of them contains long cylinders where $N = 16$. Let the number of packages be big enough to fulfill an isotropic IUR orientation of the z axes of all the packages. The maximum dimension of the length of the shortest cylinder is assumed to be large compared with the diameter d of the single cylinder. Let L_1 and L_2 be the largest diameters of a package perpendicular to the z axes and in a diagonal direction, respectively. The assumption is made that the smallest distance between two packages is greater than $Max[L_1, L_2]$ (quasi-diluted packages arrangement). The cylinders are tightly packed, inside the order range L_1. The interactions between any two packages are excluded. The volume fraction inside a package is c. Evidently, $g(s) = 0$ if $s < d$. No further special assumptions about the basic behavior of $g(s)$ are inserted and $g(s)$ describes the packing of the cylinders.

The SAS CF of the single cylinder $\gamma_0(r, d)$ results from its CLDD $A_0(r, d)$, as shown by (see Chapter 2)

$$A_0(r, d) = \begin{cases} 0 \leq r < d: \ 3r \cdot {}_2F_1\left(\frac{1}{2}, \frac{5}{2}; 3; \frac{r^2}{d^2}\right) \Big/ (4d^2) \\ d < r < \infty: 3d^3 \cdot {}_2F_1\left(\frac{1}{2}, \frac{5}{2}; 3; \frac{d^2}{r^2}\right) \Big/ (4r^4). \end{cases} \tag{4.44}$$

The actual representation is based on hypergeometric functions, i.e.,

$${}_2F_1(a, b, c, z) = 1 + \frac{abz}{c} + \frac{a(1+a)b(1+b)z^2}{2c(1+c)} + \frac{a(1+a)(2+a)b(1+b)(2+b)z^3}{6c(1+c)(2+c)} + O[z]^4.$$

The first moment of $A_0(r, d)$ equals d. Based on Eq. (4.44) with $\gamma_0(r, d) = \int_r^\infty (x - r) \cdot A_0(x, d)dx \Big/ \int_0^\infty x \cdot A_0(x, d)dx$, the formal integration of the hypergeometric function terms leads to the representation

$$
\gamma_0(r, d) =
\begin{cases}
0 \le r < d: & 1 - \dfrac{{}_2F_1\left(\frac{1}{2}, \frac{3}{2}; 3; \frac{r^2}{d^2}\right) r^3}{4d^3} - \dfrac{{}_2F_1\left(-\frac{1}{2}, \frac{3}{2}; 2; \frac{r^2}{d^2}\right) r}{d} \\[3mm]
d < r < \infty: & 1 - \dfrac{{}_2F_1\left(\frac{1}{2}, \frac{3}{2}; 3; \frac{d^2}{r^2}\right) d^2}{4r^2} - {}_2F_1\left(-\frac{1}{2}, \frac{3}{2}; 2; \frac{d^2}{r^2}\right).
\end{cases}
\tag{4.45}
$$

For the single cylinder CF defined by Eq. (4.45), the properties $\gamma_0(0, d) = 1$ and $\gamma_0(\infty, d) \to 0$ are fulfilled. In summary, for big r, where $d \ll r < \infty$, $\gamma_{0\infty}(r, d) = d^2/(8r^2) + d^4/(64r^4) + 5d^6/(1024r^6)\ldots + \ldots$ holds true.

Calculation of the CF $\gamma(r)$ of a cylinder package

Based on Eq. (4.45), the following calculations are analogous to those for ensembles of spheres [see Eqs. (4.34) and Eq. (4.36)]. The analysis of a package can be handled as a plane problem. Surprisingly, the parametric integrals obtained for this cannot be essentially simplified. Compared with the sphere case, the final CF of a cylinder package involves a more complex representation.

The starting point is Eq. (4.36), which involves the essential parameter N. This means that $\gamma(r)$ will result in terms of c, the *pair correlation distribution density* $p(s) = 4\pi s^2 g(s)/(V_a - V_0)$, $g(s)$ and $\gamma_0(r)$, the volume of the single cylinder V_0 and the volume V_a available for the single cylinder.

For the analysis of a cylinder package, Eqs. (4.35) and Eq. (4.36) (see Section 4.3.A) are modified further. First, the function $P_{AB}(r, s, d)$ is fixed for two parallel cylinders A and B (see Fig. 4.15 and compare with Fig. 4.7). A modified function P_{AB} is introduced. The modification concerns the volumes V_0 and V_a in terms of the right section of the cylinders and the functions $g(s)$ and $p(s)$, as well as with regard to the two-dimensional case.

The probability $P_{AB}(r, s, d)$ for two parallel infinitely long cylinders

The function $P_{AB}(r, s, d)$ is the geometric probability that the endpoint of a vector $\mathbf{r}$ of amount r, which starts at a point Q in cylinder A, is placed somewhere inside the other cylinder B. In analogy with Fig. 4.7, P_{AB} can be traced back to the overlapping integrals between two cylinders. The following two basic cases exist: Cylinder A overlaps the $\mathbf{r}$-translate of cylinder B, or cylinder B overlaps the $\mathbf{r}$-translate of cylinder A. In these cases, $P_{AB}(r, s, d)$ is represented as a parametric double integral operating with the direction angles ϕ and θ of vector $\mathbf{r}$ for IUR averaging in space. The integration limits depend on r, s and d. The two basic r-intervals that have to be distinguished are $s - d \le r \le s + d$ and $s + d < r < \infty$ (Fig. 4.15).

An analytic expression of P_{AB} requires an r-interval splitting for constant s and d. Therefore, the function $P_{AB} = P_{AB}(r, s, d)$ can be represented by

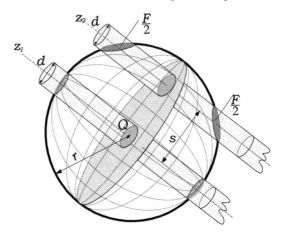

FIGURE 4.15

The probability $P_{AB}(r, s, d)$ for two infinitely long cylinders.

The parallel z axes of the cylinders are separated by the const. length s. The random point Q positioned everywhere in cylinder A is the center of a spherical shell of radius r which intersects B. The total surface area of both intersections $F = F(r, s, d)$ depends on the position of point Q. The probability P_{AB} is the averaged ratio $F/(4\pi r^2)$ when Q is placed at random positions in A. For a cylinder of finite height H, the probability P_{AB} fulfills the property $\int_{r=s-d}^{r=L} 4\pi r^2 P_{AB}(r, s, d, H)dr = \pi/4 \cdot d^2 H$, where $L = \sqrt{(s+d)^2 + H^2}$.

parametric double integrals as

$$P_{AB}(r, s, d) = \begin{cases} s - d \leq r < s + d : \dfrac{\int_0^{\cos^{-1}\left(\frac{s-d}{r}\right)} \int_0^{\phi_3} \frac{S(x,d)\cos(\theta)}{\pi/2} d\phi d\theta}{d^2 \pi/2} \\[3em] s + d < r < \infty : \dfrac{\int_{\cos^{-1}\left(\frac{d+s}{r}\right)}^{\cos^{-1}\left(\frac{s-d}{r}\right)} \int_0^{\phi_3} \frac{S(x,d)\cos(\theta)}{\pi/2} d\phi d\theta}{d^2 \pi/2} . \end{cases} \quad (4.46)$$

Here, the geometric covariogram of a circle of diameter d (the common overlapping surface area of two circles of diameter d, whose centers are separated by a distance x) is denoted by the function $S(x, d)$, $S(x, d) = d/4\left[\pi d - 2d\sin^{-1}(x/d) - 2x\sqrt{1 - x^2/d^2}\right]$. Furthermore, Eq. (4.46) uses both the abbreviations

$$x = \sqrt{s^2 - 2rs \cdot \cos(\theta) \cdot \cos(\phi) + r^2 \cdot \cos^2(\theta)}, \quad \phi_3 = \cos^{-1}\left(\frac{s^2 - d^2 + r^2\cos^2(\theta)}{2r\cos(\theta)}\right).$$

Examples of the probability $P_{AB}(r, s, d)$ for two long cylinders

Figure 4.16 demonstrates the typical behavior of $P_{AB}(r, s, d)$ in four cases. A fixed diameter $d = 1$ and the varying distance between the cylinder axes s

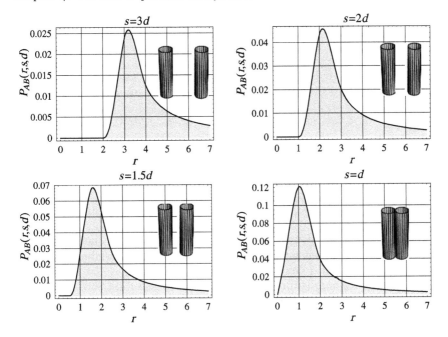

FIGURE 4.16
The case of two parallel infinitely long cylinders.
The function $P_{AB}(r, d, s)$ for cases $s = (3d, 2d, 1.5d, d)$. The smaller the ratio of s/d, the more the main part of P_{AB} shifts to relatively small r values. For analytic approximations of the probabilities $P_{AB}(r, s, d)$, see (Gille, 2004) [98].

are inserted. For the four configurations $\{s = 3d, s = 2d, s = 1.5d, s = d\}$, the total probabilities (i.e., the areas under the P_{AB} curves in Fig. 4.16) are $\{0.047, 0.097, 0.132, 0.21\}$. The closer the cylinders are, the greater the total probability.

For $d \ll r$, i.e., for $r \to \infty$, the geometric situation in Fig. 4.15 can be essentially simplified. For very big radii r of the spherical shell, i.e., for $s + d \ll r$, the surface area $F/2$ is nearly independent of the position of point Q of the center of the spherical shell. In this case, $P_{AB}(r, s, d)$ is defined by $P_{AB}(r, s, d) \approx (2 \cdot \frac{\pi}{4}d^2)/(4\pi \cdot r^2) = d^2/(8r^2)$.

If a special integral involving the function P_{AB} is considered by analogy with the sphere case, the result is

$$\int_0^L 4\pi r^2 P_{AB}dr = \int_0^{r_0} 4\pi r^2 P_{AB}dr + \int_{r_0}^L \frac{d^2}{8r^2} \cdot 4\pi r^2 dr = T_0(s, d) + \frac{\pi}{2}d^2(H - r_0).$$

Clearly, for $H \to \infty$, the integral involving the integrand function $4\pi r^2 P_{AB}(r, s, d)$ diverges.

Result of the CF of an IUR package [compare with Eq. (4.34)]

In the case $H \to \infty$, there is no difference between the volume fraction and area fraction of the cylinders in a package. Hence, Eq. (4.45) together with P_{AB} from Eq. (4.46) yield the connection between the functions $\gamma(r)$, $g(s,d)$, $P_{AB}(r,s,d)$ and $\gamma_0(r)$ (Gille, 2004) [98, p. 46], as shown by

$$\gamma(r) = \frac{1}{1-c} \cdot \left(\gamma_0(r,d) - c + \frac{c}{\frac{\pi}{4}d^2 \cdot H} \cdot \int_{\min[s-d,0]}^{\infty} H \cdot 2\pi g(s,d) P_{AB}(r,s,d) ds \right). \quad (4.47)$$

As correctly shown above, H can be reduced in Eq. (4.47). The term $\gamma_0(r,d)$ is defined by Eq. (4.45). The probability $P_{AB}(r,s,d)$ is given by Eq. (4.46). The isotropized set covariance $C(r)$ consists of two terms, as shown by

$$C(r) = c \cdot \left(\gamma_0(r,d) + \frac{c}{\frac{\pi}{4} \cdot d^2} \cdot \int_{\min[s-d,0]}^{\infty} 2\pi s \cdot g(s,d) \cdot P_{AB}(r,s,d) ds \right). \quad (4.48)$$

In analogy with the sphere case discussed in Section 4.3, there exist synonymous formulas which involve the number N of cylinders belonging to a package, which is shown below.

The cylinder volume is directly proportional to the right section area S_0, where $S_0 = \pi d^2/4$ and $c = N \cdot S_0/S_t$. Furthermore, taking into account the volume of the single cylinder and by analogy with Eq. (4.35) and Eq. (4.36), it follows that $S_t/N = (S_a - S_0)/(N - 1)$. Introducing the (mean) surface area S_a available for a circular right section S_0 into Eqs. (4.47) and (4.48) gives

$$\gamma(r) = \frac{1}{1-c} \cdot \left(\gamma_0(r,d) - c + (N-1) \int_{\min[s-d,0]}^{\infty} \left[\frac{2\pi s \cdot g(s)}{S_a - S_0} \right] \cdot P_{AB}(r,s,d) ds \right) \quad (4.49)$$

and

$$C(r) = c \cdot \left(\gamma_0(r,d) + (N-1) \int_{\min[s-d,0]}^{\infty} \left[\frac{2\pi s \cdot g(s)}{S_a - S_0} \right] \cdot P_{AB}(r,s,d) ds \right). \quad (4.50)$$

Equations (4.49) and (4.50) define $\gamma(r)$ and $C(r)$ in terms of c, $\gamma_0(r,d)$, N, S_a and S_0. The first term in the integrands of Eqs. (4.49) and (4.50) defines the pair distribution density $p(s)$ in the two-dimensional case. Hence, analytic results for $\gamma(r)$ and $C(r)$ follow based on a *proper* pair correlation function.

The special case of small volume fractions c and small r values is of practical relevance. Furthermore, this case allows some of the results to be checked: For small volume fractions, Eqs. (4.47) and (4.50) simplify. In the limiting case of relatively small r values, where $r \to 0$, the integral term disappears. Thus, $\gamma(r)$ [and $C(r)$ as well] do not depend on $g(s)$ and P_{AB}, regardless of the behavior of $g(r)$ near $r = d$. The same holds true for the first derivatives $\gamma'(r)$ and $C'(r)$.

Near the origin, $\gamma(r) \equiv \gamma(r,c) = [\gamma_0(r,d) - c]/(1-c)$ and $\gamma'(r) = \gamma_0'(r)/(1-c)$ hold. The first derivative of the set covariance $C'(r) = c \cdot \gamma_0'(r)$ follows from $C(r) = c(1-c) \cdot \gamma(r) + c^2$. These formulas hold exactly true for quasi-diluted particle ensembles (see Section 4.2). For a QDPE, a specific r-interval can be given [see Eq. (4.7)].

Touching cylinders and typical applications for $s = const.$

The behavior of $\gamma_0(r,d)$ for small r is defined by Eq. (4.45). The function $P_{AB}(r,s,d)$ disappears for $r \to 0$. It has a smooth behavior near the origin (Fig. 4.16). This also holds true for touching cylinders, where $s = d$.

In the following, Eqs. (4.49) and (4.50) are applied in three special cases (Gille, 2004) [98, p. 48].

Application 1: Two cylinders, where $N = 2$ and $c \to 0$, which are separated by $s = s_0$.

The first two applications possess only one fixed distance, i.e., $s = s_0$. Consequently, $p(s)$ collapses to the Dirac delta function $p(s) = \delta(s - s_0)$. Based on Eq. (4.49), the function $\gamma(r) = \gamma_2(r)$ (isotropized case of two isolated parallel cylinders) follows from Eq. (4.51) by inserting the $\delta(s)$ function instead of $p(r)$ and $N = 2$, which gives $\gamma_2(r, s_0, d) = \gamma_0(r,d) + (2-1) \cdot \int_0^\infty \delta(s - s_0) \cdot P_{AB}(r,s,d)ds$ and $\gamma_2(r, s_0, d) = \gamma_0(r,d) + P_{AB}(r, s_0, d)$. This special correlation function [Eq. (4.51] is in agreement with the special case of what is referred to as the infinitely diluted two-cylinder arrangement (see [76, 99]). Approximations of P_{AB} for small r, where $P_{AB} \approx P_{AB0}$, are discussed in [98].

Application 2: Three cylinders, where $N = 3$ and $c \to 0$, which are in a triangular arrangement separated by $s = s_0$.

With $N = 3$ and applying the assumed conditions,

$$\gamma_3(r, s_0, d) = \gamma_0(r,d) + (3-1) \cdot \int_0^\infty \delta(s - s_0) \cdot P_{AB}(r,s,d)ds,$$

$$\gamma_3(r, s_0, d) = \gamma_0(r,d) + 2 \cdot P_{AB}(r, s_0, d) \tag{4.51}$$

follows. Even in such simple special cases like Eqs. (4.51) and (4.52), different basic ratios for $2d/s_0$ clearly yield different results. Both of the curves given in Fig. 4.17(left) show a peak near $r = s_0$.

Application 3: A Dead Leaves model (DLm) defines $g(s,d)$ and $c \to 0.25$.

This is an application of the theory for the pair correlation of a DLm with circular grains of diameter d. Here, $c = 1/4$ is a fixed parameter. The pair correlation of such a plane DLm is

$$g(r,d) = \begin{cases} 0 \le r < d: & 0 \\ d < r \le 2d: & \frac{2}{2 - \beta_c(r,d)} \\ 2d \le r < \infty: & 1 \end{cases} \tag{4.52}$$

Here, the symbol $\beta_c(r,d)$ is an abbreviation of a transformed function of the geometric covariogram of a single circle of diameter d, as shown by

$$\beta_c(r,d) = 1 - \frac{r \cdot \sqrt{4d^2 - r^2}}{2\pi d^2} - \frac{2\arcsin[r/(2d)]}{\pi}$$

(see Stoyan & Schlather) [213]. In this case, the basic properties $\beta(0,d) = 1$, $\beta(d,d) = 2/3 - \sqrt{3}/(2\pi)$ and $\beta(2d,d) = 0$ result. Consequently, $g(2d,d) = 1$.

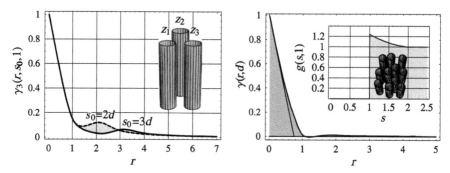

FIGURE 4.17

The SAS CF of IUR-cylinder packages for selected cases (see also the initial figure of this chapter).

The length s_0 denotes the distance between the cylinder axes, z_1, z_2, z_3, ...
Left: The function $\gamma_3(r, s_0, d)$ for a symmetrical triangular arrangement of three cylinders where $d = 1$, i.e., here $c = 0$. The special cases $s_0 = 3d$ (solid line) and $s_0 = 2d$ (dashed line) are considered. The shaded region indicates the difference between both cases.
Conclusion: For relatively small r, the parameter d can be estimated from the linear term $\gamma(r, d) \approx 1 - r/d + \dots$.
Right: The SAS CF of parallel cylinders, in which the distances between the z axes are described by a pair correlation function $g(s, d)$ resulting from a plane DLm (Stoyan & Schlather, 2001) (see insert).
In summary $g(s, d) = 2/(2 - [1 - 2/\pi \cdot \arcsin[s/(2d)] - s \cdot \sqrt{4d^2 - s^2}/(2d^2\pi)])$, where $0 \le r \le 2d$, holds true. The intermediate check values are $g(1, 1) = 1.24301$ and $g(1, 2) = 1$. The volume fraction is $c = 0.25 = const.$ The CF asymptotically approaches the axis of the abscissa. The cylinders can nearly touch. As a consequence of the non-zero volume fraction, the behavior of the CF for relatively small r is modified slightly as $\gamma(r, d) \approx 1 - r/[d(1 - c)] + \dots$.
The straight line hits the r axis at $r = 0.75 = -1/\gamma'(0) < 1$.

A graph of the pair correlation $g(r, 1)$ is included in the insert of Fig. 4.17. This function possesses a finite jump at $r = d = 1$. For Fig. 4.17 as a whole, $d = 1$ has been applied. The function $\gamma_{DLM}(r, d)$ follows in terms of $g(r, d)$ by application of Eq. (4.52). There exists a small interval where $\gamma(r) < 0$ for the disordered particle ensemble.

Near the origin, the CF is proportional to that of a single cylinder $\gamma_0(r)$. The diameter d results from $|\gamma'(0)| = |\gamma_0'(0)|/(1 - c) = (1/d)/(1 - c)$. For the three cylinder model case illustrated in Fig. 4.17(left), $d = 1/|\gamma'(0)|$ results with $c \to 0$.

4.6   Connections between SAS and WAS

Small-angle scattering (SAS) and wide-angle scattering (WAS) are interrelated (see Fig. 4.18). The interrelations between the scattering patterns
$I_{SAS}(h)$ and $I_{WAS}(h)$ will be discussed in the following sections.

4.6.1   The function $FREQ(r_k)$ describes all distances r_k

In WAS, small spheres (atoms) and their spatial positions are analyzed. A
typical size is $d = 0.1$ nm. The central part of Fig. 4.18 shows the frequency
$FREQ(r_k)$ of the distances between $N = 64$ unit sphere centers. The spheres
are touching. There exist 2016 vectors between any two sphere centers. An
investigation of the amounts of these vectors shows that there are $n = 18$
different distances r_k. These r_k stretch from $r_1 = 1$ and $r_2 = \sqrt{2}$,..., up to
$r_n = 3\sqrt{3}$. A distance $r = 0$ cannot exist for hard particles. The distance
$r_{18} = 3\sqrt{3}$ occurs 4 times, which describes the space diagonals of this sphere
matrix (cube). The function $FREQ(r_k)$ consists of a sum of 18 terms, where
$FREQ(r_k) = \sum_{k=1}^{18} w_k \cdot \delta(r - r_k)$ [see the plot $FREQ(r_k) = w_k(r_k)$].
    Obviously, a distance r_k either does not exist at all or it exists at least once.
The latter case is typical of the case of *orderless particles*, which always holds
true when r_k is very big. Consequently, $FREQ(\infty) = 1$ for $r_k \to \infty$. The
limiting property $g(\infty) = 1$ is typical of the pair correlation function $g(r_k)$ of
the atom centers of an amorphous phase or a real crystal.
    A discrete distance distribution density, denoted as the normalized function
$p(r_k)$, summarizes n probabilities. Actually, $\sum_{k=1}^{n=18} p_k(r) = 1$ holds true. By
inserting increasingly smaller spheres, the limiting case $N \to \infty$ describes a
homogeneous cube. Hence, for such a geometric "body" of volume V_0, the
function $p(r_k)$ is written $p(r_k) \to p_0(r) = 4\pi r^2 \gamma_0(r)/V_0$.
    Actually, γ_0 denotes the SAS CF of a cube of edge length a. The function
$p_0(r)$ involves the dimension $[1/m]$ and is referred to as the *distance distribution density of the body* (of the single particle *volume* of the set considered).
It involves the property of normalization $\int_0^{L_0} p_0(r)dr = 1$.
    The size, shape and spatial arrangement of such a *"huge sphere matrix
particle"* of largest diameter L_0 are the subject of SAS investigations. Since
$d \ll L_0$, in SAS the significant parts of the scattering pattern shift to very
small h, i.e., to very *small scattering angles* (small-angle scattering). Hence, in
SAS the scattering pattern of a compact, homogeneous particle is investigated
and the distances are described by the continuous distance distribution density
function $p_0(r)$. This is often referred to as *distance distribution* for short (see
Glatter, 1979–1991) [131, 133, 134, 136, 138].
    In the following, the connection between the distance distribution density
function $p_0(r)$ of a homogeneous body and $p(r_k)$ is explained in detail.

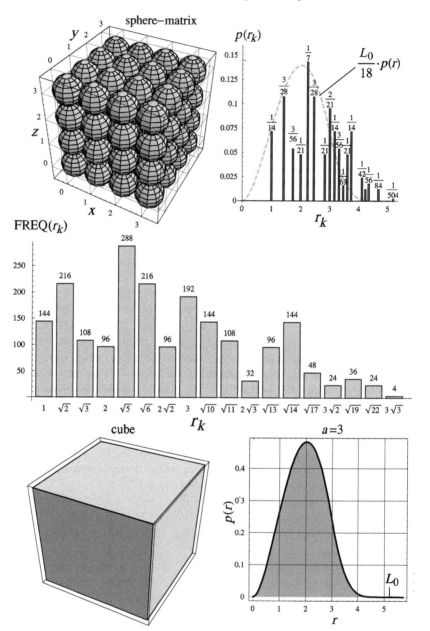

FIGURE 4.18

From WAS to SAS: WAS describes the characteristics of atom positions (actually, $N = 64$ lattice spheres). The term $FREQ(r_k)$ describes atom-atom distances. If $N \to \infty$, the homogeneous cube results, which is described by $p_0(r)$ and $L_0 = 3\sqrt{3}$.

Connection between the functions $p(r_k)$ and $p_0(r)$

These two functions cannot be directly compared. This becomes clear when comparing their dimensions: $p(r_k)$ is dimensionless, but $p_0(r)$ has a dimension $[1/m]$. In order to show that both functions are of the same origin, an additional normalization step of one of the functions is indispensable, which transforms one function (directly) into the other. The function values of $p(r_k)$ are probabilities for $n = 18$ non-overlapping classes, where $0 \leq r_k < 3\sqrt{3} = L_0$. Hence, the term factor $18/L_0 \approx 3.5\ [1/m]$ transforms the discrete values of $p(r_k)$ into the continuous distribution density $p_0(r)$.

For a sufficiently large number of distances n, $L_0/n \equiv p(r_k)/p_0(r)$ holds true (see the right upper subfigure in Fig. 4.18). The dashed line represents the function $p_k \approx (L_0/18) \cdot p(r)$. While this example is simple because the number of distances represented by $n = 18$ is small, there is still a slight difference between the terms $(L_0/18) \cdot p_0(r)$ and $p(r_k)$.

Connection between $I_{SAS}(h)$ and $I_{WAS}(Q)$: The functions $g(r)$ and $\gamma(r)$

Denoting the SAS pattern of a single sphere by $I_0(h)$, where $I_0(0) = V_0$, and that of the SAS particle ensemble by $i(h)$, $i(0) = v_c$, the Fourier transformation of Eq. (4.34) leads to

$$i(h) \cdot v_c = \frac{V_0 \cdot I_0(h)}{1 - c} \cdot \left(1 - \frac{c}{V_k} \cdot \int_0^\infty [1 - g(r)] \cdot 4\pi r^2 \cdot \frac{\sin(hr)}{hr} dr \right). \quad (4.53)$$

Here, V_k again represents the volume available for a sphere and $g(r)$ the pair correlation of the centers. Equation (4.53) has been long known in the field of SAS (Porod, 1951, 1952, 1982) [191, 192, 194].

Equation (4.53) can also be considered in the context of a WAS experiment (see Fig. 4.18). Modeling the scattering intensity of a single "atomic-sphere" by $I_0(h)$, the diameter d will be in the order of magnitude of 0.1 nm. For example, the atomic radius of a silver atom is about 0.16 nm. Hence, the essential information content of the scattering pattern extends to $Q_{max} = h_{max} = 50$ nm^{-1} and greater. This is about 20 times more than usual in the field of SAS, where a typical limit is about $h_{max} = 2.5$ nm^{-1}. In WAS, the pair correlation is of special interest. On the one hand, this function follows from an analysis of the assumed lattice structure. On the other hand, $g(r)$ explicitly results from Eq. (4.53), assuming that $i(h) = i(Q)$ represents the scattering pattern of a WAS experiment. Starting from Eq. (4.53), a formal calculation explicitly yields the term $4\pi r^2 \cdot [1 - g(r)]$ as shown by

$$4\pi r^2 \cdot [1 - g(r)] = \frac{2}{\pi} \cdot \int_0^\infty h^2 \cdot \frac{V_k}{c} \left(1 - i(h) \cdot \frac{v_c(1 - c)}{V_k I_0(h)} \right) \cdot \frac{\sin(hr)}{hr} dh. \quad (4.54)$$

Here, $i(h) \to i(Q)$ represents a WAS pattern recorded at a typical interval $0 < Q < 50$ nm^{-1}. The strategy of normalization $\lim_{Q \to \infty} \left[i(Q) \cdot \frac{v_c(1-c)}{V_k I_0(Q)} \right] = 1$ is usual. Hence, the function $g(r)$ is written

$$g(r) = 1 - \frac{1}{2\pi^2} \frac{V_k}{c} \int_0^\infty Q^2[1-i(Q)]\frac{\sin(Qr)}{Qr}dQ = 1 - \frac{\frac{1}{2\pi^2}\int_0^\infty Q^2[1-i(Q)]\frac{\sin(Qr)}{Qr}dQ}{\frac{c}{V_k}}.$$

(4.55)

Equation (4.55) automatically satisfies the relation $g(\infty) \to 1$. With respect to WAS, the term c/V_k denotes the number of atoms per volume element. With regard to WAS, this term equals the point density (particle center density) $\rho = c/V_k$. Hence, Eq. (4.55) gives

$$g(r) = 1 - \frac{\frac{1}{2\pi^2} \cdot \int_0^\infty Q^2[1-i(Q)] \cdot \frac{\sin(Qr)}{Qr}dQ}{\rho}.$$

(4.56)

The condition $g(0) = 0$ is fulfilled for the atoms considered. With respect to Eq. (4.56) this means

$$0 = 1 - \frac{\frac{1}{2\pi^2} \cdot \int_0^\infty Q^2[1-i(Q)]dQ}{\rho} \quad and \quad \frac{c}{V_k} = \rho = \frac{1}{2\pi^2} \cdot \int_0^\infty Q^2[1-i(Q)]dQ.$$

(4.57)

In the WAS case, the interrelations from Eq. (4.53) and Eqs. (4.56) to (4.58) can be summarized as

$$g(r) = 1 - \frac{\frac{1}{2\pi^2}\int_0^\infty Q^2[1-i(Q)]dQ}{\rho} = 1 - \frac{\frac{1}{2\pi^2}\int_0^\infty Q^2[1-i(Q)] \cdot \frac{\sin(Qr)}{Qr}dQ}{\frac{1}{2\pi^2}\int_0^\infty Q^2[1-i(Q)]dQ}.$$

(4.58)

In analogy with the SAS CF, which is normalized by $\gamma(0) = 1$ using the denominator term $\int_0^\infty h^2[I(h)]dh$, the condition $g(0) = 0$ is fulfilled by the normalizing term $\int_0^\infty Q^2[1-i(Q)]dQ$.

The following with regard to WAS should be noted: WAS experiments describe "spatial sphere arrays." If the arrays are limited in size, the size of their outer shape can be investigated by SAS. *The more spheres* there are and *the smaller* they are, the better this works.

The determination of the function $g(r)$ from the WAS pattern is analogous to that of the $\gamma(r)$ of an SAS pattern. There exist only the following two modifications: $i(Q \to \infty) = 1$ instead of $i(h \to 0) = 1$ and substituting $i(h)$ with $1 - i(Q)$. However, the data evaluation of WAS patterns requires consideration of the polarization factor $[1 + \cos^2(2\theta)]/2$. This term leads to a modification of $i(Q)$, especially for greater scattering angles. In SAS, such a modification can be neglected.

4.6.2   Scattering pattern of an aggregate of N spheres (AN)

Let AN be an aggregate consisting of N hard spheres of const. diameter d, the centers of which are separated by the lengths l_i. The total number of lengths $n_t = n_t(N)$ is $n_t = N(N-1)/2$. It is useful to write down these n_t

terms as $\sum_{i=1}^{n_t} \delta(l - l_i)$. In the following, the CF $\gamma_N(r,d)$ and the scattering pattern $I_N(h,d)$ for an AN are derived. This exercise is based on the functions $P_{AB}(r, l_i, d) = P_{AB}(r, l_i)$, which describe all the distances between every two spheres [see Eq. (4.29)]. In the case $N = 4$, there exist $n_t = 4 \cdot 3/2 = 6$ distances. Some distances may agree, but this is not necessarily the case.

Hence, the SAS CF $\gamma_N(r,d)$ of the AN can be pieced together from the sum

$$\gamma_N(r) = \frac{N \cdot \gamma_0(r,d) + 2 \cdot \sum_{i=1}^{n_t} P_{AB}(r, l_i)}{N} = \gamma_0(r,d) + \frac{2}{N} \cdot \sum_{i=1}^{n_t} P_{AB}(r, l_i). \quad (4.59)$$

The factor 2 in front of the sum corresponds to the fact that two spheres are always interrelated; however, P_{AB} is normalized for one sphere [see Eq. (4.28)].

In the A2 case, where $N = 2$, $n_t = 1$ results. This yields $\gamma_2(r) = \gamma_0(r) + P_{AB}(r, l_0)$. For a tetrahedral configuration A4, $N = 4$, $l_i = l_0 = const.$ and $n_t = 6$, Eq. (4.59) results in $\gamma_4(r) = \gamma_0(r,d) + \frac{2}{4} \cdot \sum_{i=1}^{6} P_{AB}(r, l_i) = \gamma_0(r,d) + 3P_{AB}(r, l_i)$.

This example shows that in many cases the sum term in Eq. (4.59) can be simplified. This is possible by taking into account the function $FREQ(l_i)$, where $FREQ(l_i) = \{(w_1, l_1), (w_2, l_2), (w_3, l_3), ..., (w_n, l_n)\}$, which completely describes all the n distances that exist in the AN in a compact form. For the example in Fig. 4.18, these frequencies are defined by a list with 18 elements $FREQ(l_i) = \{(1, 144), (\sqrt{2}, 216), ..., (4, 3\sqrt{3})\}$. The ith distance "repeats" w_i times (see Fig. 4.18). Denoting the number of different l_i by n (see last section), in most practical cases $n \ll n_t$. Without any approximation, see also case A64, where $n_t = 2016$ and $n = 18$, the sum $\sum_{i=1}^{n_t} P_{AB}(r, l_i)$ is written more shortly

$$w_1 \cdot P_{AB}(r, l_1) + w_2 \cdot P_{AB}(r, l_2) + ... + w_n \cdot P_{AB}(l_n) = \sum_{i=1}^{n} w_i \cdot P_{AB}(r, l_i). \quad (4.60)$$

Thus with Eq. (4.60), Eq. (4.59) is written as

$$\gamma_N(r) = \gamma_0(r,d) + \frac{2}{N} \cdot \sum_{i=1}^{n} w_i \cdot P_{AB}(r, l_i). \quad (4.61)$$

Equation (4.61) fixes the scattering pattern $I_N(h,d)$ in terms of $\gamma_N(r,d)$. The application of the transformation $I_N(h,d) = \int_0^L 4\pi r^2 \gamma_N(r) \cdot \frac{\sin(hr)}{hr} dr / \int_0^L 4\pi r^2 \gamma_N(r) dr$ to both terms involved in Eq. (4.61) yields the normalized scattering curve $I_N(h) = I_N(h,d)$

$$I_N(h) = \frac{\int_0^d 4\pi r^2 \gamma_0(r) \frac{\sin(hr)}{hr} dr + \frac{2}{N} \sum_{i=1}^{n} \left(w_i \int_{d-l_i}^{d+l_i} 4\pi r^2 P_{AB}(r, l_i) \frac{\sin(hr)}{hr} dr \right)}{\int_0^d 4\pi r^2 \gamma_0(r) dr + \frac{2}{N} \cdot \sum_{i=1}^{n} \left(w_i \int_{d-l_i}^{d+l_i} 4\pi r^2 P_{AB}(r, l_i) dr \right)}.$$

$$(4.62)$$

Denoting the scattering pattern of a single homogeneous sphere by $I_0(h \cdot d/2) = [3(\sin(hd/2) - h \cdot d/2 \cos(h \cdot d/2))/(h^3 \cdot (d/2)^3)]^2$ and inserting the integrals

$$\frac{1}{V_0} \cdot \int_{d-l_i}^{d+l_i} 4\pi r^2 P_{AB}(r,l) \cdot \frac{\sin(hr)}{hr} dr = I_0(h \cdot d/2) \cdot \frac{\sin(hl)}{hl},$$

$$\frac{1}{V_0} \cdot \int_{d-l_i}^{d+l_i} 4\pi r^2 P_{AB}(r,l) dr = 1, \qquad (4.63)$$

into Eq. (4.62) results in a compact formula for the AN scattering pattern. In detail, Eq. (4.62) with Eq. (4.63) yields

$$I_N(h,d) = \frac{V_0 \cdot I_0(h \cdot d/2) + \frac{2}{N} \cdot \sum_{i=1}^n \left[w_i \cdot V_0 I_0(h \cdot d/2) \cdot \frac{\sin(h \cdot l_i)}{h \cdot l_i} \right]}{V_0 + \frac{2}{N} \cdot \sum_{i=1}^n [w_i \cdot V_0]},$$

$$I_N(h,d) = \frac{I_0(h \cdot d/2) + \frac{2}{N} \cdot I_0(h \cdot d/2) \cdot \sum_{i=1}^n \left[w_i \cdot \frac{\sin(h \cdot l_i)}{h \cdot l_i} \right]}{1 + \frac{2}{N} \cdot \frac{N(N-1)}{2}}, \qquad (4.64)$$

$$I_N(h,d) = \frac{1}{N} \left[I_0(h \cdot d/2) + \frac{2}{N} \cdot I_0(h \cdot d/2) \cdot \sum_{i=1}^n \left(w_i \cdot \frac{\sin(h \cdot l_i)}{h \cdot l_i} \right) \right].$$

Equation (4.64) is a universal approach for the development of the scattering pattern on an AN. The scattering patterns of relatively complicated particle shapes can be approximated by a suitable AN, for example the modeling of macromolecules in the field of molecular biology [164]. Figure 4.19 includes the A64 result for the *sphere matrix* shown in Fig. 4.18. For $N = 2$, Eq. (4.64) collapses to the special scattering pattern

$$I_2(h,d) = I_0(h \cdot d/2) \cdot \left[1 + \sin(hl_1)/(hl_1) \right]/2,$$

which was discussed in the textbook by Guinier & Fournét (1955) [143, pp. 141–142, figure 46].

A *Mathematica* module spherematrixAN[h_, d_, freq_] [see Eqs. (4.64)] yields the scattering pattern shown in Fig. 4.19 of an AN sphere matrix:

```
(*the list freq: {{distance1,frequency1},{distance2,frequency2}...} *)
freq = {{1,144}, {Sqrt[2],216}, {Sqrt[3],108}, {2,96}, {Sqrt[5],288},
{Sqrt[6],216}, {2*Sqrt[2],96}, {3,192}, {Sqrt[10],144}, {Sqrt[11],108},
{2*Sqrt[3],32}, {Sqrt[13],96}, {Sqrt[14],144}, {Sqrt[17],48},
{3*Sqrt[2],24}, {Sqrt[19],36}, {Sqrt[22],24}, {3*Sqrt[3],4}};

(* Calculation of the scattering pattern of an AN sphere-matrix. *)
spherematrixAN[h_, d_, freq_] := Module[{n,nt,nN,x,i0,result},
    n = Length[freq]; nt = Sum[freq[[i,2]], {i, 1, n}];
    nN = (1/2)*(1 + Sqrt[1 + 8*nt]); x = h*(d/2);
    i0 = Which[x == 0, 1, True, 9*((Sin[x]-x*Cos[x])/x^3)^2];
result = (1/nN)*(i0 + (2/nN)*i0*Sum[freq[[i,2]] *
    Which[h == 0,1,True,Sin[h*freq[[i,1]]]/(h*freq[[i,1]])],{i,1,n}])];
(* This is a plot operating with the actual list freq. *)
Plot[spherematrixAN[h, 1, freq],{h, 0, 8}]
```

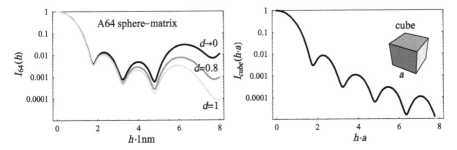

FIGURE 4.19

Approximation of a compact body (here the cube) via an ensemble of spheres (see the *Mathematica* module spherematrixAN[h_,d_,freq_]).

Left: Scattering patterns of an A64 sphere matrix (three different sphere diameters $d = 1, d = 0.8$ and $d \to 0$).

Right: Scattering pattern of a cube of edge $a = 4$. The first three minimum positions are at $h \cdot 1$ nm $= 1.7$, $h \cdot 1$ nm $= 3.2$ and $h \cdot 1$ nm $= 4.8$. For relatively small scattering vectors h, where $0 \leq h < 3$, the scattering patterns nearly agree. Therefore, the AN model works as an approximation. Naturally, the bigger h is, the greater the difference between the curves. For $h \to \infty$, the scattering of the homogeneous body decreases more rapidly. The information contained in the AN matrix for distances $r \to 0$ relates touching spheres.

Figure 4.20 illustrates another application of Eq. (4.64). Here, the symbols "SAS" embedded in an xyz coordinate system are represented via a sphere matrix array AN with $N = 52$. The single sphere diameter d can vary, as in $0 < d \leq 1$ nm. The standard case $d = 1$ nm is used in the figure. The maximum diameter of this sphere matrix array is $L \approx 16$ nm. A smaller order range L describes details, but not the whole aggregate. The solid curves in Fig. 4.20 can be approximated by the parabola (dashed line in Fig. 4.20, $I(h) = 1 - [R_g(L)]^2 h^2 / 3$, where $0 \leq h < 0.15$ nm^{-1}). The radius of gyration of the whole aggregate is $R_g = R_g(L) \approx 5.2$ nm [see also first pages of Chapter 1, details in Fig. 1.1, Fig. 1.11 and Eq. (10.24)].

Any particle shape can be approximated by use of a suitable sphere aggregate of N spheres. This technique is applied in molecular biology for modeling the scattering of macromolecules (see Damaschun, et al. 1969; Glatter & Kratky, 1982; Kratky, 1983; Feigin & Svergun, 1987) [25, 136, 164, 36]. In order to do this, a basic scattering model of the macromolecule under investigation must be known beforehand. This approach is suitable for excluding those models of the simulated scattering patterns that do not conform with the experimental data (Kratky, 1983) [164, pp. 34–41].

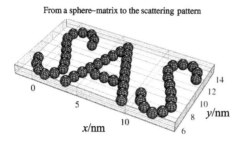

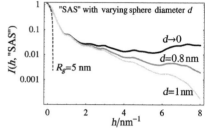

FIGURE 4.20

From a real space A52 sphere matrix, which represents the *abbreviation* "SAS", to the corresponding small-angle scattering pattern $I(h, SAS)$.

The symbols **"SAS"** are represented by $N = 52$ spheres of diameter d, where $0 \leq d \leq 1$ nm. There exist 1326 vectors between any two sphere centers. The module spherematrixAN[h, d, freq] calculates the scattering pattern $I(h, SAS)$.

The list $freq$ contains 1223 different lengths and their frequencies. The largest diameter of the sphere matrix is about $L = 15.52$ nm $+d$. This maximum length exists only once.

In analogy with Fig. 4.19, three diameters d are considered, where $d = 1$ nm, $d = 0.8$ nm and $d \rightarrow 0$. In the interval 0.005 nm$^{-1} < h < 2$ nm^{-1}, which is a typical h interval of an SAS experiment, the three curves nearly agree.

The scattering pattern is influenced by smaller distances of about 1 nm and larger distances of about (5–12) nm. However, the larger ones significantly impact $I(h)$ in $h < 2$ nm^{-1}. Hence, it is really possible to reach a fine approximation of the scattering pattern of the sphere aggregate in $h < 2$ nm^{-1}, by operating with a *point matrix* instead of a *sphere matrix*.

4.7 Chord length distributions: An alternative approach to the pair correlation function

The theory investigated in this chapter is essentially based on the pair correlation between two particles. However, it should be mentioned that there are other, frequently more effective, strategies for describing ensembles of particles.

Let $\varphi(l)$ be the chord length distribution density (CLDD) of the particle phase. Additionally [somewhat in analogy to $g(r)$], the CLDD $f(m)$ describes the distribution law of random chord lengths outside the particles (inside the matrix phase). In order to do this, a certain order range L is selected.

The functions φ, f and γ are closely interrelated (Gille, 2000) [66]. For a detailed description see the paper *Intersect distributions of a "Dead Leaves" model with spherical primary grains* [100]. One of the three functions can be determined if the other two are known. A known c simplifies this task, but it is not absolutely necessary to know c beforehand (see Chapter 8). Two special Boolean models (Bms) are analyzed in this context in Chapter 5, in which the functions γ'', φ and f are explained for *a Bm with spherical primary grains* and the *Poisson slice model* (PSm). Chapter 9 will summarize and further generalize the theory, which discusses the interrelations between the moments of the CLDD and c. The CLDD approach is briefly explained with several examples below:

It is useful to consider the DLm to start (Fig. 4.22). Based on the analytic results of Section 4.3, the function $f(m)$ is derived in terms of γ'' and φ. The particle shape fixes the CLDD $\varphi(l)$. The CLDD $f(m)$ describes "some details of the shape/size" of the homogeneous matrix phase. In other words, it characterizes random distances between "particle ends" and "particle beginnings," which always come from the *mark side* (from the matrix phase).

In many cases, the experimental data has been interpreted via the three functions γ'', φ and $f(m)$ (e.g., see applications for precipitations in Al-Zn-Mg alloys) (Gille, 1983) [48, pp. 79–83]. In this case, homogeneous, spherical precipitates of const. diameter and the function $\gamma(r)$ are inserted in order to derive $f(m)$ explicitly.

The approach has been used to analyze micropowders (see Fig. 4.23). For other examples, see (Gille, 2000) [66, pp. 77–379]. That paper also includes a study of different chord types, such as μ-, ν- and λ-chords. The meaning and some applications of ν-chords are explained in chapter 8.

The CLDD $f(m) = f_\mu(m)$ describes μ chords (typically called IUR chords) (see Chapter 1, Fig. 1.9). If the index μ is omitted, the meaning is always μ, i.e., $A(r) = A_\mu(r)$.

In each case, the functions φ and f fix the behavior of γ'' (Méring & Tchoubar, 1968) [182]. The inverse procedure, i.e., the the separation of φ and f is a complex task. In most cases, the CLDDs φ and f cannot be separated explicitly. Nevertheless, a discussion of γ'', based on the point of view that γ'' is finally *composed of* φ and f, is useful. Following this strategy, the next chapters involve examples.

The information involved in the scattering pattern of a polyvinyl chloride powder (PVC powder) (see Fig. 4.23) allowed to perform such a way of interpretation. For this, based on typical PVC micrographs, the order range $L = 20$ nm was fixed beforehand. The investigation of PVC powders is currently an obsolete procedure in materials research. Nevertheless, such investigations were standard in the 1970s, when the author was involved in the characterization of the morphology of PVC and its derivatives (mechanical and electrical properties).

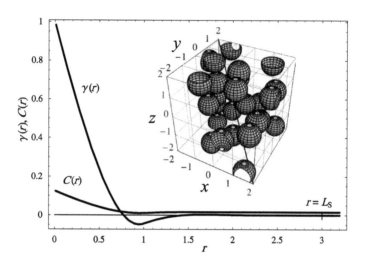

FIGURE 4.21
An ensemble of spheres $d = 1$, the centers of which are described by a specific pair correlation function [see Eq. (4.21)]. The set covariance $C(r)$ and $\gamma(r)$ follow, where $L_s = 3d$. An *infinitely long straight line* intersects many particles leading to the specific CLDDs $\varphi(l)$ and $f(m)$ (see Fig. 4.22).

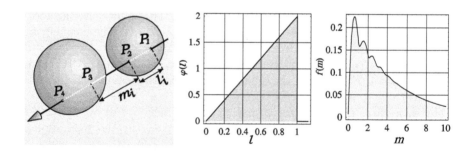

FIGURE 4.22
Analysis of a DLm with spherical grains of diameter $d = 1$ via the two CLDDs $\varphi(l)$ and $f(m)$.
The CLDD $\varphi(l)$ describes the particles, but $f(m)$ describes the chord lengths outside the spheres. Sometimes, $f(m)$ is referred to as the "*mark distribution density*." The first moments of the CLDs are $l_1 = 2d/3$ and $m_1 = 14d/3$. Based on Monte Carlo simulations it was shown that (for many hard particle cases) $f(m)$ is described by an exponential function if $m \to \infty$. This approximation breaks down for smaller arguments m (see the right part for $m < 3$).

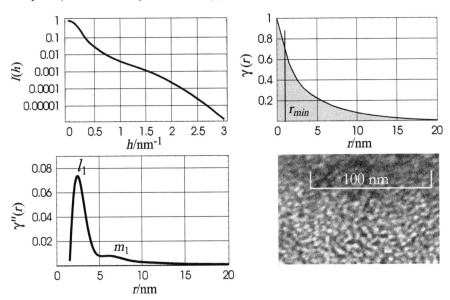

FIGURE 4.23

Investigation of a PVC micropowder via SAS and a transition electron microscope for $L = 20$ nm.

The normalized scattering pattern $I(h, L)$ fixes the resolution limit $r_{min} = \pi/3.02$ nm ≈ 1 nm, which is marked by a line in the field of the CF. The parameters estimated from the functions $\gamma(r, L)$ are the following: Porod length $l_p = 2.1$ nm; mean chord length of the primary particles $l_1 = 2.8$ nm; mean chord length of the gaps $m_1 = 7$ nm; volume fraction $c(L) = 25\%$. The estimated parameters can be verified from the transmission micrograph of this powder. The resulting mean particle diameter of about 4 nm corresponds approximately to $l_1 = 2.8$ nm for (nearly) spherical particles.

Application of the CLDD approach for a tightly packed particle ensemble of a Ni-base alloy

Experiments performed with the microfocus beamline (ID13) at the European Synchrotron Radiation Facility in Grenoble in 2001/2002 dealt with the investigation of tightly packed phases of a Ni-base alloy (see Roth et al, 2002, 2003 [198, 199] and Gille, 2012 [126]). The Figs. 4.24 and 4.25 illustrate the CLD approach. SAS was applied for determining the functions $I(h)$, $P_1(h)$, $\gamma(r)$, $C(r)$, $\gamma''(r)$ and $r \cdot \gamma''(r)$ for the specific order range $L = 150$ nm.

Approximating the scattering particles by cubes of edge length $a = 60$ nm, which nearly agrees with the data from the micrograph, the result obtained from $l_1 = 4V_0/S_0$ is $l_1 = 2a/3 \approx 40$ nm. Using the equation $1/|\gamma'(0)| = m_1 c = l_1(1 - c)$ with $\gamma'(0) \approx -1/(13$ nm$)$ results in $m_1 \approx 20$ nm. These

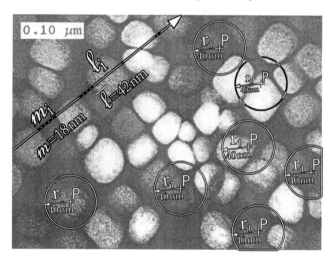

FIGURE 4.24

Micrograph of a W-rich Ni-base γ/γ' alloy (see (Roth et al., 2003) [198, 199]. These images make it possible to estimate $L \approx 150$ nm. The shape of the γ' particles can be approximated by the parallelepiped. The radius of the joined circles is identical to the first zero $r = r_1 \approx 40$ nm of the sample CF.

mean chord lengths are the starting point for a simplified *linear simulation model* (see Chapter 8) with the parameters $l_1 = 4$ and $m_1 = 2$. The model involves the property $L = l_1 + m_1 = 6$. The function $\gamma(x)$ results from the linear convolution square of the density fluctuation $\eta(x)$ (see Fig. 1.10). The analysis of the curvature of $\gamma(x)$ yields three singularities, which mark the positions of m_1, l_1 and $L = 6$. The knowledge of $c = c(L)$ makes it possible to plot the set covariance $C(r, L)$.

These parameters can also be verified by an analysis of the functions $\gamma''(r)$ or $r \cdot \gamma''(r)$ (see insert of the lower figures of Fig. 4.25). The distribution densities $\varphi(l)$ and $f(m)$ of the chord lengths l and m, respectively, are somewhat intermixed. Nevertheless, the first moments l_1 and m_1 can be estimated via the derivatives of the sample CF $\gamma(r, L)$.

The first derivative $\gamma'(0) \approx -1/(13 \text{ nm})$ is connected with the mean chord lengths $l_1 = (42 \pm 2)$ nm and $m_1 = (18 \pm 2)$ nm in $1/l_1 + 1/m_1 = -\gamma'(0) = 1/l_p$. For details, see Chapters 8 and 9. From these mean chord lengths, a volume fraction $c \approx 70\%$ of the γ' particles results via the Rosiwal relation $c = l_1/(l_1 + m_1)$. For the matrix, $1 - c = m_1/(l_1 + m_1)$ holds.

The scattering data available are not suitable for studying the internal structure of the nearly cuboid particles, i.e., for inserting $L \leq 10$ nm.

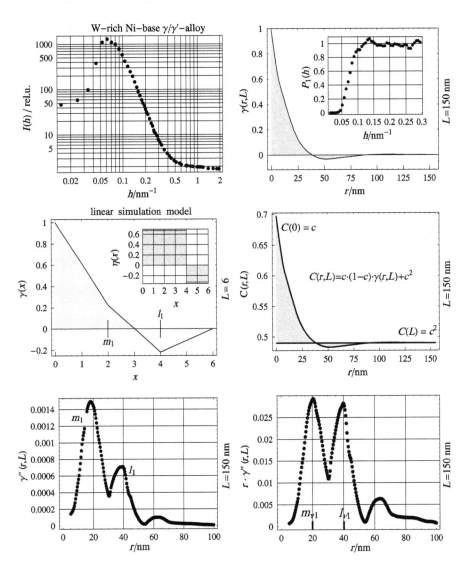

FIGURE 4.25

Investigation of γ' particles of a Ni-base alloy by use of the SAS CF $\gamma(r)$ and the set covariance $C(r)$.

The first two figures illustrate the scattering pattern and the CF for $L = 150$ nm. The data evaluation was carried out by applying $h_{max} \approx 0.5$ nm^{-1}. The parameters $l_c = 22$ nm and $f_c \approx 400$ nm^2 result from $\gamma(r, L)$. A normalized Porod plot $P_1(h)$ is a subfigure of the $\gamma(r)$ plot, where $h_{min} < h < 0.3/$nm. The CF is partly negative. The first zero is $r = r_1 \approx 40$ nm. The position of this zero is influenced by a couple of functions [12, 126].

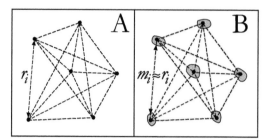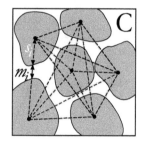

FIGURE 4.26

Similarity between the pair correlation and the ν chord length distribution density with changed normalization.

It goes without saying that g and $f_\nu = \frac{m}{m_1} \cdot f_\mu$ are different functions. Nevertheless, there is an analogy with the case of small particles (small volume fractions c of around $c < 0.2$).

In parts A and B, the lengths r_i and the chord lengths m_i have a similar meaning. The m_i give an approximate description of the distance to the next point. However, in part C this is not the case. The influence of the volume fraction on the CLDD $f_\nu(m)$ is too big. For larger r_i and m_i, the interrelation is not so clear.

Connection between the CLD approach and the pair correlation $g(r)$

In contrast to the CLDD approaches, the pair correlation function $g(r)$ describes *point-to-point distances* (and sometimes lattice distances). These distances start from (lattice) points that are not connected with a field of parallel straight lines (see Fig. 1.9). This means that the function $g(r)$ is somewhat connected with *weighted randomness chords* or ν-chords. For a single convex particle, the connection between μ and ν-chords (see Gille, 2000) [66] is written as

$$f_\nu(m) = \frac{m \cdot f_\mu(m)}{\int_0^{L_0} m \cdot f_\mu(m)dm} = \frac{m}{m_1} \cdot f_\mu(m).$$

In many cases, the relative behavior of $g(r)$ – for small r not farther than the nearest neighbor – is similar to that of the product of $m \cdot f(m)$, where $g(r) \approx r \cdot f(r)$ (see Fig. 4.26). By contrast, for relatively large chords m, i.e., for the second and third coordination spheres, the interrelation between $g(r)$ and $f(m)$ is not trivial.

Evidently, the denotation $g(r)$ [or $g(m)$] is used for two different functions: Traditionally, it is used for the pair correlation; however, it is also used for the function $g(r) = l_p \cdot \gamma''(r)$ (see Burger & Ruland, 2001) [7]. Perhaps this agreement between the denotations is supposed to be intentional because sometimes (i.e., in many practical cases), these functions are quite similar!

5

Scattering patterns and structure functions of Boolean models

The focus of Chapter 4 was mainly based on the pair correlation function. In this chapter, *Boolean models* (Bms) are discussed. The basic idea of Bms is to *neglect each kind of interrelation* between the spatial positions of the particles. Sometimes, the Bm is called the *Poissonian penetrable grain model*. The connections between the set covariance $C(r)$ and derived scattering parameters of the model are explained. The Bm is a suitable approximation for many practical applications, which has been demonstrated in several papers (see Hermann, 1991 [150] and references to other papers included therein). In this regard, the paper by Sonntag et al. (1981) [208] is remarkable.

Holes in cheese, bakery products (length scale $L = L_0 \approx 20$ cm) and many kinds of porous materials in materials science (called non-connected pores, $L = 20$ nm) that are *shaped like overlapping grains* can be described by a Bm with isotropic convex grains. This model possesses great flexibility because the size, shape and concentration of the inserted grains are free parameters. The superimposition of two Bms is again a Bm. In many cases, the Bm represents a "wild set."

The intrinsic limitations of this model (in which the interrelations between the grains are not considered for the most part and the physical germ/grain mechanism is also neglected) are known. There is a large sequence of applications. However, there are still questions of practical relevance for which there is no solution. Stoyan et al. [210, p. 82] described one such issue (see Fig. 5.1) as follows: "The interior chord lengths from Ξ_L have a complicated distribution depending on the distribution of Ξ_0 of the typical primary grain."

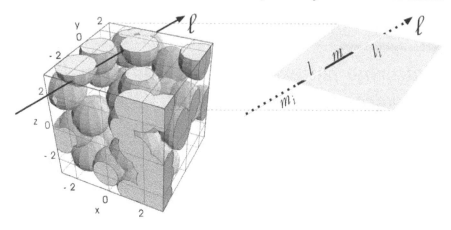

FIGURE 5.1
Simulation of a cubic section of a Bm intersected by the infinitely long straight
line ℓ (compare with the starting figure of Chapter 5).
The section of the set was constructed from 50 spherical primary grains of
diameter $d = 1$. Based on $n = 0.969$, the grain phase 1 of the infinitely large
two-phase set results in a volume fraction $c = 62\%$.

5.1   Short-order range approach for orderless systems

The Bm is a short-order range approach for modeling homogeneous ensembles of overlapping (sometimes called *soft*) particles. The starting points are arbitrarily fixed *grain prototype particles* with maximum diameter L_0 and CF $\gamma_0(r, L_0)$. A Bm is a special two-phase system of volume fraction c, where $0 \leq c < 1$. Its SAS CF $\gamma(r, c)$, where $0 \leq r \leq L_0$, is a modified exponential function term depending on $\gamma_0(r)$ and c. For smooth grains, $\gamma(r)$ fixes c and the CF of the grain prototype $\gamma_0(r)$.

The foundations of stochastic geometry are useful here (Levitz & Tchoubar, 1992) [165], (Hermann, 1991) [150], (Torquato, 2002) [225], (Gille, 2005) [100]. The Bm is a well-investigated approach (Serra, 1982) [205], (Stoyan et al., 1987) [210], (Jeulin, 2002) [156]. The scattering patterns of "wild sets" can be simulated (see the irregular particle ensemble Fig. 5.1). This simulation of a compact set is based on unit spheres $(R = 1)$ as initial grains. Other grain shapes can be inserted. Experience in simulating porous two-phase materials (Enke et al., 2001) [78] shows that the Bm is tailor-made for many applications with two-phase media (Sonntag et al., 1981) [208], (Serra, 1982) [205], (Jeulin, 2002) [156], (Torquato, 2002) [225], (Jeulin, 2005) [157].

In this section, scattering patterns, CLDDs and characteristic parameters are investigated for *spherical grains* and the *Poisson slice model* (PSm).

5.2 The Boolean model for convex grains – the set Ξ

The term *grains* is equivalent to *grain particles*. For a summary of basic formulas describing the Bm, see (Gille, 2011) [123]. The set Ξ (phase 1) results from the unconditional overlapping of primary grains Γ_i of positive volume, which have been thrown into the space "at random." "At random" means that there is no interrelation between any two grains, grain positions or spatial orientation of the grains. The grain positions result from a spatial Poisson process of const. intensity. An overlap between the Γ_i, $i = 1, 2, ...\infty$, results. The set Ξ is composed of overlapping grains (see Fig. 5.1). Clearly, there are still spatial regions (phase 2) that are not covered by any Γ_i.

For a realistic solid, it is not always clear and may even be contradictory to accept the assumption that the density of the grain phase 1 is defined to be constant. However, the basic idea of the model is to accept this assumption (which could be viewed as provocative).

Let the volume fraction of phases 1 and 2 be c and $1 - c$, respectively. The main grain volume is denoted by V_Γ. Let $\gamma_0(r)$ (which is different from zero in the interval $0 \leq r \leq L_0$) be the main CF of the Γ_i. Thus, c depends on the mean number n of grains belonging to the grain volume V_Γ. Then, $c = 1 - \exp(-n)$, where $0 \leq c < 1$, $0 < n < \infty$ (Serra, 1982) [205]. The set covariance of Ξ is written as

$$C(r, L_0) = C(r) = \begin{cases} 0 \leq r \leq L_0 : & 2c - 1 + (1 - c)^2 \cdot \exp[n \cdot \gamma_0(r)] \\ L_0 \leq r < \infty : & c^2 \end{cases} . \quad (5.1)$$

Equation (5.1) describes the isotropic set Ξ. Since the sample CF $\gamma(r)$ is connected with $C(r)$, where $C(r) = c(1 - c) \cdot \gamma(r) + c^2$, several synonymous representations exist. In particular, $\gamma(r)$ can be expressed in terms of $\gamma_0(r)$ (which stands for all the Γ_i) and n or in terms of $\gamma_0(r)$ and c [see Eq. (5.2)].

There is no long-order range (see Section 1.4). In fact, if $L_0 \leq r < \infty$, Eq. (5.1) and the property $\gamma_0(r) \equiv 0$ result in $\gamma(r) \equiv 0$. Furthermore, if $c \to 0$, it follows that $\gamma(r) \to \gamma_0(r)$. Since $\gamma_0(r)$ can vary, Ξ possesses a large flexibility. For smooth Γ_i (i.e., $\gamma_0''(0) = 0$), $c = \gamma''(0)/[\gamma'(0)]^2$ follows from Eq. (5.1) (see the *Mathematica* three-liner in Section 5.2.1). However, if a primary grain contains surface singularities (tips, corners, edges), $\gamma_0''(r) = 0$ is not fulfilled, which leads to a complex connection between c and CF $\gamma(r, c)$. Details are investigated in Section 10.3.5.

Section 5.4 explains the reconstruction of the size distribution of varying grain diameters in terms of $I(h)$ via $\gamma(r)$ (Gille, 1995) [53].

The isotropic set Ξ [see Eq. (5.1)] correlates with the isotropic scattering pattern $I(h)$, which is not necessarily a strictly monotonously decreasing function (see Section 5.3). Based on Eq. (5.1), characteristic parameters like l_p, l_c, f_c and v_c (Guinier & Fournét, 1955) [143] result. Generally,

$l_p = -1/\gamma'(0) = [\ln(1-c)] \cdot \gamma_0''(0)/c$ holds. Hence, if c is somewhat smaller than 1, l_p sensitively depends on c.

5.2.1 Connections between the functions $\gamma(r)$ and $\gamma_0(r)$ for arbitrary grains of density $N = n$

Without any restriction on grain shape, i.e., also considering more general grain shapes* with $\gamma_0''(0) > 0$, the relations

$$\gamma(r) = \frac{1-c}{c} \cdot [\exp[N \cdot \gamma_0(r)] - 1], \quad \gamma_0(r) = \frac{1}{N} \cdot \ln\left[1 + \frac{c}{1-c} \cdot \gamma(r)\right] \quad (5.2)$$

hold. Equations (5.1) and (5.2) are synonymous, but the practical relevance of Eq. (5.2) is greater. From Eq. (5.2),

$$\gamma'(r) = \frac{1-c}{c} \cdot N\gamma_0'(r)e^{N \cdot \gamma_0(r)}, \quad \gamma''(r) = \frac{1-c}{c}N\left[N\gamma_0'(r)^2 + \gamma_0''(r)\right]e^{N\gamma_0(r)} \quad (5.3)$$

follows. Let L denote the largest diameter of the largest grain described by $\gamma_0(r)$. Based on Eqs. (5.3), an investigation of $r = 0$ and $r = L$ leads to

$$\frac{\gamma'(r)}{\gamma'(L)} = \frac{1}{1-c} \cdot \frac{\gamma_0'(0)}{\gamma_0'(L)} \quad and \quad \frac{\gamma''(0)}{\gamma''(L)} = \frac{1}{1-c} \cdot \frac{\gamma_0'(0) + N\left[\gamma_0'(0)\right]^2}{\gamma_0'(L) + N\left[\gamma_0'(L)\right]^2}. \quad (5.4)$$

In fact, Eqs. (5.4) are useful to estimate c from γ, γ' and L (see Gille, 1995 [53, p. 128] and Chapter 8).

Volume fraction determination of Bm samples with smooth grains

If $\gamma_0''(0) = 0$, an analysis of the function $\Phi(r) = \gamma''(r)/[\gamma'(r)]^2$ yields c (see Fig. 5.2). The final result $c = \Phi(0)$ is based on the following assumptions for the sample CF $\gamma(r) = \gamma(r,c)$: $N = n$; $c = 1 - \exp(-n)$; $\gamma(r) = \gamma(r,c) = \gamma(r,n) = (1-c)\left[e^{n\gamma_0(r)} - 1\right]/c$ for $0 \le r \le L$ with $\gamma_0(0) = 1$, $\gamma_0(L) = 0$, $\gamma'(L) = 0$ and $\gamma_0''(0) = 0$. The following three *Mathematica* lines yield c.

```
gamma[r_,c_,n_]:=((1-c)/c)*(Exp[n*g0[r]]-1);
term =Simplify[D[gamma[r, c, n], {r, 2}]/D[gamma[r, c, n], {r, 1}]^2];
term0=term//.{r->0, g0[0]->1, Derivative[2][g0][0]->0, c->1-Exp[-n]}
```

Investigation of the ratio N/c

Another useful relation, which connects c with the terms $\gamma'(0)$ and $\gamma_0'(0)$ is $[\ln(1-c)]/c = -\gamma'(0)/\gamma_0'(0)$. Eqsuations (5.2) to (5.4) hold true for any grain shape, even if the grain is constructed from a group of separated particles (see Fig. 5.3). Generally, $N/c = N/[1 - e^{-N}] = -[\ln(1-c)]/c$ holds. Figure 5.4 illustrates the behavior of the ratio N/c as depending on c or N.

---

*The grains of a Bm do not have to be connected sets (see Fig. 5.3). For example, it is possible to have sets of discrete points. In such cases, the Bm is a Neymann-Scott point process [210, 211, 212].

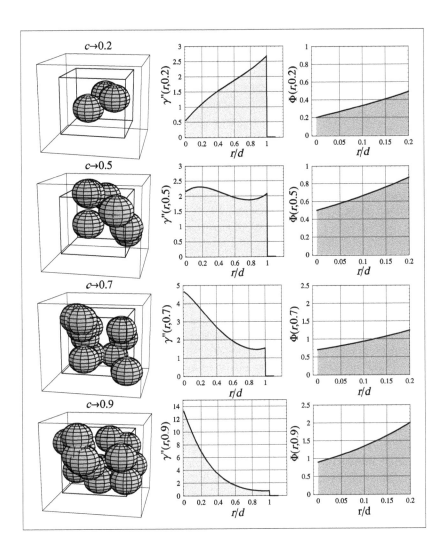

FIGURE 5.2

Connection between the sample CF $\gamma(r)$ (see $\gamma''(r)$ in the middle part) and the grain phase volume fraction c. Spherical grains of diameter $d = 1$, the centers of which are placed in an (inner) *generation box* of volume 2^3, are inserted for this illustration. The outer box volume is 3^3. The working function for estimating c is $\Phi(r)$, i.e., $\Phi(0+) = c$ (see right). This relation holds true for ensembles of grains with a grain CF $\gamma_0(r, d)$, where $\gamma_0''(0) = 0$. For differently sized grains, $\Phi(0) = c$ remains true, i.e., a grain size distribution does not influence this connection.

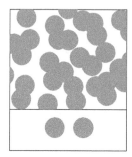 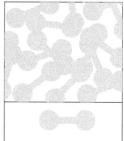 

FIGURE 5.3
The grains of a Bm are *not required to be connected sets*. The superimposition of two Bms is again a Bm (see Serra, 1982) [205]. The sections of three Bms (three different grain prototypes) illustrate this. The all-inclusive Eqs. (5.2) to (5.4) hold true. Certainly, the estimation of the function $\gamma_0(r, parameters)$ is complicated.

Some remarks about the volume fraction limiting case $c \equiv 1$ are useful. This limiting case probably has no real practical meaning. Nevertheless, if grain phase 1 occupies the whole space, a huge homogeneous sample particle results. This represents an interesting theoretical special case, which is analyzed in greater detail in various contexts in the following sections.

The chord length distributions of the Bm are analyzed in the left part of Fig. 5.1 for the following: For *any grain shape*, the CLDD of a random chord m inside the connected phase is well-known. It is an exponential distribution. For a Bm, the chord lengths (of both phases), which appear on a straight line through the set, are independent random variables. The interrelation between the second derivative of the SAS correlation function, $f(m)$, and the CLDD of the non-connected phase $\varphi(l)$ is given by a renewal process of independent events. The CLDD $\varphi(l)$ results from the grain shape and the volume fraction.

5.2.2 The chord length distributions of both phases

Both the CLDDs inherent in the Bm are analyzed. The intersection of Ξ by a line ℓ, which is expressed as $\Xi \cap \ell$ (see Fig. 5.1), yields the following two interwoven sequences of chords: The interior chords l possessing a CLDD $\varphi(l)$ and the exterior chords m possessing the CLDD $f(m)$ (Mering & Tchoubar, 1968) [182]. The ith moments of φ and f are denoted by l_i and m_i, respectively.

On the one hand, the chord length m (of the exterior chord sequence, phase 2) is exponentially distributed, which is expressed as $f(m) = 1/m_1 \cdot \exp(-m/m_1)$. The ith moment is $m_i = i! \cdot m_1{}^i$. This condition exists regardless of the shape of the Γ_i and is a well-known fact.

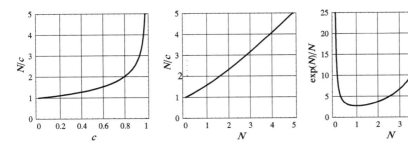

FIGURE 5.4

Volume density of the primary grains over the volume fraction of the primary grain phase for a Bm in terms of c and N.

The case $c \to 0$ corresponds to the infinitely diluted case $N \to 0$. For $c \to 1$, $N \to \infty$ is required (complete filling of the whole space). The function $\exp(N)/N$ involves a minimum at $N = 1$ (see Chapter 8).

On the other hand, carrying out an analysis of the distribution law $\varphi(l)$ of the interior chord lengths (grain phase 1) is problematic. This difficulty has existed for a long time and has been touched upon in the textbooks by Torquato (2002) [225] and Stoyan et al. (1987) [210]. The following investigation is based on the SAS-specific function $g(r) = l_p \cdot \gamma''(r)$ of the Bm. The mathematical background of $g(r)$ and the analysis of CLDDs from the corresponding scattering pattern were studied by Burger and Ruland (2001) [7]. Equation (5.1) yields

$$g(r) = -\frac{\exp[n \cdot \gamma_0(r) - 1] \cdot [n \cdot [\gamma_0'(r)]^2 + \gamma_0''(r)]}{\gamma_0'(0)}. \tag{5.5}$$

From Eq. (5.5), $g(0+) = -(n[\gamma_0'(0)]^2 + \gamma_0''(0))/\gamma_0'(0)$ follows. Hence, the left-hand limit $g(L_0-)$ results in

$$g(L_0-) = -\frac{e^{-n} \cdot \gamma_0''(L_0)}{\gamma_0'(0)} = -\frac{(1-c) \cdot \gamma_0''(L_0)}{\gamma_0'(0)}. \tag{5.6}$$

Equation (5.6) is interrelated with the leading asymptotic terms of the Porod plot $\sim h^4 \cdot I(h)$. For grains Γ_i with the properties $\gamma_0''(0) = 0$ or $\gamma_0''(L_0) = 0$, the limiting Eqs. (5.5) and (5.6) simplify.

The Fourier transformation of $g(r)$, where $Q(t) = \int_0^L g(r) \cdot e^{i \cdot t \cdot r} dr$, is connected with the (here unknown) characteristic function $p(t) = \int_0^{L_0} \varphi(r) \cdot e^{i \cdot t \cdot r} dr$ and the characteristic function $q(t)$ of $f(m)$, where $q(t) = \int_0^L f(m) \cdot e^{i \cdot t \cdot m} dm$. The latter results in the function

$$q(t) = \frac{i}{i + t \cdot m_1}. \tag{5.7}$$

The moments $m_{1,2,3}$ of $f(m)$ [connected with the derivatives of Eq. (5.7) in the origin, $q^{(n)}(0)$] are interrelated, where $m_3 = 3m_2^2/(2m_1)$. Based on the connection $Q(t) - 2 = [p(t) + q(t) - 2]/[1 - p(t)q(t)]$, which has been discussed and made use of in various papers (Mering & Tchoubar, 1968) [182], (Levitz & Tchoubar, 1992) [165], (Gille, 2005) [100], the function $p(t)$ is written as

$$p(t) = \frac{Q(t) - q(t)}{1 - 2q(t) + q(t) \cdot Q(t)}. \tag{5.8}$$

Lastly, this characteristic function fixes $\varphi(l)$. The inverse transformation of Eq. (5.8) yields[†]

$$\varphi(l) = \frac{1}{2\pi} \int_{-\infty}^{\infty} p(t) \cdot e^{-i \cdot t \cdot l} dt = \frac{1}{2\pi} \int_{-\infty}^{\infty} \frac{Q(t) - q(t)}{1 - 2q(t) + q(t) \cdot Q(t)} \cdot e^{-i \cdot t \cdot l} dt. \tag{5.9}$$

After a further analysis of the model, these interrelations and Eq. (5.9) will be verified for two special cases, namely, for *spherical grains* and for *"slice" grains*. This results in the basic parameters of SAS in terms of the CLDD parameters as follows: The first moment of $g(r)$, which is Porod's length parameter l_p, is fundamental for any tightly packed two-phase system (Burger & Ruland, 2001) [7]. The volume fraction c is given in terms of the first moments $c = l_1/(l_1 + m_1) = 1 - l_p/l_1 = l_p/m_1$. Furthermore, c is connected with the higher CLDD moments M_2, l_2 and m_2. There exist second moment interrelations between g, φ and f. With respect to unavoidable experimental errors, the analysis of still higher moments is limited.

5.2.3 Moments of the CLDD for both phases of the Bm

The first two moments of the function $g(r)$, M_1 and M_2, are interrelated with the corresponding moments of the CLDDs.[‡] In summary, the connections $M_1 = l_1 m_1/(l_1 + m_1)$ and $M_2 = l_p^2 (l_2/l_1^2 + m_2/m_1^2 - 2)$ hold true. As depending on c,

$$M_1 = \frac{(1 - c) \cdot l_1 + c \cdot m_1}{2}, \quad M_2 = (1 - c)^2 l_2 + c^2 m_2 - 2M_1^2 \tag{5.10}$$

can be confirmed (Gille, 2007) [107].

In particular, these relations must be fulfilled for small c, including the limiting case $c \to 0$. Generally, in the limiting case $c \to 0$, there exist interrelations between the moments m_1 and m_2, m_1 and m_3, m_1 and m_4, etc. The smaller c is, the larger the moments of f. Generally, for $c \to 0$, there is the interrelation $m_2 = 2m_1^2$ connecting m_1 and m_2. By way of this connection,

---

[†]The numerical analysis of Eq. (5.9) is a tricky job. It is useful to operate with modified, synonymous representations.

[‡]For more details, see Chapter 9 for analyses of more general cases.

$M_2 \rightarrow l_2$ can be confirmed for $c \rightarrow 0$ from Eq. (5.10). Thus, it must be possible to determine c in terms of all the moments involved via Eq. (5.10).

These connections simplify for the Bm. In fact, the function $f(m)$ involves the property $m_2/m_1{}^2 = 2$ [see Eq. (5.7)]. Therefore, Eq. (5.10) is written as $M_2 = (1-c)^2 \cdot l_2 = l_p{}^2 l_2/l_1{}^2$. Thus, the second moment of $\varphi(l)$ can be represented as

$$l_2 = \frac{M_2}{(1-c)^2} = \frac{l_p l_c}{(1-c)^2} = \frac{l_1 l_c}{1-c} = \frac{m_1 l_c}{c}. \tag{5.11}$$

Equation (5.11) and similar connections with the volume fraction have not yet been investigated.

5.2.4 The second moments of $\varphi(l)$ and $f(m)$ fix c; $0 \le c < 1$

The limiting case $c \equiv 1$ is a special one. This "puzzle case" must be excluded here, since it requires very special assumptions (Gille, lecture ICCMSE/2009, Rhodes/Greece) [114]. For $c \equiv 1$, Eqs. (5.9) and (5.10) can lead to contradictions. The shape of a given phase 1 is so manifold that it cannot cover the whole space. Only in exceptional cases can grain phase 1 be puzzled together.

Eliminating c from Eq. (5.10) gives the (positive) solution

$$c = \frac{l_2 - \sqrt{(m_2 - 2m_1{}^2)M_2 + (2m_1{}^2 - m_2 + M_2)l_2}}{l_2 + m_2 - 2m_1{}^2}. \tag{5.12}$$

Several CLD moments define c, where $0 \le c < 1$, via Eq. (5.12), which allows the following conclusion: On the one hand, limiting relations for very small c can be derived for any random two-phase system. For small c, both factors of the term $c^2 \cdot (m_2 - 2m_1{}^2)$ in Eq. (5.10) can be neglected. The main term $(1-c)^2 l_2$ remains on the right. Consequently, taking into account $M_2 = l_p \cdot l_c$ (which can be derived by partial integration of M_2), two special representations for c follow, which are

$$c = 1 - \sqrt{M_2/l_2}; \; (c \rightarrow 0) \; and \; c = 1 - \sqrt{l_p \cdot l_c/l_2}; \; (c \rightarrow 0). \tag{5.13}$$

On the other hand, the following conclusion can be drawn from Eq. (5.11): In the Bm case, Eq. (5.13) is not an approximation. In a certain way, Eq. (5.13) is the extrapolation of the relation $c \equiv 1 - l_p/l_1$ (Guinier & Fournét, 1955) [143] to second moments.

5.2.5 Interrelated CLDD moments and scattering patterns

Characteristic scattering parameters, CLDD moments and the asymptotic behavior of the scattering pattern are interrelated. The M_i of $g(r)$ are connected via the characteristic scattering parameters $M_1 = l_p$, $M_2 = l_p \cdot l_c$,

$M_3 = 3l_p f_c/\pi$ and $M_4 = 3l_p v_c/\pi$. The derivatives of $Q(t)$ in the origin define the M_i, $M_0 = Q(0)$, $M_1 = Q'(0)/i$, $M_2 = -Q''(0)$ and $M_3 = -Q^{(3)}(0)/i$. Thus, the Taylor series at $t = 0$, which fixes all the characteristic scattering parameters, is

$$Q(t) = 1 + il_p \cdot t - \frac{l_c l_p}{2} \cdot t^2 - i\frac{3l_p f_c}{\pi} \cdot \frac{t^3}{6} + \frac{3l_p v_c}{\pi} \cdot \frac{t^4}{24} + \cdots \qquad (5.14)$$

The asymptotic behavior of the scattering intensity $I(h)$, where $I(0) = 1$, for large h is the sum of the following several terms: the Porod term, the Kirste-Porod term plus oscillating terms, which are all multiplied by non-positive powers of h. These terms depend on the behavior of $g(r)$ on the abscissas $r = 0+$ and $r = L_0-$ [see Eq. (5.5)]. A detailed investigation shows that the function $g(r)$ possesses a finite jump at $r = L_0$ and starts with a positive value at $r = 0$. An analysis of the left-hand derivatives $g^{(n)}(L_0-)$ and right-hand derivatives $g^{(n)}(0+)$, where $n = 0, 1, 2, ...$, yields the asymptotic behavior of $I(h)$ for $h \to \infty$ (Ciccariello, 1993) [15]. Normalizing the elementary Porod term to 1, the dimensionless Porod term is written as $P_{1\infty}(h) = v_c l_p h^4 \cdot I(h)/(8\pi) = M_4 \cdot h^4 I(h)/24$. Based on this, Eq. (5.5) results in the leading asymptotic terms

$$P_1(h) = 1 - \frac{2\gamma'''(0)}{\gamma'(0)} \cdot \frac{1}{h^2} + \frac{L_0 \cdot g(L_0) \cdot \cos(hL_0)}{2}$$
$$- \frac{[L_0{}^2 \cdot g'(L_0) + 3L_0 \cdot g(L_0)]}{2} \cdot \frac{\sin(hL_0)}{hL_0} \cdots \qquad (5.15)$$

Another representation of the Kirste-Porod term is $-2\gamma'''(0)/[\gamma'(0) \cdot h^2] = 2g(0)/h^2$. Equation (5.15) demonstrates the meaning of the parameter L_0, regardless of the grain shape. The quality of the approximating Eq. (5.15) mainly depends on the actual volume fraction c. For small c, Eq. (5.15) is already a high-level approximation if $3 < hL_0$ holds. If $c \to 1$, it is not surprising that the approximation can fail even for $1\,000 < hL_0$. In other words, in such cases the chance to find any ordered interfaces is very small. The disorder is immense and Ξ is described by $I(hL_0) \approx 1$. The introduction of further terms into Eq. (5.15) is more or less useless.

Detection of the whole grain correlation function γ_0

The Fourier sine transformation of $h \cdot I(h)$ yields $r \cdot \gamma(r)$, but not $r \cdot \gamma_0(r, L_0)$, which is defined by the grain shape. For $c \to 1$, the grain shape (especially the maximum diameter of the Γ_i) can only be estimated by analyzing the image material. The manifold examples of images in the textbook by Serra (1982) [205] demonstrate this. From the point of view of a scattering experiment, the actual grain shapes cannot be detected at all. However, the structure functions of the Bm are closely connected with the grain CF $\gamma_0(r, L_0)$. Hence, it is not surprising that a complete reconstruction of the grain CF is quite possible. See examples in (Weil, 2004) [231], (Torquato, 2005) [226] and (Gille, 1995) [53]. A simultaneous analysis of Eqs. (5.1) and (5.5) allows L_0, $M_{1,2,3,4}$ and c, n to be detected in the first step. In this respect, smooth grains considerably simplify the considerations.

5.3   Inserting spherical grains of constant diameter

There exists a broad spectrum of possible grain prototypes, for which the SAS correlation function $\gamma_0(r)$ is known analytically (see Chapter 2). Furthermore, a superimposition of different grain shapes has been well investigated. This basic knowledge allows analytic expressions for the functions $\gamma(r)$, $g(r)$, $q(t)$, $Q(t)$ and finally for $I(h)$ or $\varphi(l)$ to be obtained [Eqs. (5.1) to (5.9)].

A series of trials shows that the actual integrals of type $\int e^{f(x)} \cdot e^{-i \cdot t \cdot x} dx$, where $f(x)$ is at least a polynomial of third degree, can be traced back to an implicit solution. Corresponding expressions are elaborate and involve very special functions of mathematical physics (Wolfram Research, 2005) [234]. Therefore, numerical integration has been applied for obtaining Figs. 5.1 to 5.7, which restrict to spherical grains of constant diameter d.

With the correlation function of a single sphere $\gamma_0(r,d)$, Eqs. (5.1) and (5.5) give $M_1 = l_p = -1/\gamma'(0) = -2cd/[3\ln(1-c)]$. Furthermore, l_c, f_c, v_c and $M_{2,3,4}$ result. Numerical integration is by far the simplest way to specialize Eq. (5.14). By applying $g(L_0) = 2(1-c)/d$ and $g'(L_0) = 2(1-c)/d^2$, as well as $\gamma'''(0) = [3\ln(1-c) \cdot (9[\ln(1-c)]^2 - 8)]/(8cd^3)$ [obtained from Eq. (5.5)], Eq. (5.15) takes the special shape

$$P_{1\infty}(h) = 1 - \frac{9[\ln(1-c)]^2 - 8}{2 \cdot h^2 d^2} + (1-c) \cdot \cos(hd) - 4(1-c) \cdot \frac{\sin(hd)}{hd}. \quad (5.16)$$

The term $g(0+) = -3 \cdot \ln(1-c)/(2d)$, which is useful for estimating c, does not directly influence Eq. (5.16). The quality of the approximation [Eq. (5.16)] depends on c, where $0 \le c < 1$. This is illustrated in Fig. 5.5. If c is somewhat smaller than 1, the whole asymptotic approximation collapses to the pure Porod asymptote $P_{1\infty}(h) = 1$ for $h \to \infty$. In doing this, the Kirste-Porod term disappears since $\lim_{x \to 0} x^2 \cdot \ln(x) = 0$. Thus, $[\ln(1-c)]^2/h^2 \to 0$ for { $c \to 1, h \to \infty$ } must be taken into account. Adding further oscillating terms to Eq. (5.16) is not effective. Figure 5.6 contradicts the widespread opinion that the $I(h)$ of a Bm is a monotonously decreasing function. A sequence of c_i values ($c = 0 \ldots 1$) have been inserted—from the single sphere case to a compact block of material. The greater c is, the smoother $I(h)$. The change in the maximum positions is insignificant compared with the single sphere case.

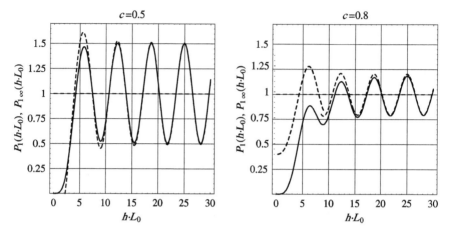

FIGURE 5.5

Scattering behavior of a Bm with spherical grains $d = L_0 = 1$.
Normalized Porod plot $P_1(hL_0)$ (solid line) and the asymptotic approximation $P_{1\infty}(hL_0)$ (dashed line) for selected volume fractions $c = 0.5$ and $c = 0.8$ and operating with Eqs. (5.15) and (5.16). The smaller c is, the better the approximation [see the discussion for Eq. (5.15)].

Obviously, for monodisperse grains, the scattering intensity can involve weak maximums for relatively small c only. Otherwise, a relatively smooth $I(h)$ is obtained from the manifold shape of the resulting grain phase 1. All the more reason, this fact holds true for polydisperse grains, where still smoother scattering patterns are expected. Maximums occur only rarely (see Fig. 5.6 for an example of smooth curves, which exclude extremely small volume fractions c). This behavior results from the missing order of the model. Throwing more and more grains Γ_i into the space *ignores any law of combination* of the position between any two grains. A long-order range cannot be produced.

These properties are reflected in the analytic representation of the scattering intensity for large h. Any oscillations of the leading asymptotic terms [Eqs. (5.15) and (5.16)] disappear for $c \to 1$. This means that the pure Porod term $P_{1\infty}(h) = 1$ is the only term which describes the asymptotic behavior.

The special case of spherical grains discussed in Figs. 5.1 to 5.6 has been extended to Fig. 5.7. Here, the functions $g(r)$ and $\varphi(l)$ [see Eq. (5.9)] are plotted for varying c.

The function $\gamma_0(r)$ can be reconstructed from $\gamma(r)$ and $g(r)$. The central function of these investigations is $g(r)$. An estimation of the whole model parameter set c, n, L_0, $\gamma_0'(0)$ and $\gamma_0''(0)$ can be performed by operating with $\gamma(r)$ or $g(r)$. After fixing c, the grain CF $\gamma_0(r)$ can be reconstructed. Using

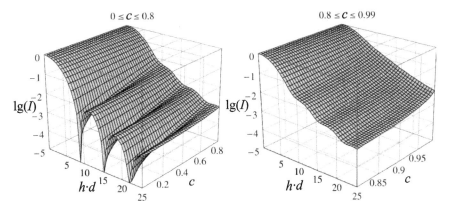

FIGURE 5.6

Logarithmic representation of the scattering patterns $lg(I) = lg[I(h \cdot d, c)]$ of a Bm with spherical grains, which are based on the CF defined by Eq. (5.2) for $0 < h \cdot d < 25$.

The functions do not intersect. Different curves $I(h \cdot d, c)$ result for different c. The volume fraction is split into the two intervals $0 \le c \le 0.8$ and $0.8 \le c \le 0.99$. In the limiting case of an infinitely large isotropic object $(c \to 1)$, the identity $I(h \cdot d, c) \equiv 1$ holds. The more c approaches 1, the smoother the scattering pattern. Applications for cases in which $0.99 < c < 1$ are rare.

Eq. (5.1) results in

$$\gamma_0(r) = -\frac{\ln\left[1 + \frac{c}{1-c} \cdot \gamma(r)\right]}{\ln(1-c)} \approx 1 - \frac{c\gamma'(0)}{\ln(1-c)} \cdot r + \frac{c(c[\gamma'(0)]^2 - \gamma''(0))}{2\ln(1-c)} \cdot r^2 + O[r]^3.$$

$$(5.17)$$

In detail, the derivatives of the functions $\gamma_0(r)$ and $\gamma(r)$ in the origin are connected via

$$\gamma_0'(0) = -\frac{c \cdot \gamma'(0)}{\ln(1-c)}, \quad \gamma_0''(0) = \frac{c(c[\gamma'(0)]^2 - \gamma''(0))}{\ln(1-c)} = \frac{[c \cdot \gamma'(0)]^2}{\ln(1-c)} - \frac{c \cdot \gamma''(0)}{\ln(1-c)}.$$

$$(5.18)$$

In certain special cases, Eq. (5.5) can be interpreted as a differential equation for $\gamma_0(r)$, which also allows a complete reconstruction of the function γ_0 if c is known.

Contrary to the simple interrelations $\gamma(r) \leftrightarrow \gamma_0(r)$, like Eqs. (5.17) and (5.18), the construction of any hard particle model requires a well-selected pair correlation function $g(r, c)$. Exact analytic expressions for $g(r, c)$ functions are seldom available (see Stoyan & Stoyan [211, p. 276] and Chapter 4). The pair correlation function of the Γ_i centers of the Bm equals unity *for all pair distances*.

More intrinsic properties of the CLDDs $\varphi(l)$ and $f(m)$

It is of interest to investigate the CLDDs $\varphi(l)$ and $f(m)$ (of the grain-phase 1 and complementary phase 2) for different volume fractions c (see Fig. 5.7). Here, the (only) experimental length variable r is inserted for all CLDDs. The continuous part of the result $\varphi(l) \equiv \varphi(r)$ fulfills $g(0) = \varphi(0)$. It seems to be a paradox that the maximum chord lengths of both phases are unlimited if the case $c \equiv 0$ is excluded. However, the probability of detecting such huge chord lengths is small. The cases where $c \to 0$ and $c \to 1$ are discussed in Fig. 5.7.

There exists a set of useful relations between c and the moments of the functions $g(r)$, $\varphi(l)$ and $f(m)$ [Eqs. (5.10) to (5.13)]. However, taking $\varphi(l)$ alone does not allow c to be ascertained, whereas the information content of the $f(m)$ CLDD moments of the "connected phase" is much larger (see Fig. 9.5 in Chapter 9). This agrees with the fact that the function $f(m)$ bears a certain similarity to the pair correlation function (see Fig. 4.26 and the final part of Chapter 4).

In contrast, the function $\varphi(l)$ characterizes the *non-connected particle phase 1* of the models, but nothing more. The study of particles does not involve information about the particle-to-particle distances, c, etc. In this regard, the CLDDs $f(m)$ contain more information than $\varphi(l)$. With respect to the Bm, further conclusions can be drawn.

Information content of the function $f(m)$

For grains of mean chord length l_1, the CLDD $f(m) = f(m, c, l_1)$ is written as

$$f(m) = \frac{-\ln(1-c)}{l_1} \exp\left[m \frac{\ln(1-c)}{l_1}\right] = -\frac{(1-c)^{m/l_1}\ln(1-c)}{l_1}, \quad 0 \leq m < \infty. \tag{5.19}$$

The function f is shown in Fig. 5.7 for the case $d = 1$, i.e., for $l_1 = 2d/3$. Regardless of whether $c < 1$, there exist chords of length $m \to \infty$. The first three moments of $f(m)$, given as

$$m_1 = -\frac{2d}{3 \cdot \ln(1-c)}, \quad m_2 = \frac{8d^2}{9 \cdot [\ln(1-c)]^2}, \quad m_3 = -\frac{16d^3}{9 \cdot [\ln(1-c)]^3}, \tag{5.20}$$

decrease for $c \to 1$ (i.e., if more and more space is taken from the grain phase 1). In the special case $m = l_1$, Eq. (5.19) yields $f(c) = -(1-c) \cdot \ln(1-c)/l_1$. There is a maximum of $f(c)$ at $c = 1 - 1/e \approx 0.632$. This property is typical of the CLDD $f(r,c)$ of a Bm (see Fig. 5.7).

For $r = 0$, but with increasing c, the values $f(r,c)$ increase, i.e., $f(0,c) \to \infty$. Bigger values of c result in extremely small chord lengths $m_i \to 0$. In contrast to this, the relation $0 < r$ leads to $f(r,c) \to 0$ if $c \to 1$ (see Fig. 5.7).

Summary: For all chord lengths m, where $0 < m < \infty$, $f(m) \equiv 0$ holds true in both cases $c \equiv 0$ and $c \equiv 1$.

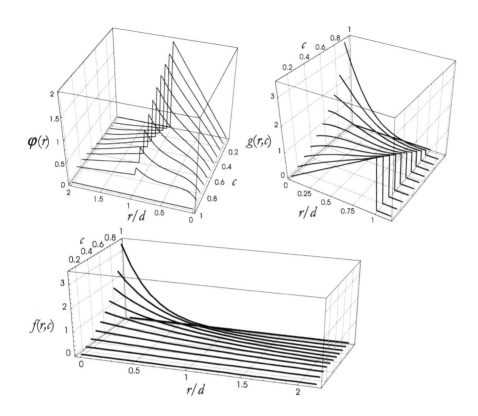

FIGURE 5.7
The function $g(r) = g(r,c)$ and the CLDDS $\varphi(r) = \varphi(r,c)$ and $f(r) = f(r,c)$ (normalized functions) for a Bm with spherical grains $d = 1$, depending on the volume fraction c of grain phase 1, where $0 \leq c \leq 1$.
The order range is $L = d$, i.e., the function $g(r,c)$ possesses the property $g(r,c) \equiv 0$ if $d < r < \infty$.
Several limiting cases are of interest: In the single sphere case (i.e., for $c \to 0$), the result is $g(r) = 2r/d^2$, where $0 \leq r < d$. On the other hand, if $c \to 1$, $g(r) \equiv \delta(r)$ is expected. If the case $c \equiv 1$ is excluded, $g(r,c)$ involves a finite jump at $r/d = 1$.
The interpretation of the CLDDs φ and f involves (at least) two limiting cases: In the single sphere case $c = 0$, the result is $\varphi(r) = 2r/d^2$ where $0 \leq r < d$ and $f(r) \equiv 0$ where $0 < r < \infty$. The limiting case of an extremely filled system $c \to 1$ or even $c \equiv 1$ is characterized by $\varphi(r) \equiv 0$ and $f(r) = \delta(r)$. Such details cannot be directly recognized from these plots. The representation of f emphasizes the existence of many extremely long chord lengths $m \to \infty$.

5.4   Size distribution of spherical grains

This section deals with a problem which does not seem to be tackled very often.[§] If there is a size distribution of equally shaped grains, its distribution law can be determined from the scattering pattern (see the article entitled *"Diameter Distribution of Spherical Primary Grains in the Boolean Model from SAS"* [53]). In the following, selected results from that article will be summarized.

Let $f(D)$ be the distribution density of the random grain diameter D, where $0 \le D < L$. Let c be the volume fraction of the grain phase defined by the intensity λ of the point Poisson process, $c = 1-\exp(-\mu \cdot \lambda) = 1-\exp(-n)$. The parameter $\lambda = n/\mu$ is connected with $f(D)$ via $\lambda = n / \int_0^L \pi D^3/6 \cdot f(D) dD$. The CF, which is defined by the scattering pattern $I(h) = I(h, c, f(D))$ of the Bm constructed from those grains, is denoted by $\gamma(r) = \gamma(r, f(D), c)$. Finally, denoting the CF of a single spherical grain of diameter D by $\gamma_0(r, D)$, all of these terms are interrelated via the starting equation [see Eq. (5.2)].

$$\frac{\gamma(r) \cdot c}{1-c} + 1 = \exp\left[\lambda \cdot \int_{D=r}^L \frac{\pi D^3}{6} \cdot \gamma_0(r, D) \cdot f(D) dD\right], \quad 0 \le r < L, \ 0 \le c < 1.$$

(5.21)

Equation (5.21), which is a specific generalization of Eq. (5.1), describes all the geometric details of the model. The mean grain volume is $\mu = \pi \overline{D^3}/6$. The properties $\gamma(0) = 1$ and $\gamma(L) = 0$ are fulfilled. Furthermore, the well-investigated limiting case $c \to 0$ is involved, which leads to the explicit result $f(D) = -\overline{D^3}/3 \cdot (\gamma''(D)/D)'$, where $\overline{D^3}$ is the third moment of $f(D)$.

In the following lines, the volume fraction c is assumed to be known. Below, c will be detected in terms of the sample CF. The abbreviation term $T(r)$, where $T(r) = 1 - \ln[\gamma(r) \cdot c + (1-c)]/\ln[1-c]$, is defined. By use of this term and after some arithmetic, Eq. (5.21) gives

$$\frac{T''(r)}{r} = \frac{\pi}{2\mu} \cdot \int_{D=r}^L f(D) dD, \quad 0 \le r < L, \ 0 \le \mu < \frac{\pi}{6} L^3.$$

(5.22)

The distribution density $f(D)$ results from Eq. (5.22). Differentiation with respect to r leads to the explicit representation in terms of $\gamma(r)$ and c, where $0 \le r < L$, as shown by

$$f(D) = -\frac{2\mu}{\pi} \cdot \frac{d}{dr}\left[\frac{T''(r)}{r}\right] = -\frac{2\mu}{\pi} \cdot \frac{d}{dr}\left[\frac{(1 - \ln[\gamma(r) \cdot c + (1-c)]/\ln[1-c])''}{r}\right].$$

(5.23)

Equation (5.23) still contains the volume fraction as an unknown quantity. This parameter results in terms of $\gamma(r)$ in the following way:

[§] In any case, the author is not aware of a detailed consideration for a case with grains of fixed shape that have a certain size distribution.

Volume fraction of grain phase 1 in terms of the sample CF $\gamma(r)$

For the random closed set considered, $\gamma(r) = (1-c)/c \cdot \left[e^{n \cdot \overline{\gamma_0(r)}} - 1\right]$ holds true [see Eq. (5.1)]. The symbol $\overline{\gamma_0(r)}$ denotes the mean volume-averaged CF of the grains [see Eq. (5.21) for spherical grains].

In order to eliminate this term for smooth grains [which means $\gamma_0''(0) = 0$ and $\overline{\gamma_0''(0)} = 0$], at least the two differentiation steps $\gamma'(r)$ and $\gamma''(r)$ must be taken into account. Altogether, this once again leads to the surprisingly simple connection (see also Section 5.2.1) (i.e., here under the assumption of an existing polydispersity)

$$c = \lim_{r \to 0+} \frac{\gamma''(r)}{[\gamma'(r)]^2} = \lim_{r \to 0+} \Phi(r) = \frac{\gamma''(0+)}{[\gamma'(0+)]^2} = \Phi(0+). \qquad (5.24)$$

Equation (5.24) holds true for any Bms constructed from polydisperse, smooth grains. It contains the first and second derivatives of the sample CF and unambiguously defines c. There exist other approaches to fix c in terms of two scattering patterns, e.g., $I_{1,2}(h)$ (see Damaschun & Mueller, 1967) [24].

Consequently, $f(D)$ results from Eqs. (5.23) and (5.24) (see the section *experimental results* in [53, p. 130]). In that paper, the Bm approach has been applied for the analysis of a ceramic γ-Al$_2$O$_3$ micropowder. Spherical grains of diameter D, which are described by $f(D)$, have been inserted. The data evaluation was based on the assumption that $L = 30$ nm (see Chapter 1). Fixing the length parameter L beforehand stabilizes all the steps of the data evaluation. Without this assumption, essential transformations like $I(h) \to \gamma(r) \to c$, $\gamma(r) \to T(r)$ and $T(r) \to f(D)$ are "ill-posed."

Application: Grain size distribution $f(D)$ of a ceramic micropowder

Based on the theory of Section 5.4 and the previous sections, a ceramic micropowder was investigated. The results demonstrate the meaning and importance of the parameter L (see Chapter 1). According to the micrograph information in Fig. 5.8, $L = 600$ nm is inserted for the data evaluation. From this micrograph, it follows that the primary grains themselves consist of considerably smaller globular particles possessing a nearly constant diameter of about 20 nm. A detailed investigation of this particle type (inserting $L_1 = 20$ nm) would require an extension of measurement and data evaluation to include larger h values of at least $h_{max} = 0.2$ nm^{-1}.

The structure functions explained in Fig. 5.9 correlate with a typical micrograph of this powder (see Fig. 5.8). This image was obtained with a transmission electron microscope (working voltage $U = 80$ KV, original magnification $V = 6\,300$) in the physical laboratory complex of *Ceramic Works Hermsdorf* (KWH) in 1989.¶

¶In 1957, the *first bipolar transistor* in the world was developed here. However, later electronic research at this place took another direction.

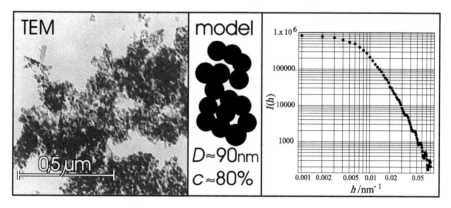

FIGURE 5.8

Micrograph, model and scattering pattern of a ceramic γ-Al$_2$O$_3$ micropowder.
Left: From the image it can be derived that $L = 0.6$ µm $= 600$ nm is a suitable length to describe the microparticle ensemble using a Boolean model with nearly spherical grains (see the model figure in the middle). As a consequence of the overlapping packing of the "clusters," their volume fraction c is much bigger than that of tightly packed hard spheres.
Right: Scattering pattern (relative measurement) of γ-aluminum oxide (Tonerde) from a relative measurement with a Kratky plant (recorded h-interval $h_{min} = 0.005$/nm$< h < 0.09$/nm).
Based on the sampling point theorem (SPT), the coordinates of the first 4 points were added by extrapolation. According to the SPT, the resolution limits $r_{min} = 30$ nm and $r_{max} = 630$ nm result. The sample thickness of this ceramic powder was about 0.7 mm. The significant part of the recorded scattering curve consists of 81 points of measurement $[h_k, I(h_k)]$. According to the large electron density differences, a strong scattering effect was observed. There exist another volume fractions $c(L)$ of this material. An estimation of $c(L_1)$ of the clusters requires $L_1 \to 5\,000$ nm and another model.

In fact, this and similar experiments always lead to an isotropic scattering pattern. There is no interrelation between the material grains.

Multiple scattering effects, which frequently exist for such powders, can be neglected. This is confirmed by the nearly perfect Porod law behavior of the scattering pattern at relatively large h. A discussion of these experiments by introducing *fractal parameter sets* does not seem to be useful here (at least not for the actual scattering pattern).

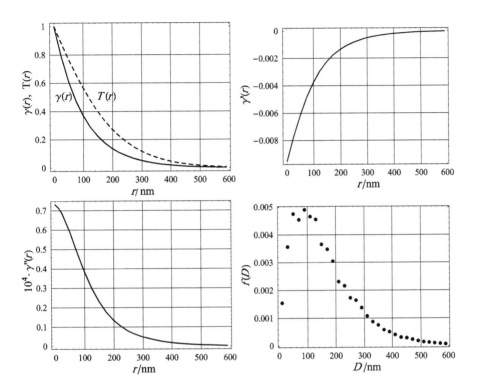

FIGURE 5.9

Structure functions obtained from the scattering pattern of the micropowder after inserting $L = 600$ nm.

A Boolean model with spherical primary grains of diameter distribution density $f(D)$ was applied to describe the powder. Operating with the functions $\gamma'(r)$ and $\gamma''(r)$ results in a volume fraction of nearly 80% for the grain phase [Eq. (5.24)].

The function $T(r)$, where $T(r) = 1 - \ln[\gamma(r) \cdot c + (1 - c)]/\ln[1 - c]$, is represented by the dashed line. Based on Eq. (5.23), a highly non-symmetric distribution density $f(D)$ of the grains results. The number of smaller grains contained in the sample is much bigger than that of larger ones. There is a maximum at about $D \approx 90$ nm. Furthermore, $n = 1.625$, $\lambda \approx 2.5 \cdot 10^{-7}$ nm^{-3} and the mean diameter of the grains is $\overline{D} \approx 160$ nm. A mean grain volume of $V \approx 6 \cdot 10^6$ nm^3 follows.

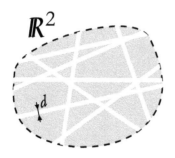
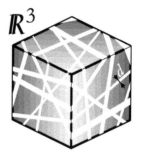

FIGURE 5.10

In the Poisson slice model (PSm), a block of material is intersected by layers of thickness d. The parameter c, where $1 - c = 1 - \exp(-2\rho d)$, denotes the volume fraction of phase 1, which represents the remaining block of material not occupied by a layer. The analytic representation of $\gamma(r)$ requires an interval splitting in the variable r as follows: $0 \leq r \leq d$ and $d < r < \infty$. The chord lengths m_i in grain phase 2 are limited by $d \leq m_i < \infty$.

5.5 Chord length distributions of the Poisson slice model

The Poisson slice model (PSm) is another special case of a Boolean model. Its characteristic scattering pattern $I(h)$ and real-space structure functions, will be derived. This approach is useful for the characterization of porous materials. Actually, the primary grains are infinitely long layers of constant thickness d. According to the Babinet theorem, the layers can be substituted by slices without changing the CF of the model. The layers (or slices) are uniformly arranged in space in an isotropic, random manner. Additionally, Fig. 5.10 explains the plane case. The layer (or slice) density is interrelated with the volume fraction $1 - c$ of the layer and can be expressed by a sequence of interrelated parameters. This complicates matters.

The denotations are as follows: The layer (or slice) primary grains represent phase 2, which has the volume fraction $1 - c$. Let $f(m)$ be the CLDD of phase 2. The characteristic function of $f(m)$ is $q(t)$. Phase 1 (volume fraction c, CLDD $\varphi(l)$ and the characteristic function $p(t)$; see the darker regions in Fig. 5.10) is limited by planes originating from the union set of the slices of const. thickness d.

In the following, the spatial PSm set is analyzed. The density of the layers $n = \lambda \cdot V_0$ (see the considerations in Section 5.2) is described by a parameter ρ operating with the abbreviations $n = \lambda \cdot V_0 = n/(d \cdot S_0) \cdot (S_0 d) = 2\rho d$, i.e., $\rho = n/(2d)$, where $0 < \rho < \infty$. The dimension of ρ is 1/length. Consequently, $c = \exp(-n) = \exp(-2\rho d)$ results and $0 < c < 1$. Based on the grain CF of the single lamella, the formulation of which requires the analysis of two intervals ($0 \leq r < d$ and $d < r < \infty$), the CF $\gamma(r)$ of the whole tightly packed two-phase system can be formulated in terms of r, d, ρ or in terms of r, d, c, which works equally as well. The latter, i.e., $\gamma(r, d, c)$ will be favored. Based on $\gamma_0(r, d)$ and $\gamma(r) = [C(r)/c - c]/(1 - c)$ (see Chapter 1), the CF of the PSm results in[||]

$$\gamma(r, d, c) = \begin{cases} 0 \leq r \leq d: \ (c^{\frac{r}{2d}} - c)/(1 - c) \\ d < r < \infty: \ (c[c^{-\frac{d}{2r}} - 1])/(1 - c) \end{cases}. \tag{5.25}$$

While $\gamma(r) = \gamma(r, d, c)$ is continuous, $g(r) = l_p \cdot \gamma''(r)$ involves a finite jump of size $2\sqrt{c}/d$ at $r = d$ (see Fig. 5.11). From the functions γ and g, the parameters $l_p = 2(1 - c) \cdot d/\ln(c)$ and $l_1 = l_p/(1 - c)$, $m_1 = l_p/c$ result. The CLDD $\varphi(l) = \varphi(l, d, c)$ is an exponential function of the variable l. The representations $\varphi(l, l_1) = 1/l_1 \cdot \exp(-l/l_1)$ and $\varphi(l, \rho) = \rho \cdot \exp(-l \cdot \rho)$ are equivalent (see Fig. 5.12). It holds that $\rho = 1/l_1$.

In contrast, determining the CLDD $f(m)$ [CLDD of the grain phase 2 of the PSm] is a complicated matter and requires some effort. However, in analogy with (Gille, 2007) [107] and (Gille, 2009) [115], it can be handled analytically. Again, the approach is based on the fundamental connection

$$q(t, d, \rho) = q(t) = \frac{Q(t) - p(t)}{1 - p(t) \cdot [2 - Q(t)]} \approx 1 + Q'(0) \cdot t + O(t)^2, \quad q(0, d, \rho) = 1 \tag{5.26}$$

between the (unknown) characteristic function $q(t)$ and the known terms $Q(t) = Q(t, d, \rho) = \int_0^\infty g(r, d, c)e^{itr}dr$ and $p(t, d, \rho) = i\rho/(t + i\rho)$. From $q(t)$, the function $f(m)$ results (see Fig. 5.12, right). The function $f(m)$ was determined by a *Mathematica* program involving the sequence of three Fourier transformations by use of the function *SequenceLimit* (see Gille, 2007) [107].

The moments of f, φ and g are defined in terms of the derivatives of the corresponding characteristic functions [see Eq. (5.26)]. Important extensions of Eq. (5.26) are discussed in Chapter 9. This includes the analysis and generalization of the connection

$$-\gamma'(0) = \frac{1}{l_p} = \frac{1}{l_1} + \frac{1}{m_1} = \rho + \frac{1}{m_1}, \tag{5.27}$$

which was discovered by the great Porod as early as 1950, [191, 192, 194]. Equation (5.27) is involved in all Bm cases, especially in Eq. (5.26).

The asymptotic representation of the scattering pattern is investigated in Fig. 5.13. The normalized Porod plot $P_1(h)$ follows from the CF, as well as from the CLDDs $\varphi(l)$ and $f(m)$ (Levitz & Tchoubar, 1992) [165]. Comparing the theoretical asymptotic approximation $P_{1\infty}(h)$ with experiments allows the model parameters d and ρ to be detected.

[||]In the limiting case $c \to 1$, the CF $\gamma_0(r, d)$ of the single layer of thickness d results in $\gamma_1(r, d) = 1 - r/(2d)$, if $0 \leq r < d$ and $\gamma_2(r, d) = d/(2r)$, if $d < r < \infty$.

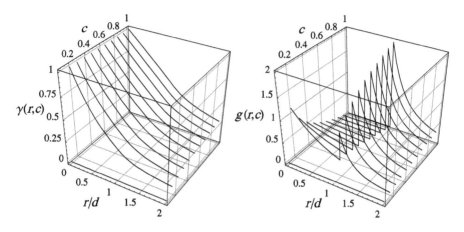

FIGURE 5.11

The functions $\gamma(r, 1, c)$ and $g(r, 1, c)$ of the PSm (see Fig. 5.10) for $c = 0(0.1)1$. The case $c \to 1$ corresponds to the single lamella case, where $\rho \to 0$. If $c \to 0$, which corresponds to $\rho \to \infty$, the space is completely occupied by slices.

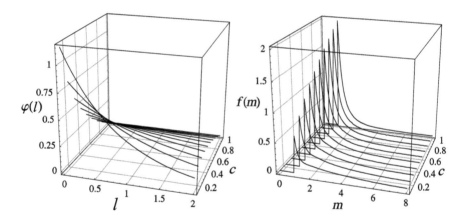

FIGURE 5.12

The CLDDs of both phases of the PSm illustrated by a parametric representation of the functions $\varphi(l, 1, c)$ and $f(m, 1, c)$ for varying $c = 0(0.1)1$ (see Fig. 5.11).

The exponential function φ does not depend on d, since $\rho = n/(2d) = 1/l_1$. The smaller c is, the greater the first moment m_1, where $m_1 = l_p/c = 2(1 - c) \cdot d/[c \cdot \ln(c)]$ of $\varphi(l)$. In the limiting case $c \to 0$, when the space is completely occupied by the overlapping slices (see Fig. 5.10), $\varphi(l) \to \delta(l)$ and $f(m) \to 0$. On the other hand, if the number of slices is very small, $\varphi(l) \to 0$ for $c \to 1$. Hence, $f(m)$ approaches the CLDD of a single lamella and Porod's length parameter $l_p = 2d$ follows.

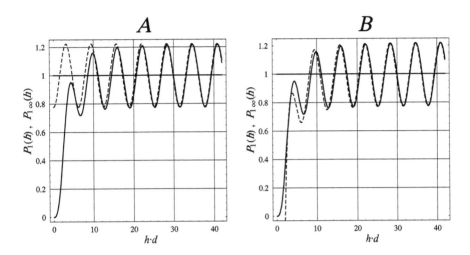

FIGURE 5.13

Illustration of the scattering pattern of the PSm in the special case $d = 1$, $\rho = 3/2$ and $c \approx 0.05$ for the interval $0 \leq h \cdot d \leq 40$.

Because of $v_c \to \infty$ and $f_c \to \infty$, $I(0) \to \infty$ results. Nevertheless, similar to the infinitely long cylinder case (see Chapter 3), the $P_1(h)$ plot is well-suited for inspecting the asymptotic behavior of $I(h)$.

The exact curve $P_1(h) = l_p \cdot h^4 I(h)/(8\pi)$ (solid line) is compared with two asymptotic approximations $P_{1\infty}(h)$ (see the dashed lines), which are shown as steps A (left) and B (right).

Step A (left): $P_{1\infty A}(h) = 1 - \cos(hd)/(2d^2\rho^2)$ and

Step B (right): $P_{1\infty B}(h) = 1 - 2\rho^2/h^2 - \cos(hd)/(2d^2\rho^2)$.

The more general case of approximation B, which involves a term h^{-2} and a cosine term, is based on

$$P_{1\infty}(h) \approx 1 - 2\frac{\gamma'''(0+)}{\gamma'(0+) \cdot h^2} - \frac{\cos(hd)}{2d^2\rho^2}$$

[see Eqs. (1.39) to (1.41) and Fig. 1.6].

The term $-2\gamma'''(0+)/[\gamma'(0+) \cdot h^2] = -2\rho^2/h^2$ essentially improves the approximation for relatively small h. However, this modified Kirste-Porod term is not important for larger h. The behavior of $P_1(h)$ for $h \to \infty$ is analogous to that of a Bm with spherical grains (compare with Fig. 5.5).

More general formulas for asymptotic approximations of $I(h)$ were given by (Ciccariello, 2002) [18].

5.6 Practical relevance of Boolean models

The Bm, which was first developed in the field of stochastic geometry, has been applied in many fields and is also one of the most widely investigated models of stochastic geometry. The model, which is simple yet of a large flexibility, involves two basic independent parameters, which are c (or n) and the function $\gamma_0(r)$ with L_0. Extensions for varying grain shapes of a fixed size (Sonntag et al., 1981) [208] or for a fixed grain shape possessing a certain size distribution (Gille, 1995) [53] are feasible and do not require great effort. Due to the existence of the parameter L_0, which is the maximum diameter of the biggest grain particle, this type of model can never describe a compact sample, where $0 \leq r < L$. The *short-order range model* restricts to the superimposition of any two grains, which are placed at the germ positions.

Applying the approach derived in Section 1.5.2, the maximum grain diameter L_0 can be studied. Based on an (all-including) scattering pattern $I(h)$, the relative characteristic volume $v_{c,rel}(L)$ fixes the parameter L_0 [see Eq. (1.72)]. This theory includes the case of differently sized grains. In such a case, the interpretation of $v_{c,rel}$ is a complicated matter. Superimpositions of the scattering for different L_i yield smooth curves $v_{c,rel}(L)$. In limiting cases, this does not allow different lengths L_i to be separated.

In Section 5.5, spherical and plate-like grains were considered. Similar considerations follow for a large variety of compact grain shapes, which characterize $\gamma_0(r)$. In all of these cases, the scattering pattern is smoother if more germs per volume are inserted and more grains per volume contribute to the scattering pattern (see Fig. 5.6).

Nevertheless, the moments of both the CLDDs and c can be detected from the pattern $I(h)$ for all the grain shape parameters inserted. For an analysis of the interrelations between the CLDD moments, see Chapter 9. In fact, the volume fraction is interrelated with higher CLDD moments. An extension of Rosiwal's linear integration principle $c = l_1/(l_1 + m_1)$ to higher moments has practical relevance for the data evaluation of astronomic occulation experiments and in astrophysics for the estimation of volume fractions in the universe.

Another essential model of stochastic geometry, the *Dead Leaves model*, is briefly explained in the next chapter. As already mentioned in Chapter 4, this model even allows analytic expressions to be constructed for pair correlation functions.

Furthermore, Chapter 6 provides another model for constructing random particle shapes. Here, the idea is to start from a tessellation of space and not from overlapping grains, as was done for the Boolean model in this chapter.

6

The "Dead Leaves" model

Besides the Boolean model (Bm), the "Dead Leaves" model (DLm) is a fundamental and well-investigated model of stochastic geometry. It has two modifications, which are the "covered DLm" (random tessellations of the space) (see Section 6.1) and the "uncovered DLm" (tightly packed hard particles) (see Section 6.2). Both modifications have practical relevance.

Natural phenomena that can be well described by a "Dead Leaves" model.
The DLm in stochastic geometry was developed by the French mathematician George Matheron [174]. Its applications include random sequential adsorption processes, as well as the description of tessellations, puzzles and collections of separated randomly shaped particles [205]. The puzzle interlayer model (PIM), which has recently been introduced in scattering theory, is based on these foundations (Gille, 2007) [107].

There are mainly two variations of this type of model. In pointing out this fact, the mathematician Serra [205] says that "*it (the DLm) has the dual advantage that it provides us with a tessellation of the space as well as a model for non-overlapping particles.*" Sections 6.1 and 6.2 explain both of these aspects in the context of the scattering pattern of a particle arrangement.

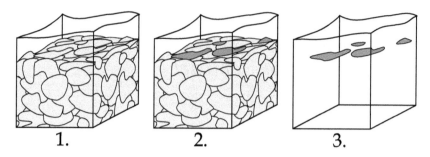

FIGURE 6.1

The term *"Dead Leaves"* model (German: *"Laubmodell"*) goes back to a time sequence process of falling leaves, which cover a plane completely (1) and is accompanied by a process of selection of the upper leaves (2) and the decomposition of the covered leaves (1,2,3). Substituting the term *leaf* with *particle grain* or simply with *grain* (G), the last grain partly covers one or more of the grains lying below [see (1) and (2)]. Four leaves survive.

Figure 6.1 shows a semi-two-dimensional illustration of the DLm. Establishing practicable models of random media (see Jeulin, 2002) [156] mainly includes two aspects: The description of ensembles of microparticles with suitable functions and parameters, as well as the generation of random shapes and sizes of structuring elements, e.g., single particles of random shape. Both aspects are involved in the DLm, (Matheron, 1968, 1975) [174], (Filipescu, et al.) [37] and (Serra, 1982) [205]. For the field of physical scattering methods, the books by (Stoyan et al., 1987) [210] and (Hermann, 1991) [150] are important. The physicist Porod (1951) was not able to apply these fundamental principles, however he was able to successfully use his own "homemade" considerations [191, 192].

The DLm provides us with a tessellation of space (see Serra, 1982) [205, pp. 508–509]. In Figs. 6.1 and 6.2, grains have been placed at random positions in space beginning at time $t = -\infty$. As long as no overlapping happens, the grains survive at fixed places. If a "new" grain G_n at time $t = t_n$ overlaps with "old" ones, G_n survives. The "old" intersected grains are eliminated. For spherical grains of const. diameter, this process results in the volume fraction $1 - c = 7/8$ of the connected phase (outside the grains) after an endless time at $t = 0$. Then, the volume fraction c of remaining grains is $c = 1/8$.

The pair distribution function $g(r)$ (see part (3) of Fig. 6.1, but not Fig. 6.2), the volume fraction and the set covariance of the remaining grains were the subject of intensive investigations (Stoyan & Schlather, 2000) [213]. In contrast, the subject of Section 6.1 is the *the single cells* of the DLm tessellation, i.e., the isolated prototype of the DLm cells. For these *puzzle cells*, the denotation PC is used in this chapter. In Chapter 7, PP (for puzzle particle) or FP (for fragment particle) is used instead in connection to the puzzles. The

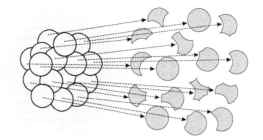

FIGURE 6.2
DLm tessellation construction (covered DLm) with circular primary grains (two-dimensional case).
A falling "leaf" (a new grain at a random position) covers the part below its own surface area. This tessellation consists of 16 cells. After shifting these cells (indicated by the given arrows) to another position (possibly followed by a random rotation of the cell), a puzzle results. The scattering pattern of the cells can be recorded (quasi-diluted) "one by one." In this chapter, the abbreviation PC is used for the prototype of an (isolated) puzzle cell. It is important to note that the mathematical description of the randomly shaped PCs is surprizingly simple. The real-space structure functions of the PC can be traced back to the autocorrelation function of the grain.

shape and size of the primary grain fix the shape and size of the PC (Fig. 6.2). The covariance and linear erosion of the PCs are defined in terms of the grain CF. Stereological parameters up to the scattering pattern $I(h)$ of the PCs result. Such PCs are the patches of dried soil, which are shown in the upper left part of the introductory figure of this chapter.

6.1   Structure functions and scattering pattern of a PC

Figure 6.2 explains the connection between starting mosaic and tessellation cells. The PC shape(s) result(s) from covered circles, which define random shapes of PCs. The investigation of the real-space SAS structure functions and the isotropic scattering pattern of the typical isolated PC is simple.*
This holds regardless of the grain shape inserted. For the sake of simplicity, however, most of the following illustrations use spherical primary grains of constant diameter.

---

*In 1982, this fact caused the mathematician Serra to limit the representation to only a few pages (Serra, 1982) [205].

Model parameters of independently scattering PCs

The DLm considered produces isotropic PCs and is isotropic. Clearly, one single PC in a fixed position is anisotropic. The typical size and shape of the PCs are defined by the fixed size and shape of the grain(s)(G). No interaction between any two PCs is assumed. Sufficiently small L, i.e., sufficiently large scattering vector amounts h, are considered. A quasi-diluted scattering pattern of PCs is the sum of the scattering intensities of the single PCs. An averaged isotropic band limited $I(h, L)$ results [194], which is defined by a Fourier transformation of the CF $\gamma_{PC}(r, L)$. Indeed, $\gamma_{PC}(r, L)$ is known. According to a proof (see Serra, 1982) [205, p. 509] for any grain shape corresponding to a grain CF $\gamma_G(r)$, the CF of the PC is

$$\gamma_{PC}(r) = \frac{\gamma_G(r)}{2 - \gamma_G(r)}. \tag{6.1}$$

Equation (6.1) is not an approximation. Its meaning is immense. It holds for any CF $\gamma_G(r)$. The coefficients of the series expansion of $\gamma_{PC}(r)$ are given in terms of the derivatives of $\gamma_G(r)$ in the origin as

$$\gamma_{PC}(r) = 1 + 2\gamma_G'(0) \cdot r + \left[2\gamma_G'(0)^2 + \gamma_G''(0)\right] \cdot r^2 + O\left(r^3\right). \tag{6.2}$$

Equations (6.1) and (6.2) are general for compact grains and the resulting compact PCs. Substituting $\gamma_G(r)$ by the CF of a single sphere of diameter d into Eq. (6.1) gives

$$\gamma_{PC}(r, d) = \begin{cases} \frac{(d-r)^2(2d+r)}{(d+r)^2(2d-r)} & , \quad if \quad 0 \le r \le d, \\ 0 & , \quad if \quad d < r < \infty. \end{cases} \tag{6.3}$$

The first three derivatives in the origin are $\gamma_{PC}'(0) = -3/d$, $\gamma_{PC}''(0) = 9/d^2$, $\gamma_{PC}'''(0) = -69/(2d^3)$. Consequently, the mean average chord length for IUR chords of the (typically) non-convex PC is $\bar{l}_{PC} = d/3$. This is half the mean chord length of the grain $l_{1G} = \bar{l}_G = 2d/3$. The intrinsic properties of the CLDD of the PC are dealt with later starting with Eq. (6.13).

Mean average parameters of a PC particle: The PC prototype can be characterized by its radius of gyration R_g and the basic integral parameters v_c, f_c and l_c (Porod, 1951) [191, 192]. Operating with Eqs. (6.3) and (10.24), the analysis of the moments of $\gamma_{PC}(r, d)$ leads to Eqs. (6.4) to (6.7), starting with

$$R_g^2 = \frac{\int_0^d r^4 \gamma_{PC}(r) dr}{2 \int_0^d r^2 \gamma_{PC}(r) dr} = \frac{69 - 100 \ln(2)}{10[4 \ln(2) - 3]} \cdot d^2 \approx 0.1384 \cdot d^2. \tag{6.4}$$

Thus, $R_g \approx 0.372 \cdot d$. Substituting $t = 3 - 4 \cdot \ln(2)$, the characteristic volume

$$v_c = \int_0^d 4\pi r^2 \cdot \gamma_{PC}(r) dr = \frac{4}{9}\pi t \cdot d^3 = 0.3175 \cdot d^3 \tag{6.5}$$

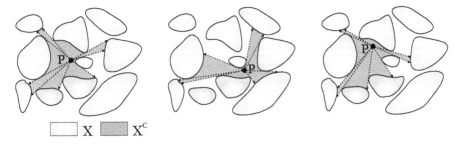

$\square$ X $\blacksquare$ X^c

FIGURE 6.3

Explanation of the *star* of a non-convex set in the two-dimensional case (for details see Serra, 1982) [205, p. 332]. The point **P** sweeps across the whole region outside the (hard, non-transparent) particles X (illustration of three positions of **P**). The boundary of the field of view defines a region X^c for any position of **P**. The "lines of sight-limit" depend on the position of **P**. The *star* is the average of all these contributions to the field of view.

In analogy with this, the *star* of a single particle is defined (see Fig. 6.4). Here, the total boundary of the field of view results in terms of the (inner) border of the particle. For a convex particle of volume V_0, the mean "region of total sight-limit" equals V_0, i.e., *star* = V_0.

follows. This volume v_c is 61% that of the primary sphere. Furthermore,

$$f_c = \int_0^d 2\pi r \cdot \gamma_{PC}(r)dr = \frac{\pi d^2 \cdot [32\ln(2) - 21]}{9} \approx 0.4121 \cdot d^2, \qquad (6.6)$$

$$l_c = 2\int_0^d \gamma_{PC}(r)dr = \frac{2d \cdot [8\ln(2) - 3]}{9} \approx 0.5656 \cdot d. \qquad (6.7)$$

The star parameter and the characteristic volume: The so-called *star* belongs to the basic parameters defined in stochastic geometry. It is useful to compare v_c with the *star*, denoted by st_3, which is also a volume parameter. It is given in terms of the linear erosion $P_{PC}(r)$ (Serra, 1982) [205, p. 332] (see Fig. 6.3 and Fig. 6.4).

For a single PC, the parameter st_3 has the following meaning: Imagine a transparent PC prototype and consider a random point **P** inside. It follows then that a certain zone volume of the PC can be directly seen from **R**. The star st_3 is the average volume of this zone when the point **P** is evenly distributed inside the PC (when it sweeps across the whole PC).

For the tessellation model considered (for the DLm), the linear erosion function $P_{PC}(r)$ connected with the grain via $\gamma_G(r)$, as given by

$$P_{PC}(r) = \frac{\gamma_G(r)}{1 - r \cdot \gamma_G'(0)}, \qquad (6.8)$$

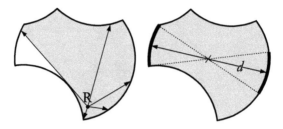

FIGURE 6.4
Geometric properties of a puzzle cell (PC) in a two-dimensional case.
Left: Illustration of the parameter st_2 for a typical PC. Since there are regions inside the PC, which cannot be reached from point **R** by a straight line, st_3 can never be greater than the PC volume. The white region(s) do not contribute to the star (volume) for the fixed position of point **R**.
Right: This PC has a *limited* parallel surface area (indicated by the bold lines). This limited parallelism is the reason for the *restricted* elliptic oscillation term of the normalized Porod plot $P_1(h, d)$ (see Gille, 2001, 2003) [74, 89].

must be inserted into Eq. (6.5) instead of $\gamma_{PC}(r)$. Equation (6.8) is the most important equation in the DLm approach. Equations (6.8) and (6.5) result in

$$st_3 = \int_0^d 4\pi r^2 \cdot P_{PC}(r, d)dr \approx 0.3036 \cdot d^3 \approx 0.956 \cdot v_c. \qquad (6.9)$$

For a convex particle, there is no difference between st_3 and v_c. However, the majority of the PCs are non-convex. Therefore, the star is somewhat smaller than v_c, where $st_3 < v_c$.

Furthermore, the linear erosion is connected with the linear size distribution density $f_{PC}(l)$ of the PC. The relation for any compact particle is $f_{PC}(r) = P_{PC}''(r)/ \mid P_{PC}'(0) \mid$ (see the corresponding curve f_{PC} in Fig. 6.7). However, with a convex particle, the function $f_{PC}(l)$ is the distribution density for IUR chords l_i of the particle. The actual case includes non-convex PCs. Hence, careful consideration shows that the function $U_{PC}(r) = \gamma_G''(r)/ \mid \gamma_G'(0) \mid$ (see Fig. 6.7) is not a pure CLDD of *one* random length variable.

More interesting details about typical second-order characteristics of a PC are summarized in the three parts of Fig. 6.7.

The surface area of the PC is not smooth; consequently $0 < P_{PC}''(0)$ and $U_{PC}(0, d) = 3/d > 0$. Furthermore, the PC is characterized by the distance distribution density $p(r, d) = 4\pi r^2 \cdot \gamma_{PC}(r)/v_c$ and the transformed correlation function γ_{TPC} of the PCs, which is defined in terms of γ_{PC} via $\gamma_{TPC}(r, d) = 2/\pi \cdot \arcsin(1 - [\gamma_{PC}(r, d)]^{1/3})$, where $0 \leq r < d$ [66, 77].

The transformed CF (see dashed line in Fig. 6.7) fixes the largest diameter (here $r = d$) of the largest PC.

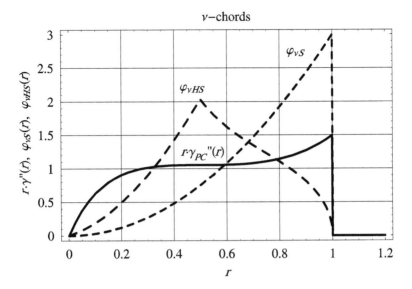

FIGURE 6.5

Analysis of PCs constructed from spherical grains of diameter $d = L_0 = 1$ and other particles via ν-chords.

The full curve represents the function $r \cdot \gamma''_{PC}(r)$ of a prototype puzzle cell. The dashed lines represent the CLDDs for ν-chords of a single sphere $\varphi_{\nu S}(r) = 3r^2/d^3$, where $0 \le r < d$ and of a hemisphere $\varphi_{\nu HS}(r) = r \cdot A_{HS}(r)/l_1$ [see Eqs. (2.47) and (2.32)]. For the typical PC, the number of small ν-chords is mostly increased (compared with the sphere). The curves for the hemisphere and the PC share a certain similarity (see also the scattering patterns in Fig. 6.6). The parameter $d = L_0$, or actually $d = 1$, can be fixed from these functions.

The scattering pattern of quasi-diluted PCs

The scattering intensity function I of a tessellation broken into puzzle pieces results from Eq. (1.33) with Eq. (6.3). With the CF of a single sphere $\gamma_G(r) = \gamma_G(r, d)$ and using the abbreviations $y = h \cdot d$, $Si(x)$ for the sine integral function and $Ci(x)$ for the cosine integral function, $I(y) = I(h \cdot d)$ is

$$I(y) = \frac{1}{t \cdot y^3} \{ 2y^2 [(2Ci(y) - 2Ci(2y)) \cdot (2\sin(y) - 2\sin(2y) + 3y \cdot \cos(y))$$
$$+ (2Si(2y) - 2Si(y)) \cdot (2\cos(y) - 2\cos(2y) - 3y \cdot \sin(y)) + 3\sin(y)]$$
$$- 9(\sin(y) - y \cdot \cos(y)) \}. \tag{6.10}$$

By use of a *Mathematica* program, some useful plots of the scattering behavior of the PCs result [see Eq. (6.10) and Fig. 6.6].

```
(* scattering pattern i[h, d], asymptotic scattering pattern ias[h,d],
   characteristic volume vc[d] and asymptotic Porod Plot Plas[h,d] for PCs *)

i[h_, d_] :=(y = h*d; (1/(y^3*(-3 + Log[16])))*
              (9*((-d)*h*Cos[y] + Sin[y]) + 2*y^2*(-3*Sin[y]+
              2*(-CosIntegral[y] + CosIntegral[2*y])*
              (3*d*h*Cos[y]+2*Sin[y]-2*Sin[2*y])+
              2*(SinIntegral[y] - SinIntegral[2*y])*
              (2*Cos[y]-2*Cos[2*y]-3*y*Sin[y])))));

vc[d_]      :=(-(4/9))*d^3*Pi*(Log[16]  - 3);
ias[h_,d _]:=-9*((-552*(Pi/(d^3*h^6))+24*(Pi/(d*h^4)))/
              (4Pi d^3*(-3 + Log[16])));
Plas[h_,d_]:=(d/3)*vc[d]*h^4*(i[h, d]/(8*Pi));

Needs["Graphics`Graphics`"];
LogPlot[{1, i[h, 1]}, {h, 1/100, 20}];
Plot[{1, Plas[h, 1], (1/3)*vc[1]*h^4*(ias[h, 1]/(8*Pi))}, {h, 10, 50}];
```

Equation (6.10) gives the power series

$$I(y) = 1 + a_2 \cdot y^2 + a_4 \cdot y^4 + a_6 \cdot y^6 + a_8 \cdot y^8 + O\left(y^{10}\right),$$

$$a_2 = -\frac{100 \cdot \ln(2) - 69}{30t}, \quad a_4 = \frac{329 \cdot \ln(2) - 228}{210t},$$

$$a_6 = -\frac{466 \cdot \ln(2) - 323}{2520t}, \quad a_8 = \frac{437800 \cdot \ln(2) - 303459}{39916800t}. \quad (6.11)$$

The coefficients a_i reflect the moments of $\gamma_{PC}(r, d)$ [see Eqs. (6.4) and (6.5)]. The coefficient a_2 of the quadratic term is connected with the SAS parameter R_g (see Table 1.1).

Equations. (6.10) and (6.11) fulfill the normalization $I(0, d) = I(0) = 1$. As a check, the invariant of the standard scattering intensity $2\pi^2$ [see Eq. (1.38) and Eq. (1.70)] (Porod, 1951) [191] results from Eq. (6.10). In this case [see Eq. (6.2)], there do not exist logarithmic singularities in the behavior of $\gamma_{PC}(r, d)$; a related discussion can be fpund in (Ciccariello, 1993 & 1995) [15, 16] and (Gille, 2000) [65]. This can be traced back to the simple shape of the grains and the resulting random PC shapes. The asymptotic behavior $I_\infty(h)$ can be approximated by a Porod term plus a Kirste-Porod term. Operating with Eqs. (6.1) and (6.2), the formula $v_c \cdot I_\infty(h) = -8\pi\gamma'(0)/h^4 + 16\pi\gamma'''(0)/h^6$ yields the asymptotic approximation $I_\infty(h, d)$ as

$$v_c \cdot I_\infty(h, d) = \frac{24\pi}{d \cdot h^4} - \frac{552\pi}{d^3 \cdot h^6}. \quad (6.12)$$

The normalized Porod plot $P_1(h, d) = d \cdot v_c \cdot h^4 \cdot I(h, d)/(24\pi)$ compares the exact oscillating curve Eq. (6.10) with the Kirste-Porod approximation Eq. (6.12) in the interval $5 < h \cdot d < 50$. A *Mathematica* program based on Eq. (6.10) is given for these functions.

Linear erosion and SAS correlation function of DLm PCs

The correlation function $\gamma(r)$ and the isotropized set covariance $C(r)$ go back to the same structure element, i.e., two points A and B, separated by a distance r.

In contrast, the structure element of the linear erosion is the line $\overline{AB}$ of length r (see Serra 1982 [205], section B1 *Linear erosion*).

Furthermore, for a compact *convex* set (convex particle), the second derivative $\gamma''(r)$ is proportional to the chord length distribution density for IUR chords. For any compact set, the linear size distribution density $f(r)$ is proportional to the second derivative $P''(r)$ (see Gille et al., 2005) [102]. By use of the normalization $\gamma(0) = P(0) = 1$, the behavior of $\gamma(r)$ and $P(r)$ near the origin is identical and the normalization factors are given by the same length. This normalization term is the mean chord length $l_1 = 1/\mid \gamma'(0)\mid = 1/\mid P'(0)\mid$. This length is given for any compact set by $l_1 = 4V/S$ (Gille et al., 2005) [102]. Here, S is the whole surface area (all surface parts must be included) of the particle. In fact, the series expansion of Eq. (6.8), which is written as

$$P_{PC}(r) = 1 + 2\gamma_G'(0) \cdot r + \left[2\gamma_G'(0)^2 + \frac{\gamma_G''(0)}{2}\right] \cdot r^2 + O\left(r^3\right) \quad (6.13)$$

contains the same linear term as Eq. (6.2) for any compact grains considered.

Furthermore, if all the grains have sufficiently smooth surfaces, i.e., $\gamma_G''(0) \to 0$, then Eq. (6.13) does not differ from Eq. (6.2). Both Eq. (6.2) and Eq. (6.13) yield $f_{PC}(0,d) \equiv U_{PC}(0,d) = 2 \cdot |\gamma_G'(0)| = 3/d$. Based on Eq. (6.8), for the single cells of the puzzle, the functions $P_{PC}(r,d)$ and $f_{PC}(r,d)$ (Fig. 6.7) give

$$P_{PC}(r,d) = \begin{cases} \frac{(d-r)^2(2d+r)}{d^2(2d+3r)}, & if \quad 0 \le r < d, \\ 0, & if \quad d < r < \infty \end{cases}, \quad (6.14)$$

$$f_{PC}(r,d) = \begin{cases} \frac{2}{9d} \cdot \left[1 + 100\left(\frac{d}{2d+3r}\right)^3\right], & if \quad 0 \le r < d \\ 0, & if \quad d < r < \infty \end{cases}. \quad (6.15)$$

Thus, for spherical grains, $f(0+,d) = 3/d$ and $f(d-,d) = 2/(5d)$. Furthermore, the function $T_{PC}(r,d) = 4\pi r^2 \cdot P_{PC}(r)/st_3$ is included in Fig. 6.7. This function T_{PC} is very similar to the SAS distance distribution density function $p_{PC}(r,d)$ [compare with Eqs. (6.5) and (6.9)].

Figure 6.7 summarizes the important properties of the structure functions of a PC. The functions given in the top and middle graphs are of different origin. Nevertheless, their behavior is quite similar and the differences $[\gamma_{PC}(r) - P_{PC}(r), p_{PC}(r) - T_{PC}(r)$ and $U_{PC}(r) - f_{PC}(r)$, see bottom graph] between them are minimal. The deviations depend on the size and shape of the grain of the DLm.

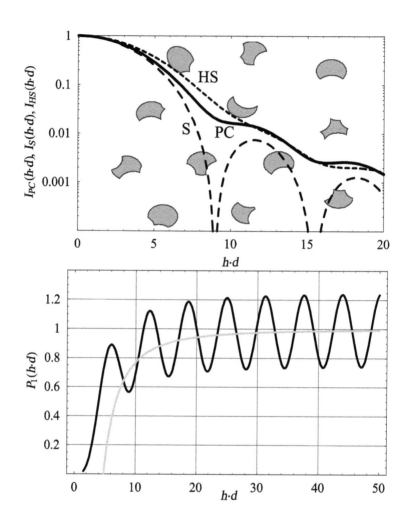

FIGURE 6.6

Analysis of the pattern $I_{PC}(h \cdot d)$ of a typical PC, where $0 < h \cdot d < 20$.

Top figure: An ensemble of PCs of a DLm tessellation, constructed from spherical grains of constant diameter d, produces the solid line scattering pattern I_{PC} (see typical PC shapes in the background). The dashed lines correspond to the pattern of a single sphere I_S and to a hemisphere I_{HS}. As is to be expected, the cases HS and PC are similar.

Bottom figure: Normalized Porod plot $P_1(h \cdot d)$ and Kiste-Porod approximation of the scattering of the PCs. Compare with Figs. 6.2 and 6.4, which illustrate two-dimensional PCs.

On the one hand, the functions given in the top graph, $[\gamma_{PC}(r), p_{PC}(r)$ and $U_{PC}(r)]$, can be detected from SAS experiments and describe the prototype of a PC. These functions can be completely traced back to the CF of the inserted grain model $\gamma_G(r, d)$ [see Eqs. (6.2) and (6.3)]. The function γ_{TPC} (dashed curve) is the transformed correlation function of the PC.

On the other hand, the linear erosion $P(r)$ of the PC, i.e., $P_{PC}(r)$ [see Eq. (6.8)], fixes the stereological structure functions $f_{PC}(r)$, $P_{PC}(r)$ and $T_{PC}(r)$ in terms of the grain shape [see Eqs. (6.14) and (6.15)]. The normalization of $T_{PC}(r)$ is traced back to the *star* [see Eq. (6.9)].

Since the PC particles do not have an equal shape or size, the impact of non-convexity is minimal. As already shown in Fig. 6.6 and Fig. 6.5, there exists a certain similarity to the case of a hemisphere (see Figs. 6.6 and 6.5). This can be investigated in detail starting with the hemisphere CLDD [see Eq. (2.47) and Eq. (2.32)].

DLm puzzle cells in the context of SAS

The scattering properties and SAS parameters of an ensemble of the PCs are fully described by the two fundamental equations Eq. (6.1) and (6.8). The SAS CF (isotropic set covariance) is useful for describing random particle models (Hermann, 1991) [150].

The one aspect of a simple DLm that has been studied in detail with regard to scattering experiments is *the tessellation of space*. The randomly shaped, predominantly non-convex PCs are created from a tessellation process. An analysis of the mode of single isolated PCs yields an analytic expression for the scattering pattern $I(h)$. Clear expressions involving only the parameter $d = L_0$ (the diameter of the primary spheres) result for the typical real-space structure functions of a PC. The simplicity of these terms favors the DLm particle model for checking data evaluation procedures (Feigin & Svergun, 1987; Svergun, 1991) [36, 218].

The asymptotic approximation of the scattering pattern does not contain (highly) oscillating terms that touch the axis of the abscissa.[†] Indeed, the two-term approximation via Eq. (6.12) is perfect (see Fig. 6.6). This behavior can be traced back to the specific shape of the PCs. The oscillations are restricted to a smaller amplitude because the number of parallel interfaces is strictly limited. Obviously, there are only a few PCs that have parallel surface areas (distance d) (see Figs. 6.2 to 6.4).

The functions $\gamma_{PC}, p_{PC}, U_{PC}, \gamma_{PCT}$ and finally $I(h)$, as well as P_{PC}, U_{PC} and f_{PC}, can be traced back to the grain CF $\gamma_G(r)$. The actual explanation is restricted to spherical primary grains. However, Eq. (6.1) and Eq. (6.8) hold true for any compact primary grain shape. These circumstances favor

---

[†]Such oscillations are well-known from the single sphere case, where $h^4 I(h) \sim P_1(hd) = [1 - cos(hd)]$.

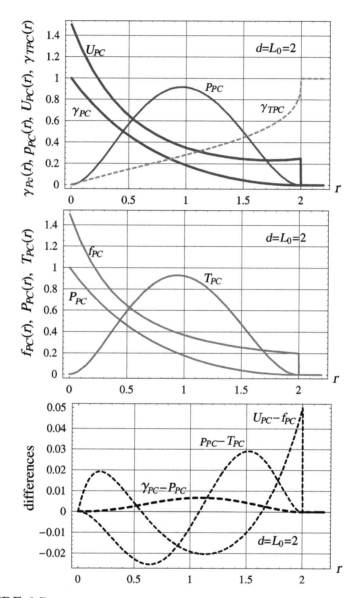

FIGURE 6.7

Behavior of SAS and stereological structure functions of the PC prototype. All of the functions in the top graph are based on the SAS CF $\gamma_{PC}(r)$ of the PC. The functions in the middle graph are based on the structure element of the linear erosion. The bottom graph illustrates the (actually small) differences between the pairs of functions.

the DLm for its application in materials science and for the identification of tessellation cells.

Further analysis of tessellation cells - relation to Chapter 7

The inverse step, i.e., the determination of the grain correlation function $\gamma_G(r)$ in terms of one of the structure functions given (Figs. 6.4 to 6.7), is also possible. However, *this back step can only be successful if the puzzle cells fit together completely* as in Fig. 6.2. In this context, it makes sense to introduce terms other than PC, like *puzzle particle* or *fragment particle*. Evidently, each of the PC structure functions and parameters analyzed in Eqs. (6.4) to (6.12) contains information about the origin of the puzzle pieces. With regard to larger length scales, some areas of application include archeology, cell biology and astrophysics.

In Chapter 7, the considerations and denotations will be inverted as follows: The prototype of a PC, i.e., typical single puzzle cells (PCs), will be considered to be fragment particles, which could *possibly* represent the *pieces of a puzzle*. The abbreviation PP will be used for the puzzle particle. In reality, Eq. (6.1) can be fulfilled for any given PPs in selected special cases. Starting from a scattering pattern I_{PP} which yields γ_{PP}, the relation $\gamma_{PP}(r, L) = \gamma_{PC}(r, L)$ does not have to be true.

6.2   The uncovered "Dead Leaves" model

The approach of an uncovered DLm is illustrated in part 3 of Fig. 6.8.

As already pointed out by J. Serra (1982) [205], the DLm provides us with a model for the spatial distribution of hard particles (see Chapter 4), which indirectly results in a special pair correlation of the DLm. Figure 6.8 illustrates a two-dimensional case, where the grains do not possess a unique shape. This aspect has absolutely nothing to do with Section 6.1.

There are several papers in which the analytic expression of the pair correlation function $g_{DLm}(r)$ of the *uncovered leaves* is used (Stoyan & Schlather, 2000) [213].

There exist analytic expressions for the set covariance of a collection of equal spheres, the center positions of which are described by *this specific pair correlation* $g_{DLm}(r)$. The function $C(r)$ is directly given in terms of $g_{DLm}(r)$ and the sphere diameter d_0 for a const. particle volume fraction $c = c_0 = 1/8$ (see Eq. (4.21) and (Gille, 2002) [80]).

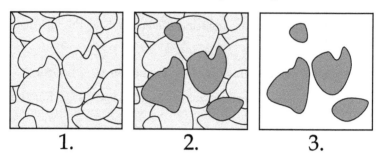

FIGURE 6.8
An ensemble of non-overlapping particles (phase 3) can be obtained from "Dead Leaves". The basic idea is simple: A grain (a particle, a leaf) that is not covered by another one "survives". Phases 1, 2 and 3 of the process are illustrated. The final result is phase 3, which especially shows the pair correlation of the intact grains. The intact grains never touch. The pair correlation of the (centers) of the intact grains was subject of intense investigations (see Stoyan & Schlather, 2000) [213].

The scattering pattern of such a model differs from the one resulting from the Percus-Yevick approximation (see [188]) for the special $c = c_0$. The reader can study some details in Section 10.3.2. Furthermore, knowledge of an analytic expression for the pair correlation makes it possible to detect the CLDD $f(m)$ of the connected phase of a hard particle "Dead Leaves" model, i.e., for hard spherical grains $d = d_0$ and $c = c_0$. The function $f(m)$ involves the properties $f(0) = f(\infty) = 0$ and clearly reflects a sequence of coordination shells of the spheres. Important formulas are derived in the paper by Gille (2005) [100] entitled *"Intersect distributions of a 'Dead Leaves' model with spherical primary grains"*. Hence, the second aspect of the DLm is also of practical relevance.

In Chapter 7, the tessellation aspect of the model is investigated in detail. An analysis of *puzzle fragment particles* via scattering patterns, the SAS structure functions and the set covariance are also included.

7

Tessellations, fragment particles and puzzles

In many areas, ranging from just a few nanometers to thousands of kilometers, the destruction of a tessellation results in single, independent objects. Obviously, these objects can be considered to be the pieces of a puzzle (see The Brown University S.H.A.P.E. website, www.lems.brown.edu/shape/). This site includes a summary of the mathematical procedures (e.g., applications, projects, specific references) for the reconstruction of archeological findings. Applications for such *geometric puzzles* occur in many areas.

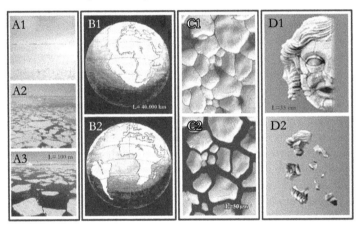

Examples of natural puzzles of different sizes (particle ensembles) that more or less fit together. The ensembles have different order ranges L, e.g., 50 µm$<$ $L < 40\,000$ km. The fragment particles and the intact tessellations are shown. These considerations require L to be fixed beforehand. The examples can be extended; e.g., for microphases in aluminum alloys, L is in the nm region.
A1/A2/A3: From an intact ice area to ice floes in the Arctic Ocean.
B1/B2: *Antonio Snider-Pellegrini Opening of the Atlantic*
(www.de.wikipedia.org/wiki/Kontinentaldrift).
C1/C2: Micrograph of a barium titanate system. Precision measurements were performed with a JSM-6480LV high performance scanning electron microscope, JEOL (2005) (see www.scan.si).
D1/D2: Shape Archeology multidisciplinary projects dedicated to solving the

puzzle problem for the automated reconstruction of archeological findings (www.lems.brown.edu/shape/).

Chapter 7 explains such connections in detail for different types of tessellation (Sections 7.2, 7.3 and 7.4). Any random tessellation of space can be the starting point for the construction of randomly shaped particles (see Figs. 7.1 and 7.2). Introducing here the particle volume fraction c as a free parameter, all cases $0 < c \leq 1$ are possible. For such ensembles of hard particles, the particle shape(s) are nor fixed, but fulfill a certain basic *law of shape*. With respect to practical application, it is worthwhile trying to investigate the scattering patterns of such particle ensembles as a function of c.

Puzzle problems in different fields

Puzzle problems, where more or less randomly shaped puzzle pieces have to be put together (or at least the possibility of a fit is uncertain), exist in many different fields, e.g., materials science, geography, biology and archeology. The puzzle pieces are separate, homogeneous particles with a maximum diameter L_0, where $L_0 \leq L$ (see Chapter 1). With respect to its analysis, the puzzle fitting function $\Phi(r, L_0)$ can be determined using elastic scattering experiments of (electromagnetic) waves of a certain suitable wavelength.

In the second step, the behavior of the function $\Phi(r, L_0)$ near the origin is important. The latter yields either $\Phi(0, L_0) < 1$ or $\Phi(0, L_0) \geq 1$. If the second relation is fulfilled, the origin of the particles can be traced back to a destroyed tessellation, which is based on a "Dead Leaves" model (DLm) of a certain grain shape (see Chapter 6). This approach, which is explained in detail in Section 7.2. can be seen as part of the broad field of automatic image analysis, even though no direct use is made of any image material. The analysis of the DLm puzzle case requires one scattering experiment, which reflects the fragment particle shapes. This chapter will explain how *the tessellation information* is considered in such an experiment.

While a great variety of other puzzle types exist, DLm tessellation puzzles are a special case. The fitting-together-ability of a Punch Matrix/Particle puzzle (PMP puzzle) (see Section 7.3) can be investigated with two independent scattering experiments. A PMP puzzle consists of a large homogeneous matrix piece (the punch matrix) and N single homogeneous fragment particles. A Fourier transformation of the scattering curve of the ensemble (matrix piece and fragment particles) yields function $g_{1N}(r)$. Additionally, the CLDD $\varphi(r)$ of the separate fragment pieces $\varphi(r)$ is assumed to be known near the origin $r \to 0$. The ratio between both terms yields the function $\Phi_{1N}(r) = g_{1N}(r)/\varphi(r)$. The behavior of $\Phi_{1N}(r)$ is significant for $r \to 0+$, i.e., *fitting-together-ability* is characterized by $\Phi_{1N}(0+) = 1$. Section 7.3 explains examples of $\Phi_{1N}(r)$. Another approach, which is applicable to fragment particles without holes, is introduced in Section 7.4. These fragments do not have to originate from a DLm. The data consists only of the whole CLDD $\varphi_P(l)$ (where $0 \leq l \leq L$) of the compact fragments.

7.1 Tessellations: original state and destroyed state

The basic denotations and functions for investigating whether given pieces can fit together have been introduced (see [108]). The scattering theory [36], the theory of chord length distributions [89, 156] and stochastic geometry [212, 225] have been applied in order to analyze this question.

Based on Chapter 1, the working function is the SAS CF $\gamma(r, L)$ (see Fig. 1.10 and Section 1.3.1). The isotropized set covariance $C(r, L)$, where $[C(0, L) = c, C(\infty, L) = c^2]$, is related to Eq. 1.46. By use of $Z(r) = C(r)/c$, the connection $\gamma(r, L) = [C(r, L)/c - c]/(1 - c)$, where $0 \leq r \leq L$ and $0 \leq c \leq 1$, can be formulated.

These functions are defined for a certain order range L of the sample. In the following equations, the length parameter L is not always explicitly involved; however, $\gamma(r) \equiv \gamma(r, L)$. For a single particle (actually for a grain or a fragment of volume V_0 and surface area S_0), the CF possesses the properties $\gamma_0(0) = 1$, $|\gamma_0'(0)| = S_0/(4 \cdot V_0) = 1/\overline{l_1} = 1/l_1$ and $\gamma_0(r) = 0$ if $r > L_0$ (see [175, 102]).

The analysis of such arrangements of puzzle particles requires the following assumptions regarding the background (origin) of the tessellation pieces (fragment particles): Based on the assumption that the origin of the random particle shapes goes back to a DLm [205], the puzzle fitting function $\Phi(r) = \Phi_{DLm}(r)$ (see Section 7.2) was introduced (see the initial paper [108]). There is no connection between the size of the puzzle fragments and their geometric shapes. If this independence does not exist, the assumption of a DLm is an approximation. The DLm has been thoroughly checked using stereological methods. Recently, scattering intensities of this type of model were determined in the special case of a puzzle interlayer model (PIM) [107] (see Section 7.6).

The DLm was considered in Chapter 6. In this context, the connections between the structure functions of grains, fragments and the corresponding scattering patterns have been explained. Considerations about the CLDD of DLm fragments have been discussed and an analysis of its scattering pattern (see Figs. 6.4 to 6.7) carried out. Based on these fundamentals, it is now possible to ask the following: *When does a puzzle fit together?* The mysteries of puzzles and their fragments will be analyzed. An analysis of randomly shaped tessellation pieces will now be considered in close connection with their scattering pattern.

It must be emphasized that the description *randomly shaped puzzle pieces* excludes trivial puzzles with a very small number of pieces. Obviously, two equally sized hemispheres fit together to form a sphere. However, such "types of puzzles", where the orientation of the pieces can be arbitrarily fixed, are not part of these investigations.

7.2   Puzzle particles resulting from DLm tessellations

There exists a large spectrum of particle shapes in nature. The DLm (see Chapter 6) describes a special class of particle shapes. Overlapping primary particles (primary grains) characterized by a grain CF $\gamma_G(r)$ result in a tessellation. Separating the pieces of the tessellation yields randomly shaped fragment particles possessing the CF $\gamma_P(r)$. These fragment particles (surface area S_P and volume V_P) can be considered to be the pieces of a puzzle. The term puzzle pieces (PP) is used instead of fragment particles in the following analyses. An investigation of the behavior of the CLDD $\varphi(r)$ of the PPs for isotropic uniform random chords yields simple results if these pieces originate from a DLm in the manner described.

A well-defined order range L is considered. For smooth primary grains, where $\gamma_G''(0, L) = 0$, the result $g(0, L) = \varphi(0, L) = 1/l_1 = S_p/(4V_P)$ holds. The function $g(r, L)$ is the superimposition of the CLDD of the random two-phase particle ensemble. A puzzle fitting for separate, i.e., for non- touching PPs, can be introduced. It is defined by

$$\Phi(r, L) = \Phi_P(r, L) = \frac{\gamma_P''(r, L)}{[\gamma_P'(r, L)]^2}, \quad 0 \leq r < r_0, \tag{7.1}$$

(derivatives with respect to r). Equation (7.1) correlates with an investigation of an ensemble of non-touching PPs. The term $\Phi(0, L)$ is the product of the CLDD of the single particle in the origin multiplied by its mean chord length l_1. Thus, $\Phi(0+, L) = \varphi(0+, L) \cdot l_1 = \varphi(0, L) \cdot 4V_P/S_P = \gamma_P''(0, L)/[\gamma_P'(0, L) \cdot \gamma_P'(0, L)]$, which resulted in the definition Eq. (7.1). Simplified denotations not involving L are frequently used. Based on the fitting property $\Phi(0+) = 1$, physical apparatuses which detect the second-order characteristics of PPs (e.g., example elastic scattering methods of electromagnetic waves) make it possible to analyze the *fitting-together-ability* of given (non-touching) PPs.

Equation (7.1) defines Φ in terms of the (mean) CF of the PPs for small r, $r \to 0$. An extension to (still non-touching PPs) possessing a certain volume fraction c is possible. Besides the CF $\gamma(r) = \gamma(r, c)$, other structure functions, defined in the field of *image analysis* and stochastic geometry, can be applied: the isotropized set covariance $C(h) \to C(r)$, the geometric covariogram $K(h) \to K(r)$, where $\gamma_P(r) = K_P(r)/V_P = K_P(r)/(c \cdot V_t)$, and the function of occupancy $Z(r)$. With the restriction $0 \leq r < r_0$, $\gamma(r) = [Z(r) - c]/(1 - c) = [C(r)/c - c]/[1 - c] = [C(r) - c^2]/[c(1 - c)]$ holds for a specific L, where $r_0 < L$. Therefore, Eq. (7.1) is written as

$$\Phi(r, L) \equiv \frac{\gamma_P''(r)}{[\gamma_P'(r)]^2} = \frac{\gamma''(r)}{(1 - c) \cdot [\gamma'(r)]^2} \equiv \frac{c \cdot C''(r)}{[C'(r)]^2} \equiv \frac{Z''(r)}{[Z'(r)]^2}. \tag{7.2}$$

The function $\Phi_{DLm}(r)$ for grains with surface singularities

The function $\Phi(r, L_0)$ was studied in several cases of DLm tessellations with smooth and non-smooth grains. For grains involving surface singularities (see the case of hemispherical grains [110]), $\Phi(r)$ is written in terms of $\gamma_P(r)$ as

$$\Phi(r) = \frac{\gamma_P''(r)}{[\gamma_P'(r)]^2} + \left(\frac{\gamma_P'(r)}{\gamma_P(r) + 1} - \frac{\gamma_P''(r)}{2\gamma_P(r)} \right). \tag{7.3}$$

For $\gamma_G''(0) \to 0$, the additional term in brackets disappears [see Eq. (6.1)]. Thus, $\Phi(r) = \gamma_P''(r)/[\gamma_P'(r)]^2$ and $\Phi(0) \equiv 1$. In the general case, i.e., if $\gamma_G''(0) > 0$, Eq. (7.3) gives $\Phi(0) = 1 - \gamma_G''(0)/(2 \cdot [\gamma_G'(0)]^2) > 1$. Thus, in addition to the relation $2 \cdot \gamma_G'(0) = \gamma_P'(0)$, the connection $\gamma_G''(0) = -[1 - \Phi(0)] \cdot \gamma_P'(0) = [1 - \Phi(0)]/\bar{l}_P$ results. By doing this, both the grain parameters $\gamma_G'(0)$ and $\gamma_G''(0)$ can be derived in terms of $\gamma_P(r)$, i.e., $\gamma_P'(0)$ and $\Phi(0)$. Equation (7.3) will now be verified for spherical and hemispherical grains.

Even in the simplest case of spherical grains with diameter $d = const.$ (Gille, 2003) [89, fig. 1], most of the fragments have complicated shapes. It follows that

$$\Phi(r) = \frac{1}{3} - \frac{r^3}{3d^3} + \frac{5r}{6d} + \frac{2d^2}{3(r-d)^2} \approx 1 + \frac{13r}{6d} + \frac{2r^2}{d^2} + \frac{7r^3}{3d^3} + O[r]^4. \tag{7.4}$$

Regardless of the shape, the condition $d = L_0$ is fulfilled. Equation (7.4) has the fitting-together-property $\Phi(0+) = 1$. Near the origin, hemispherical grains of radius R are described by the CF (see [61]),

$$\gamma_G(r) = 1 - \frac{9r}{8R} + \frac{r^2}{2\pi R^2} + \frac{r^3}{8R^3} - \frac{r^4}{80\pi R^4} + O[r]^6. \tag{7.5}$$

This yields the CF of the fragment particle

$$\gamma_P(r) = 1 - \frac{9r}{4R} + \frac{(32 + 81\pi)r^2}{32\pi R^2} - \frac{(576 + 665\pi)r^3}{256\pi R^3} + \\ + \frac{(5120 + 38624\pi + 27045\pi^2)r^4}{10240\pi^2 R^4} + O[r]^5. \tag{7.6}$$

Based on Eqs. (7.6) and (7.3), the term Φ for hemispherical grains given by

$$\Phi(r) = \left(1 + \frac{32}{81\pi} \right) + \frac{(4096 + 5184\pi + 8289\pi^2)r}{5832\pi^2 R} + O[r]^2 \tag{7.7}$$

is obtained (see Fig. 7.1). The term $\Phi(0) \approx 1.13$ *does not depend* on R. The latter case of Eq. (7.7) has been explained in detail (see [108, 112]). The PPs resulting from ellipsoidal grains have also been considered [110]. The next section summarizes the basic properties of DLm-PPs.

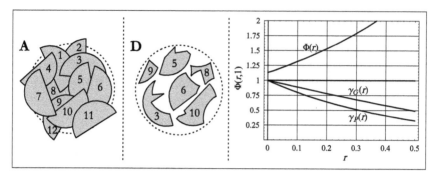

FIGURE 7.1

Analysis of the *fitting-together-ability* of homogeneous puzzle fragments via the puzzle fitting function $\Phi(r, L_0)$ in the case of hemispherical grains of constant diameter $d = 2R$.

Actually, 12 grains are inserted into a DLm model which define the specific tessellation (A). Separating the main (the central) pieces of this tessellation (separation of border effects) yields 6 isolated, typical puzzle fragment particles (PPs) (see D). These can be described by the (mean) CF function $\gamma_P(r, L_0)$ or by a CLDD. These functions result from the scattering pattern of a quasi-diluted ensemble of the PPs, i.e., via the scattering pattern of particles 3, 5, 6, 8, 9, 10 ....

Relations for non-touching PPs of volume fraction c

Let N non-touching PPs be investigated by a scattering experiment. Thus, for small r, where $0 \leq r < r_{min}$ (see Section 4.2), the sample CF $\gamma(r)$ of the PP ensemble results in terms of the (mean) CF of the single fragments $\gamma(r) = \frac{\gamma_P(r) - c}{1 - c}$, i.e., $\gamma_P(r) = c + (1 - c) \cdot \gamma(r)$. Hence, the derivatives $\gamma'_P(r) = (1 - c) \cdot \gamma'(r)$ and $\gamma''_P(r) = (1 - c) \cdot \gamma''(r)$ result. The puzzle fitting function $\Phi(r) = \Phi(r, c)$ then is written

$$\Phi(r, c) = \frac{\gamma''_P(r)}{\gamma'_P(r)^2} = \frac{\gamma''(r)}{(1 - c)[\gamma'(r)]^2}. \tag{7.8}$$

The function $\Phi(r)$ can be defined both in terms of γ_P (single PP) and in terms of c plus γ (more tightly packed ensemble of fragment particles). Expressing $\Phi(r)$ in terms of the covariance $C(r)$ or in terms of Porod's function of occupancy $Z(r) = C(r)/c$ yields

$$\Phi(r, c) = c \frac{C''(r)}{[C'(r)]^2} = \frac{Z''(r)}{[Z'(r)]^2}. \tag{7.9}$$

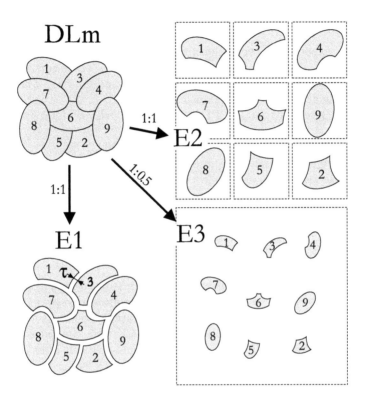

FIGURE 7.2

A tessellation of space (DLm) leads to randomly shaped particles. Assuming that these puzzle cells (PCs) form particle ensembles like E1, E2 and E3, i.e., two particles that never touch, simple analytic expressions result, which define the CF of the puzzle pieces.

Analysis of the SAS pattern of randomly shaped, homogeneous (but non-touching) "puzzle-cell" particles originating from a DLm of smooth grains: Combining these geometric assumptions with the first principles of SAS results in a sequence of structure parameters in terms of the sample CF $\gamma(r)$. The volume fraction c of the particle phase follows from Eq. (7.8), where $c = 1 - \gamma''(0)/[\gamma'(0)]^2$. This result does not depend on the initial grain shape(s) of the tessellation.

Operating with the (mean) geometric covariogram $K_P(r)$ of the single PPs, the following holds true.

$$\Phi(r,c) = \frac{1}{cV_t} \cdot \frac{K_P''(r)}{[K_P'(r)]^2}. \tag{7.10}$$

For specific experimental conditions, these relationships [especially Eqs. (7.8) to (Eqs. (7.10)] allow Φ to be determined for small r. These connections hold even more for a quasi-diluted ensemble of PPs (see Fig. 7.2). The following three particle ensembles are explained in Fig. 7.2:

E1: Puzzle interlayer model (PIM) with an interlayer thickness τ (Gille, 2007) [107].

E2: An infinitely diluted ensemble of PCs (Gille, 2003) [89].

E3: For the representation of a quasi-diluted particle ensemble (see Chapters 4 and 8), a size factor 0.5 was applied.

Both phases are characterized by their mean chord length l_1 (phase 1) and m_1 (phase 2). The specific surface area is denoted by S_V. Thus,

$$\frac{1}{|\gamma'(0)|} = l_p = \frac{4Vc(1-c)}{S} = 1 \Big/ \left(\frac{1}{l_1} + \frac{1}{m_1}\right) = l_1(1-c) = m_1 c.$$

These identities are a pivotal point for random two-phase systems (Guinier & Fournét, 1955) [143, p. 80]. The Rosiwal relation (Rosiwal, 1898) [197] $c = l_1/(l_1 + m_1)$ is included. However, these identities do not fix the structure parameters S_V, l_1, m_1 and c in terms of the CF $\gamma(r)$. In order to do this, further information is indispensable. The parameters result after introducing the *DLm tessellation assumption* about the particle shapes. The specific equation

$$g(0) = \frac{\gamma''(0)}{|\gamma'(0)|} = (1-c) \cdot |\gamma'(0)| = \frac{1}{l_1}$$

makes this possible. Two examples given for the PIM (see (Gille, 2007) [107, pp. 693–694, Fig. 4 and Fig. 6] illustrate the behavior of the function $g(r) = l_p \cdot \gamma''(r)$ and confirm this equation, i.e., $g(0) \equiv 1/l_1$.

This also holds true if several (different) grain shapes are used for the tessellation construction. Thus, the CF of the grains of the tessellation is a weighted function. The weights are the different grain volumes [see Eq. (1.35)].

7.3 Punch-matrix/particle puzzles

The most well-known elementary puzzle type is the punch-matrix/particle puzzle (PMP puzzle). See Fig. 7.3. The results obtained in Section 7.2 will be extended to another class of puzzles called PMPs. This type of puzzle consists

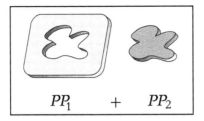

FIGURE 7.3
The simplest PMP puzzle consists of one hole embedded in a huge matrix piece (PP_1), plus the corresponding single particle (PP_2). If the size and shape of both agree, both of the scattering curves agree also.

of $1 + N$ pieces made up of one compact, huge connected punch matrix piece and N isolated puzzle pieces (see a plane PMP puzzle in Fig. 7.4). The spatial PMPs considered here do not allow any puzzling by moving the PPs. Nevertheless, the differences between two-dimensional and three-dimensional PMPs are small. Since the scattering pattern of the homogeneous matrix piece PP_1 and the particle PP_2 agree (see Fig. 7.3), an analysis of the second-order characteristics of these parts [set covariance $C(r)$, SAS correlation function $\gamma(r)$] is the starting point of the following theory.

Theory of a $1 + N$ puzzle (punch-matrix/particle puzzle)

Corresponding to Eqs. (7.1) and (7.2), a suitable puzzle fitting function $\Phi_{1N}(r)$ based on the following assumptions and denotations is introduced again: The huge matrix piece 1 with N holes and the N single PPs are simultaneously investigated by one scattering experiment.

Such an "all-inclusive" experiment yields a scattering intensity $I_{1N}(h)$ and a CF $\gamma_{1N}(r)$. This is emphasized by encircling all of Fig. 7.4 with dashed lines, regardless of whether one or more PPs are (already) inserted at the right spot of the huge matrix. Let r_{min} be the shortest distance between the borders of two holes in matrix piece 1. These holes, which are a tightly packed ensemble of N non-touching *particles*, involve the volume fraction c, where $0 \leq c < 1$. For piece 1 there exists a (sample) CF $\gamma(r)$. Let $\gamma_s(r)$ be the mean CF of a single hole inside piece 1. Regardless of the shape of the holes with the largest diameter L_{max}, for small r the functions γ and γ_s are interrelated. It holds that

$$\gamma(r) = \frac{\gamma_s(r) - c}{1 - c}, \quad 0 \leq r < \min(r_{min}, L_{max}). \tag{7.11}$$

Equation (7.11) describes the huge matrix piece in terms of c and γ_s.

The second component of the puzzle consists of a diluted arrangement of N (not necessarily equal) PPs, PP_1, PP_2,...PP_N, with volumes V_i and surface areas S_i. Let γ_i be the ith CF of PP_i. The first derivatives $\gamma_i'(0)$ are defined by

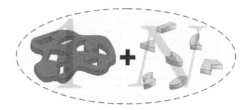

FIGURE 7.4

The punch matrix/particle puzzle case. Here, N single particles fit into the sequence of N holes one by one. The $1 + N$ puzzle consists of matrix piece 1 (N tightly packed holes) plus N single pieces (diluted arrangement). A puzzle fitting function $\Phi_{1N}(r)$ [see Eq. (7.15) and text] results for a fixed L.

$\gamma_i{}'(0) = -S_i/V_i$. Let $\gamma_s(r)$ be the mean correlation function of the single PPs. Based on the normalization $\gamma_i(0) = 1$, this mean value requires the averaging weights V_i. The mean CF is $\gamma_s(r) = \sum_{i=1}^{N} \left[V_i \cdot \gamma_i(r) \right] \Big/ \sum_{i=1}^{N} V_i$. The mean averaged chord length $\bar{l} = \bar{l_1}$ of all the N PPs is $\bar{l_1} = 4 \sum_{i=1}^{N} V_i \Big/ \sum_{i=1}^{N} S_i$ [175].

Let $\varphi(r)$ be the averaged CLDD of all the PP_i. The largest chord length (the largest diameter of the largest piece) is L_{max}. Let $\varphi_i(r)$ be the CLDD of PP_i (for IUR chords). Consequently, by averaging with a weight S_i, the mean CLDD $\varphi(r)$ of the N PPs results, which is expressed as $\varphi(r) = \sum_{i=1}^{N} [S_i \cdot \varphi_i(r)] \Big/ \sum_{i=1}^{N} S_i$.

The correlation function $\gamma_{1N}(r)$ of the ensemble of $1 + N$ pieces

Let the shape and size of the holes in piece 1 (one by one) agree with the PP_i. Only then and based on these assumptions and denotations for r relatively close to the origin, i.e., $0 \le r < \min(r_{min}, L_{max})$, the all-inclusive CF

$$\gamma_{1N}(r) = \frac{(1-c)\gamma(r) + c\gamma_s(r)}{(1-c) + c} \equiv [\gamma_s(r) - c] + c\gamma_s(r) = (1+c)\gamma_s(r) - c \quad (7.12)$$

results. The identity in Eq. (7.12) results by inserting $\gamma(r)$ from Eq. (7.11). Equation (7.12) represents $\gamma_{1N}(r)$ in terms of the parameters of the puzzle. From Eq. (7.12), the derivatives $\gamma'_{1N}(r)$ and $\gamma''_{1N}(r)$ and the ratios between them result. In detail,

$$\frac{\gamma''_{1N}(r)}{\gamma'_{1N}(r)} = \frac{(1+c) \cdot \gamma''_s(r)}{(1+c) \cdot \gamma'_s(r)} = \frac{\gamma''_s(r)}{\gamma'_s(r)}, \quad 0 \le r < \min(r_{min}, L_{max}) \quad (7.13)$$

results regardless of the value c of the holes of the huge punch matrix piece 1. In fact, for the relatively small r assumed, Eq. (7.13) contains the valuable

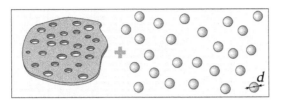

FIGURE 7.5
A PMP puzzle. The matrix piece has N spherical holes. In addition, there exist N equally sized single spheres (PPs) of diameter d (here $N = 25$). Next, the puzzle is analyzed to find out if the N PPs fit together with the matrix.

information given by the ratio

$$\frac{\gamma_{1N}''(0+)}{\gamma_{1N}'(0+)} = \frac{\gamma_s''(0+)}{\gamma_s'(0+)} = -\varphi(0), \;\; 0 \le r < \min(r_{min}, L_{max}). \tag{7.14}$$

Operating with a $g(r)$ function, where $g_{1N}(r) = \gamma_{1N}''(r)/|\gamma_{1N}'(0)| = l_P \cdot \gamma_{1N}''(r)$, Eq. (7.14) results in $g(0+)/\varphi(0+) = 1$. Actually, this connection does not require the assumption of a DLm [see Eq. (7.1)]. In this section, the shape of the PPs is arbitrary. The fitting information consists of $g(0+) = \varphi(0+)$. Thus, taking the data of both g and φ together yields the more general puzzle fitting function $\Phi_{1N}(r)$, which again possesses the property $\Phi_{1N}(0+) = 1$,

$$\Phi_{1N}(r) = \frac{g(r)}{\varphi(r)} = \frac{\gamma_{1N}''(r)}{|\gamma_{1N}'(r)| \cdot \varphi(r)} \approx \frac{\gamma_{1N}''(r)/|\gamma_{1N}'(r)|}{\gamma_s''(r)/|\gamma_s'(r)|}, \;\; 0 \le r < \min(r_{min}, L_{max}). \tag{7.15}$$

In order to apply Eq. (7.15), a maximum of two independent experiments, which define g and φ, is required. If $\varphi(r)$ is not available, it can be obtained from an additional scattering experiment via $\gamma_0''(0+)/|\gamma_0'(0+)| = \varphi(0+)$. Furthermore, if c is known a priori, $1/|\gamma_{1N}'(0)| = l_p = \overline{l}_1 \cdot (1 - c)$ results in

$$\Phi_{1N}(r) = \frac{\overline{l}_1(1 - c) \cdot \gamma_{1N}''(r)}{\varphi(r)}, \;\; 0 \le r < \min(r_{min}, L_{max}). \tag{7.16}$$

These connections will be explained for two model cases. The numerical aspects and details (for reducing truncation errors) for determining the terms $g(0+)$ and $\varphi(0+)$ from scattering patterns $I(h)$ are discussed in Section 7.3.

The function $\Phi_{1N}(r)$ for spherical holes plus spheres

The matrix piece involves N spherical holes of diameter d. In addition, there exist N equally sized single spheres (PPs) of diameter d.

The problem is determining whether the N PPs fit together with piece 1 (Fig. 7.5). Since $d = const.$, no size averaging is required and $L_{max} = d$ holds. Let the mean CF of the N PPs $\gamma_s(r) = 1 - 3r/(2d) + r^3/(2d^3)$, where $0 \le r \le d$, and the volume fraction c of the holes be given. The CLDD of all PPs and

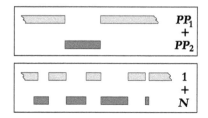

FIGURE 7.6
Two different linear PMP puzzles (line fraction c). As in the two-dimensional or three-dimensional case, there exists one matrix puzzle particle (involving N different vacancy lengths) and the corresponding N single PPs. The PPs do not necessarily fit into these holes.

holes is $\varphi(r, d) = 2r/d^2$, where $0 \leq r < d$. A certain minimum distance r_{min} between the holes in the matrix piece is assumed. From Eq. (7.12), the function $\gamma_{1N}(r, c, d)$ of the $1 + N$ puzzle gives

$$\gamma_{1N}(r, c, d) = \left(1 - \frac{3r}{2d} + \frac{r^3}{2d^3}\right) \cdot (1 + c) - c, \quad 0 \leq r < r_{min}. \tag{7.17}$$

Equation (7.17) involves the parameter c and allows the scattering pattern $I_{1N}(h)$ of the $1 + N$ particle ensemble for relatively large h to be determined. From Eq. (7.15), the function $\Phi_{1N}(r)$ results in

$$\Phi_{1N}(r, d) = \frac{\gamma_{1N}''(r)}{|\gamma_{1N}'(r)|} \cdot \frac{1}{\varphi(r)} = \frac{d^2}{d^2 - r^2} \approx 1 + \frac{r^2}{d^2} + \frac{r^4}{d^4} + \dots \tag{7.18}$$

The fitting-together-property $\Phi_{1N}(0+) = 1$ is fulfilled. In more detail, the ratio $\gamma_{1N}''(r)/|\gamma_{1N}'(r)| = 2r/(d^2 - r^2)$ results. Analogous results are obtained for other elementary PPs shapes, whereas the analytic expressions for the CF and the CLDD $\varphi(r)$ to be inserted restrict to the first r-interval near the origin.

The linear segment case (rods of fixed length d) and $\Phi_{1N}(r)$

A linear PMP puzzle is obtained by extrapolating the three-dimensional case in Fig. 7.5 to one dimension (see Fig. 7.6) for $N = 1$ and $N = 3$ with one matrix piece each. For a const. hole size d, a one-dimensional single segment CF $\alpha_s(r) = 1 - r/d$, where $0 \leq r \leq d$, results. If Eq. (7.12) is applied analogously, $\alpha_{1N}(r, c, d) = (1 - r/d) \cdot (1 - c) - c$ describes the linear puzzle. If $c \to 0$, the simple segment CF results. The CLDD of a segment of length d is $\varphi(r, d) = \delta(r - d)$. It is useful to write down the corresponding distribution function $F(r)$ of $\varphi(r)$, where $F'(r) = \varphi(r)$. The basic properties of $F(r)$ are $1 - F(r) = 1$ if $0 \leq r < d$ and $1 - F(r) = 0$ if $d < r < \infty$. A representation of Eq. (7.15) (here $'$ denotes differentiation with respect to r) is

$$\Phi_{1N}(r) = \frac{\alpha_{1N}''(r)}{|\alpha_{1N}'(r)|} \cdot \frac{1}{\varphi(r)} = \frac{\alpha_{1N}''(r)}{\alpha_{1N}'(r)} \cdot \frac{1}{[1 - F(r)]'} = \frac{(1 - r/d)''}{(1 - r/d)'} \cdot \frac{1}{(1)'}. \quad (7.19)$$

After splitting the common factor $(-1/d)$, which exists both in the numerator and denominator of Eq. (7.19), $\Phi_{1N}(r) \equiv 1$ results for $0 \le r < d$. As is to be expected, the linear PMP puzzle considered fulfills $\Phi_{1N}(0+) = 1$.

Analysis of $g(0+)$ and $\varphi(0+)$ for a selected L in terms of $I(h)$

In this subsection, the only function symbol considered is g since the numerical procedures for approximating g and φ agree (for a fixed order range L). The term $g(0+)$ results from $g(0) = 2\delta(0) + l_p \cdot \gamma''(0)$, where $l_p = 1/|\gamma'(0)|$. Starting from

$$\gamma(r) = \frac{\int_0^\infty h^2 I(h) \cdot \sin(h \cdot r)/(h \cdot r) dh}{\int_0^\infty h^2 I(h) dh} \quad (7.20)$$

and operating with the representation

$$\delta(r) = \frac{1}{\pi} \cdot \int_0^\infty \cos(h \cdot r) dh, \quad (7.21)$$

the connection

$$g(r) = \int_0^\infty \left[\frac{2}{\pi} \cos(h \cdot r) + \frac{l_p h^2 \cdot I(h)}{\int_0^\infty h^2 I(h) dh} \cdot \frac{(2 - h^2 r^2) \sin(hr) - 2hr \cos(hr)}{hr^3} \right] dh \quad (7.22)$$

follows. A scattering pattern is a band-limited function. Let $h_{max} \approx 2$ nm^{-1} be the maximum abscissa of the SAS experiment. The parameter h_{max} defines the resolution limit of the experiment r_{min}, where $r_{min} = \pi/h_{max}$ [see Eq. (7.17)]. Using the function $P_1(h)$ (see [74] and Chapter 1) gives

$$P_1 = \frac{\pi}{4} l_p \cdot \frac{h^4 I(h)}{\int_0^\infty h^2 I(h) dh}. \quad (7.23)$$

For $r \to 0$, Eq. (7.22) can be written in the following more stable form as

$$g(r) = \int_0^\infty \left[\frac{2}{\pi} \cos(hr) + \frac{4}{\pi} P_1(h) \frac{(2 - h^2 r^2) \sin(hr) - 2hr \cos(hr)}{h^3 r^3} \right] dh. \quad (7.24)$$

Based on the scattering data, i.e., inserting a truncation limit h_{max} into Eqs. (7.20) to (7.24), these relations are not yet suitable for a numerical determination of the limit $g(0+)$. A highly oscillating behavior of $g(r)$ near the origin can be expected. To obtain a stable result, it is useful to fix the special abscissas r_k down to r_{min}, where $r_k = k\pi/h_{max}$ with $(k = \ldots 4, 3, 2, 1)$. Based on the table $[r_k, g(r_k)]$, an approximation for $g(0+)$ can be calculated. A reliable method is to estimate a limit of the sequence $g(r_k)$ by use of the *Mathematica* function *SequenceLimit*. In complicated cases, the variation of the *Degree* option of this field-tested approach (Gille, 2007) [111] leads to correct limits.

7.4   Analysis of nearly arbitrary fragment particles via their CLDD

In this section *fragment particles* (*FPs*) are investigated. In this case detailed assumptions about the origin of the *FPs* are not required (e.g., *that they originate from a DLm or that they are PPs*). Much weaker assumptions are needed. The analysis of the *FPs* will be based on the CLDD of an ensemble (E) of homogeneous, hard, compact, randomly shaped fragments.

An investigation of whether such *FPs* can fit together like the pieces of a puzzle can be carried out based on the experimental information contained in the scattering pattern of E, which consists of many separate *FPs*. Let L_0 be the maximum diameter of the largest *FP*. The *one-by-one investigation* of FP_1, FP_2, FP_3 ... in a quasi-diluted arrangement (or better still in a separate state) yields the characteristic scattering pattern of E. This fixes the mean CLDD of the *FPs*.

The idea is to also construct a 50% volume fraction sample from the *PPs* and to introduce the fitting function $\Phi_{1/2}(r, L_0)$, where $0 \leq r \ll L_0$. Again, the limiting case $r \to 0+$ is of interest. If $\Phi_{1/2}(0+, 2 \cdot L_0) = 1$, the origin of the *FPs* is any destroyed mosaic. In fact, "Dead Leaves" tessellations are a special case of the approach. The connection is $\Phi_{1/2}(r, L_0) \Rightarrow \Phi(r, L_0)$.

Extending the DLm tessellation assumption to arbitrary fragments

Independent of a detailed reconstruction, the question again is whether certain randomly shaped hard FPs fit together. In other words: *Are the given FPs the pieces of* any *initial mosaic?* The idea is to separate the FPs and investigate them one by one, by recording the elastic scattering of the quasi-diluted pieces and deriving the distribution law of the length of random chords. All of these steps can be handled even without detailed image material of the FPs, i.e., based on their diffraction pattern.

In the DLm case, the mean CF $\gamma_P(r)$ of the single FP was used to explain the puzzle fitting function $\Phi_{DLm}(r) = \gamma_P''(r)/[\gamma_P'(r)]^2$. Based on certain assumptions about the primordial DLm tessellation, the limit $\Phi_{DLm}(0+) = 1$ guarantees that the given PFs fit together. For this, it was absolutely necessary to make the following detailed assumptions about the type of the initial (primordial) tessellation: Indicated by the acronym DLm, the function $\Phi_{DLm}(r)$ analyzes the FPs (the PPs in this case) resulting from a "Dead Leaves" mosaic with smooth, homogeneous primary grains.

Theory and basic assumptions about FPs in a more general case

The FPs are hard, homogeneous, randomly shaped, compact and relatively smooth connected particles without any holes or disruptive internal structure.

Let their size be detectable by an isotropic elastic scattering experiment. The cases of plane and spatial FPs are similar. Let L_0 be the largest diameter of the largest FP. The function $\varphi(l)$ denotes the CLDD for IUR chords of the typical FP with a first moment l_1.

The CF $\gamma_P(r)$ and $l_1 = 1/|\gamma_P'(0)|$ are fixed by the SAS intensity $I_P(h)$ of quasi-isolated FPs. Randomly shaped FPs lead to alternating segment lengths of l_i and m_i on any straight line ℓ passing through the ensemble E, which are independent random variables (see Fig. 7.7, parts T and T_ℓ).

Let $\gamma(r)$ be the sample CF of E with particle volume fraction c, where $0 \le c < 1$. In the actual case of independent segments l_i and m_i along ℓ, the function $g(r) = l_p \cdot \gamma''(r)$ characterizes E. The density functions $\varphi(l)$ and $f(m)$ describe the particles and intermediate space, respectively. The corresponding characteristic functions $p(t)$, $q(t)$ and the function $Q(t)$ are defined for a certain order range L, $p(t) = \int_0^{L_0} \varphi(l) \cdot e^{i\cdot t\cdot l} dl$, $q(t) = \int_0^{L} f(r) \cdot e^{i\cdot t\cdot r} dr$ and $Q(t) = \int_0^{L} g(r) \cdot e^{i\cdot t\cdot r} dr$. The last three functions are connected via $Q(t) = p(t) + q(t) - 2p(t)q(t)/[1 - p(t)q(t)]$, where $p(0) = q(0) = Q(0) = 1$ (see Chapters 6 and 9). Many modifications of this connection exist. One such modification entails inserting two of the three functions p, q and Q and determining the third one. Such an approach is the puzzle interlayer model, where Q results in terms of selected φ and f functions (see Section 7.6). In the following, Q, φ and f are used, but without explicitly defining any one of these functions.

Check of the fitting-together-ability of the FPs

If the FPs (can) fit together, the tessellation **T** can be reconstructed from the FPs without any gaps (Fig. 7.7, far left). The mosaic **T** has a 100% particle volume fraction (phase 1). Substituting each second fragment along line ℓ with a hole of equal shape (phase 2) results in a particle hole arrangement where $c = 50\%$ along ℓ. There is no difference between the CLDDs of phases 1 and 2. For fitting together FPs along ℓ, an alternating sequence of black (l_i) and white (m_i) FPs reflects the intrinsic properties of the particle ensemble. The order range L is twice the largest fragment diameter L_0. Besides $c = 1/2$, both of the CLDDs are identical, i.e., $\varphi(r) \equiv f(r)$. Thus, $p(t)$ and $q(t)$ agree as well and

$$Q(t) = 2p(t)/[1 + p(t)] = 2q(t)/[1 + q(t)], \quad -\infty < t < \infty \qquad (7.25)$$

results. Equation (7.25) reflects the significant properties of $\varphi(l)$ for $0 \le l \le L_0$, which describe the FPs. The function $Q(t)$ defines a scattering pattern $I(h)$ of E. However, the construction of Eq. (7.25) is much more than just the "invention" or "thoughtless formulation" of an arbitrary function $Q(t)$. For FPs that do not fit together, $Q(t)$ involves paradoxical properties that contradict the following trivial geometric requirements: volume fractions outside the interval $0 \le c \le 1$, negative scattering intensities $I(h) < 0$, negative

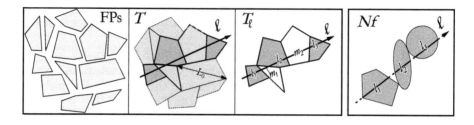

FIGURE 7.7
Typical part (12 pieces) of tessellation (T) and corresponding FPs. By intersecting T with arbitrary lines ℓ, the typical subfigure T_ℓ follows, which describes the chord sequence l_i and m_i. Along ℓ there is an alternating sequence of FPs [black (phase 1) and white (phase 2)].
FPs that do not fit together are shown on the far right (Nf). There are still gaps between the FPs along ℓ. Thus, the order relations $l_1 \neq m_1$ and $\varphi(l) \neq f(m)$ hold true. The functions p and q cannot agree and $Q(t)$ cannot be traced back to the one CLDD φ. However, starting from Eq. (7.25), contradictions (interrelated with the fitting-together-ability of the FPs) in the derivation of characteristic scattering parameters arise.

$g(r)$ moments and negative characteristic parameters of the scattering pattern $I(h)$ or theoretical scattering patterns that contradict the general rules. The function φ defines Q via the inverse Fourier transformation of $p(t)$ and Eq. (7.25). From this $Q(t)$ [in terms of $\varphi(l)$], $g(r)$ results in

$$g(r) = \frac{1}{2\pi} \int_{-\infty}^{\infty} Q(t) \cdot e^{-i \cdot t \cdot r} dt. \tag{7.26}$$

The first two moments of $g(r)$, i.e., M_1 and M_2, are defined in terms of the first and second moments of φ and f (see Chapters 6 and 9). In this case, the more general relations $M_1 = l_1 \cdot m_1/(l_1 + m_1)$ and $M_2 = (c-1)^2 \cdot l_2 + c^2 \cdot m_2 - 2M_1^2$ simplify. In the special case $l_1 = m_1$ and $l_2 = m_2$, it follows that $M_1 = l_1/2$ and $M_2 = (l_2 - l_1^2)/2$. For the first moment of $g(r)$, the denotation l_p is common. It follows that $M_1 = l_p = l_1/2$. This is all in agreement with the relation between l_1, l_p, c and $\gamma'(0)$, $l_p = l_1(1 - c) = m_1 c = 1/|\gamma'(0)|$. The functions φ and g are compared in some cases (Fig. 7.8). The property $g(0+) = 2 \cdot \varphi(0+)$ results from the following analysis of the 50% model constructed for φ and g.

An investigation of Eq. (7.26) for relatively small r shows that only big arguments t of $Q(t)$ contribute to $g(r \to 0+)$. Since $p(t) \to 0$ for $t \to \infty$, the denominator term in Eq. (7.25) simplifies to 1. Thus, for $r \to 0+$, Eqs. (7.25) and (7.26) result in

$$g(r) = \frac{1}{2\pi} \int_{-\infty}^{\infty} \frac{2p(t)}{1 + p(t)} \cdot e^{-i \cdot t \cdot r} dt \approx 2 \left(\frac{1}{2\pi} \int_{-\infty}^{\infty} p(t) e^{-i \cdot t \cdot r} dt \right) = 2\varphi(r). \quad (7.27)$$

Equation (7.27) contains the relation $g(0+) = 2\varphi(0+)$. This is confirmed by a numerical analysis of typical cases (see Fig. 7.8). Equation (7.27) yields $g(0+) = l_p \gamma_P''(0+) = 2l_1 \cdot \gamma_P''(0+)$. In contrast, cases S and H in Fig. 7.8 will prove to be "non-proper ones." A study of the invariants of the corresponding scattering patterns $I(h, L)$ [see Eq. (1.36)] will show this.

Scattering pattern and invariant of the 50% model case

The reality checks for the 50% model case are the non-negative intensity $I(h) = I(h, L)$ for the specific order range L and Porod's invariant. The parameter L and the function $g(r)$ fix $I(h, L)$ [see Eq. (1.34)] as shown by

$$I(h, L) = \frac{4\pi}{v_c \cdot l_p} \int_0^L g(r) \cdot \frac{2 - 2\cos(hr) - hr \sin(hr)}{h^4} dr, \quad I(0, L) = 1. \quad (7.28)$$

Equation (7.28) is related to the characteristic volume $v_c = \int_0^L 4\pi r^2 \gamma(r) dr$ and to the invariant inv, where $inv = v_c \cdot \int_0^L h^2 \cdot I(h) dh = 2\pi^2$ (see [143] and Table 1.3, which includes the parameter L). From these equations

$$inv = \frac{4\pi}{l_p} \int_0^\infty h^2 \cdot \left[\int_0^L g(r) \cdot \frac{2 - 2\cos(hr) - hr \sin(hr)}{h^4} dr \right] dh \quad (7.29)$$

follows. After changing the order of integration in Eq. (7.29), formal integration with respect to h and using (Wolfram *Mathematica*, 2009) [234] yields

$$\int_0^\infty \frac{2 - 2\cos(hr) - hr \sin(hr)}{h^2} dh = \frac{\pi \cdot r}{2}. \quad (7.30)$$

Finally, from Eqs. (7.29) and (7.30) the invariant of the 50% model case is

$$inv = \frac{2\pi^2}{l_p} \int_0^L r \cdot g(r) dr = \frac{2\pi^2}{l_p} \cdot l_p[model]. \quad (7.31)$$

On the right, Eq. (7.31) includes two versions for a representation of Porod's length parameter l_p: The terms $l_p[model]$ and l_p result from different *starting assumptions*, which do not necessarily harmonize with each other. The parameters $l_p[model]$ and l_p do not agree for FPs that are not able to fit together. Both l_p terms will be analyzed; in order to do this, two conditions must be taken into account. On the one hand, $l_p \equiv l_1/2$ always follows from $c = 1/2$ and $1/l_p = 1/m_1 + 1/l_1$. Otherwise, the integral $l_p[model]$ in Eq. (7.31) depends on L and $g(r)$. The relation $l_p[model] = l_1/2$ proves to be correct *for certain special cases only*.

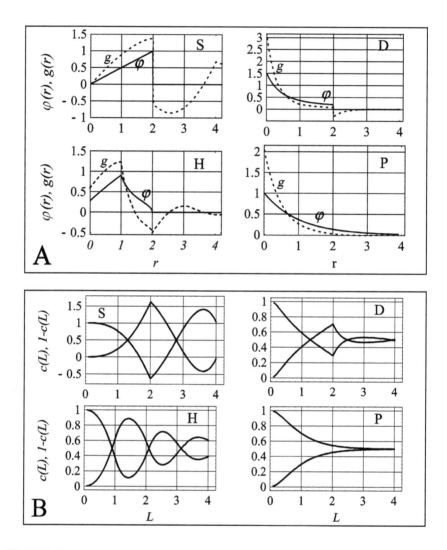

FIGURE 7.8

The functions $\varphi(r)$ and $g(r)$ (see part A) and the volume fraction c for varying L (see part B) of 4 selected types of fragment particles (FPs).

S and H: Sphere and Hemisphere (const. diameter $d = L_0 = 1$); D: DLm with spherical grains (const. diameter $d = L_0 = 1$); P: Poisson plane mosaic ($l_1 = 1$).

The functions $c(L) = 1 - l_p(L)/l_1$ and $1 - c(L) = l_p(L)/l_1$ are applied as indicator functions in all these cases. For the *fitting-together-cases* D and P, $c = 1 - c \rightarrow 0.5$ if $L \rightarrow 4$ results; in contrast, the 50% model case is not realistic in cases S and H.

The term $l_p[model]$ as depending on the FPs F_1, F_2, F_3 ... is defined by the parametric integral $l_p[model] = \int_0^L r \cdot g(r)dr$ and can be traced back to $L = 2L_0$ and $Q(t)$. This means that $l_p[model] = l_p[Q(t), 2L_0]$, whereas Eq. (7.25) traces $Q(t)$ back to $p(t)$. Furthermore, $p(t)$, the characteristic function of $\varphi(l)$ characterizing the FPs, is experimentally known. For convex FPs, the whole function $\varphi(l)$ is proportional to $\gamma_P''(r)$ of the FPs, i.e., $\varphi(r) = \gamma_P''(r)/|\gamma_P'(0)|$, which is defined in terms of their scattering pattern $I_P(h)$. The connection between the parameters l_p, L and $Q(t)$ (see [107]) is

$$l_p[Q(t), L] = \frac{1}{2\pi} \int_{-\infty}^{\infty} \frac{-1 + e^{-iLt}(1 + iLt)}{t^2} \cdot Q(t)dt. \qquad (7.32)$$

Definition of a generalized "puzzle fitting function" for FPs

For FPs that do not fit together, the invariant relation Eq. (7.31) or the derived condition $l_p = l_1/2 = l_p[Q(t), 2L_0]$ is not fulfilled. Here, the misfit of the FPs (see part Nf of Fig. 7.7) leads to an unrealistic function $Q(t)$, which corresponds to a particle ensemble which cannot exist. Thus, the function $l_p[Q(t), 2L_0]$ defined by Eq. (7.32) can be used as an "indicator of fitting-together ability."[*] In the following, checks of Eq. (7.32) in terms of $\varphi(l)$ are performed. From the assumptions given, all the information involved in the interval $0 \leq l, r \leq L_0$ for the functions $\varphi(l)$, $\gamma_P(r)$, $\gamma_P'(r)$ and $\gamma_P''(r)$ fixes $l_p[model]$ via Eq. (7.32).

Based on the parameter $l_p[Q(t), L]$, the following approach can be realized. This makes it possible to perform a simple fitting together test. Together with the mean chord length $l_1 = m_1$, the parameter $l_p[model]$ defines the particle volume fraction c, where $c = c(L) = 1 - l_p[Q, L]/l_1$. Consequently, the relation $1/2 = l_p[Q, 2L_0]/l_1$ must be fulfilled for $L = 2L_0$ and fitting together FPs. Extending this idea somewhat, a simultaneous investigation of the behavior of the two functions $c(L)$ and $1 - c(L)$ is a sensitive indicator (Fig. 7.8). For fitting together FPs, both terms must indicate $c(2L_0) = 1/2$. However, for FPs that are not able to fit together, a gap between both functions will occur (see parts S and H of Fig. 7.8). In the following, this particular feature is used to define a *generalized puzzle fitting function*.

The term puzzle fitting function denoted by $\Phi(r, L)$ was mentioned under more limited assumptions in the previous sections. From the current results [Eqs. (7.27) and (7.32)], the function $\Phi_{1/2}(r)$ (where the index $1/2$ implies the 50% volume fraction approach) can be introduced as

$$\Phi_{1/2}(r, L) = l_1 \cdot \varphi(r)/[l_p(L) \cdot g(0+)], \qquad (7.33)$$

---

[*]Analogous considerations can also be performed for the characteristic parameters $f_c[model]$ (characteristic surface) and $v_c[model]$ (characteristic volume). Higher moments of the function $g(r)$ are connected with these parameters and L. For example, v_c can be traced back to the fourth moment of $g(r)$ via $v_c = \frac{\pi}{3} \cdot \frac{\int_0^L r^4 g(r)dr}{l_p}$. Analogously to Eq. (7.32), the analysis of the v_c integral (see [107]) leads to further "indicator conditions."

where $0 \leq r < \epsilon \ll L$ and $L = 2L_0$. For fitting-together FPs based on Eqs. (7.27) and (7.32), Eq. (7.33) involves the property $\Phi_{1/2}(0+) = 1$.

Equation (7.33) corresponds to the *puzzle fitting function* $\Phi_{DLm}(r, L)$ of DLm mosaics. This function was defined in Section 7.2 for a collection of non-touching FPs arranged with a volume fraction c, where $0 \leq c < 1$ [110, 112]. Thus, the relationship

$$\Phi_{DLm}(0+) = \gamma''(0+)/[(1-c) \cdot [\gamma'(0+)]^2] = 1 \qquad (7.34)$$

holds true [see also Eq. (7.2)]. For isolated FPs, i.e., for $c \to 0$ and $\gamma(r) = \gamma_P(r)$, Eq. (7.34) specializes to $\Phi_{DLm}(r) = \gamma_P''(r)/[\gamma_P'(r)]^2$. Near the origin $0 \leq r \ll L_0$, the function $\Phi_{DLm}(r)$ has a strictly monotonously increasing behavior. For fitting-together FPs (DLm mosaic based on smooth grains), $\Phi_{DLm}(0+) = 1$. This limiting case of Eq. (7.34) is a special case of the extended definition. For diluted FPs of a DLm mosaic, the limit 1 also results.

Equation (7.33) can be further specialized. Instead of $c = 1/2$, $c \to 0$ now results in the the simplifying connections $l_p(L) = l_p(L_0) = l_p \to l_1$ and $g(0) \to \varphi(0)$. For the DLm considered, $\varphi(0) \equiv 1/l_1$ holds true [89]. The denominator of Eq. (7.33) simplifies to 1. For very small r, the remaining numerator is $\Phi_{1/2}(0) = l_1 \varphi(0+)$. With the correlation function γ_P of the fragments and $l_1 = 1/|\gamma_P'(0+)|$ and $\varphi(0) = l_1 \cdot \gamma_P''(0)$,

$$\Phi_{1/2}(0+) \Rightarrow l_1 \cdot \varphi(0) = l_1^2 \cdot \gamma_P''(0+) = \gamma_P''(0+)/[\gamma_P'(0+)]^2 \qquad (7.35)$$

results. Equation (7.35) is the limiting case of $\Phi_{DLm}(r) = \gamma_P''(r)/[\gamma_P'(r)]^2$ for $r \to 0+$. In this regard, Eq. (7.33) is a more general definition of a fitting-together function for the FPs.[†]

The approach represented consists in "comparing unknown shapes"

Actually, FPs of largest diameter L_0, which are characterized by the CLDD $\varphi(l)$, are given. The function $\varphi(l)$ is connected with the scattering pattern $I_P(h)$. This information allows an unambiguous determination of whether the FPs fit together. The derivation is based on the first principles of SAS (Gille, 2012) [128].

It could be objected that the shape of isotropically arranged homogeneous particles (actually, the shape of the FPs) cannot be detected from the corresponding isotropic scattering pattern, their CF or their CLDD!

Nevertheless, special cases have been studied by (Ciccariello, 2002). By recording scattering data with a two-dimensional detector, the particle shape of right cylinders can be reconstructed from the asymptotic SAS intensity [18].

[†]In the field of SAS, there is no unique transformation between $I_P(h)$ and $\varphi(l)$ for non-convex FPs. This fact is not important for the analysis of FPs originating from a DLm. This is a consequence of scrapping (leaving out) the restricting assumption of a DLm mosaic. Under the more general assumptions made, it seemed inevitable to start with a complete CLDD φ. The CLDDs for IUR chords can be recorded using different experimental procedures. SAS is only one of them.

Superficially considered, the current approach seems to contradict some basic principles of SAS. However, this is not the case. In this section, no attempt has been made to determine the shape of the FPs or to reconstruct the original mosaic. In fact, *such a goal* would require a determination of particle shapes. The actual approach consists in *comparing unknown shapes*, which are always the unknown shapes of two FPs. One of them remains a particle, however, for the purpose of comparing the shapes; the other one is converted into a hole of the same size and shape.

The sequence of working functions/parameters ranges from L_0, $\varphi(l)$, l_1, $p(t)$ and $Q(t)$, Eq. (7.25), the analysis of $g(r)$ and Eq. (7.26). Inserting $L = 2L_0$, the scattering pattern, invariant, volume fraction and characteristic scattering parameters follow for a linear 50% two-phase system. The investigation of Porod's length parameter $l_p[Q, 2L_0]$ yields

$$l_p[Q, 2L_0] = \frac{l_1}{2} \equiv \frac{m_1}{2} \equiv \frac{1}{2|\gamma_P'(0)|} \equiv \frac{1}{2} \int_0^{L_0} l\varphi(l)dl \equiv \frac{1}{2} \int_0^{L_0} mf(m)dm.$$
(7.36)

Contradictions arise for FPs that do not fit together. Equation (7.36) and the fundamental lemmas of SAS theory are not fulfilled. If Eq. (7.36) holds true, the FPs fit together and originate from some destroyed tessellation consisting of compact particles. The function $\Phi_{1/2}$ summarizes Eqs. (7.36) and (7.27) in one statement.

Applications are possible for any order range L_0, such as in archeology (e.g., $L_0 \approx 1$ m, astronomy (e.g., 10^6 m $< L_0$) and cell biology (say 10^{-6} m $\approx L_0$). Equations (7.25) to (7.32) are based on the knowledge of the function $\varphi(l)$ in the interval $0 \leq l < L_0$.

7.5 Predicting the fitting ability of fragments from SAS

Up to now, three different cases of puzzles have been studied in Chapter 7. In these examples, the image material is unimportant. The CF of homogeneous particles is the working function for inspecting the fragment particles.

1. DLm tessellation puzzle pieces (PPs): The CF of the pieces of a DLm tessellation is used.

2. Punch-matrix/particle (PMP) puzzles: Two separate CFs from two experiments are used.

3. More general case of fragment particles (FPs): The (whole) CLDDs of the FPs are incorporated (see Section 7.4).

In Section 7.2 and Chapter 6, the puzzles resulting from a DLm tessellation are considered. In this regard, two special cases are differentiated: For smooth grains, $\Phi(0) = 1$; however, in grains with surface singularities, $\Phi(0) > 1$ (see Fig. 7.1).

Section 7.3 analyzes punch-matrix/particle puzzles. For this, two scattering patterns leading to γ_{1N} and φ are inserted. The results Eqs. (7.12) to (7.16) and Φ_{1N} have nothing in common with the DLm pieces for which a puzzle fitting function Φ was initially defined in Section 7.2. However, in comparing the PMP case with the DLm-PPs case, a certain similarity between Φ and Φ_{1N} is evident: The PMP case starts from much weaker assumptions for the particle shape than the DLm case. Certainly, here $\varphi(r)$ is an independent function and must be known a priori, at least near to the origin. For these weaker assumptions, an investigation of PMP puzzles requires more information about the second-order characteristics of (both) the pieces. A simple technical apparatus can be constructed for predicting fitting ability.

In none of the figures and illustrations shown up to this point has a three-dimensional PMP puzzle been given. Obviously, this would be useless. In most three-dimensional cases, recognizing, moving and inserting the PPs is not possible. The transparency of matrix piece 1 represents another problem. Nevertheless, three-dimensional PMP puzzles (see Fig. 7.9) can be analyzed via the approach shown. The Fourier transformation of the scattering pattern yields the second-order characteristics required. The assumption of a fixed L is an important step for data evaluation (see Chapter 1). Examples in which representations of $\varphi(r)$ and $\gamma_{1N}(r)$ are available (i.e., tetrahedral, ellipsoidal, cylindrical and hemispherical PPs) have been simulated [113].

Section 7.4 studies the predictability of the fitting ability of randomly shaped FPs. The parameter L_0 and the whole CLDD of the fragments are inserted. The generalized function $\Phi_{1/2}$ was introduced.

Remarks about puzzles in different dimensions

Most simple puzzle games are two dimensional. Although the puzzle pieces are interlocked, they can be separated. The person performing the separation uses the "third dimension" for translating, rotating and fitting the PPs together (or working backwards: destroying, separating and mixing the pieces of the tessellation). For example, see the three-dimensional highly "interlocked" PMP puzzle consisting of two pieces in Fig. 7.9.

In such a case, the possible fit of the PPs cannot be checked by "trivial puzzling" by moving the PPs together. However, the PMP approach (compare with Fig. 7.3), can be applied to check the fit of the PPs. According to the Babinet theorem, there are no geometric limits, regardless of the shape of the pieces. The fitting property can be investigated by comparing the scattering patterns of PP_1 and PP_2. Suitable physical devices determine the second-order characteristics of PPs for a large spectrum of particle sizes.

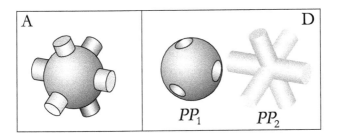

FIGURE 7.9
Example of a more complicated PMP case: Fitted state (A) and separated
puzzle-particle state (D) of two "strongly interlocked" puzzle pieces.
According to the geometric shapes of PP_1 and PP_2, it is not possible to
disaggregate the tessellation $(A \to D)$ or to fit the pieces together $(D \to A)$.

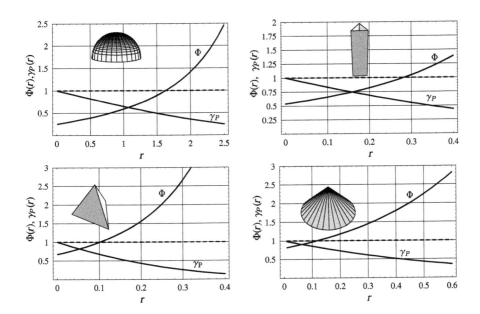

FIGURE 7.10
Puzzle fitting functions for basic cases of fragments that do not fit together:
Hemisphere: $R = 3$, $\Phi(0) = 64/(81\pi)$; triangular rod: $a = 1$, $\Phi(0) = 2(6\sqrt{3} + 4\pi)/(27\pi)$; tetrahedron: $a = 1$, $\Phi(0) = (8\sqrt{2} + 4\pi - 4\arccos(1/3))/(9\pi) \approx 0.67$; cone: $H = 2$ and $R = 1$ (see [63, 112]). The parameters $\Phi(0+)$ are
summarized in Fig. 7.11.

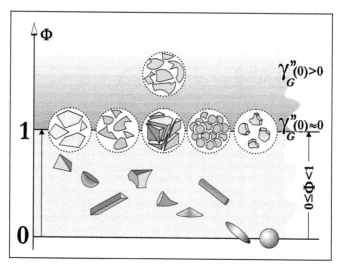

FIGURE 7.11

Analysis of elementary geometric shapes and DLm puzzle pieces via the behavior of function $\Phi_{DLm}(r)$ near the origin. The three situations $0 \leq \Phi < 1$, $\Phi \equiv 1$ and $\Phi > 1$ must be distinguished. For the DLm cases, the latter requires the grain property $\gamma_G''(0) > 0$ (see Section 7.2).

Puzzle fitting function Φ for elementary bodies

It is useful to apply the strategy of a puzzle fitting function to *special single particle shapes* where the CF is known (compare with the cases considered in Chapter 2). Four "non-fitting together cases" are selected in Fig. 7.10. An overview of this concept is shown in Fig. 7.11.

7.6 Porous materials as "drifted apart tessellations"

Finally, an application of Dead Leaves tessellations (see Section 7.2) for the characterization of porous materials is presented. In 2006, the author described the SAS patterns of isotropic arrangements of tightly packed, hard, homogeneous particles [107]. These particles have the shape of the pieces of a DLm tessellation. The model referred to as the Puzzle Interlayer Model (PIM) is based on the simple approach explained below.

The starting point is a two-dimensional or three-dimensional *initial puzzle* P_0 (see part A of Fig. 7.12). These *puzzle pieces* (PPs) fill the space completely so there are no gaps. The volume fraction c of the PPs is $c = 1$.

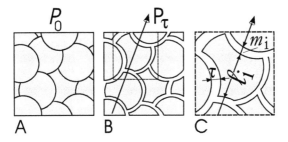

FIGURE 7.12
Illustration of a PIM as a special example with parameter τ (idealization of a plane case): The random shape of the PPs is defined by a DLm with circular primary grains.
The initial tessellation P_0 (A) and magnified part (B). Modification of the initial tessellation by a shifting process leads to τ with small intermediate layers τ (C). The parameter τ varies a bit. Other parameters of the model are the grain type of the tessellation leading to $\varphi(l)$ and l_1, $\tau = m_1/2$, $f(m)$, c and the specific surface area S/V.

Translating all the individual pieces of the initial tessellation slightly with a certain length τ produces the modified tessellation (B). Here, τ is relatively small compared to the maximum size of the PPs. Hence, the resulting particle arrangements (B, C) involve a connected intermediate region (a new phase 2), in which the PPs are embedded. The larger τ is, the larger the volume fraction $1 - c$ of phase 2.

The scattering pattern of such an ensemble of PPs (B, C) can be investigated in terms of the CLDDs along some test line ℓ, which alternatingly intersects phases 1 and 2 of the material.

For the PIM, the approximation $f(m) = 2\tau/m^3$, where ($\tau < m < \infty$ and $m_1 = 2\tau$) is inherent. A huge spectrum of initial tessellations can be inserted, especially the tessellation type resulting from a DLm of any convex grain shape. The scattering pattern $I_{PIM}(h)$ is similar to that of a layer of thickness $2\tau e$ [107]. The terms $\gamma(r)$, $\gamma'(0+)$, $\gamma''(0+)$ and $g(0+)$ are interrelated with the model parameters of the PIM,

$$g(0+) = \frac{1}{l_1}; \quad c = \frac{l_1}{l_1 + 2\tau}, \quad 1 - c = \frac{2\tau}{l_1 + 2\tau} = 2\tau \cdot \frac{S}{4V}, \tag{7.37}$$

$$|\gamma'(0)| = \frac{S}{4Vc \cdot (1 - c)} = \frac{S_V}{4c(1 - c)} = \frac{1}{2\tau \cdot c} = \frac{1}{l_1 \cdot (1 - c)}. \tag{7.38}$$

Equations (7.37) and (7.38) explicitly define the model parameters c, l_1, $m_1 = 2\tau$ and S/V in terms of the sample CF $\gamma(r)$ (Table 7.1).

It can be concluded that $\gamma''(0+)/[(1 - c) \cdot [\gamma'(0)]^2] = 1$. This fact makes it possible to define the function $\Phi_{PIM}(r, c)$ (see Table 7.1) possessing the property $\Phi(0+, c) \to 1$.

Application of the approach for a controlled macroporous glass

Figure 7.13 involves the details. It is advisable to investigate the development of the real-space PIM parameters in terms of r. In investigating terms like $\gamma'(r)$, $\gamma''(r)$ is better than restricting to $\gamma'(0+)$, $\gamma''(0+)$. Plots like $c = c(r)$ involve the limiting case $c \to 0$. This results in stable parameters. The last line of Table 7.1 defines an expression for the *puzzle fitting function* $\Phi_{PIM}(r,c)$ [see also Eq. (7.8)].

TABLE 7.1
Summary of the parameter set of a PIM.

parameter	formula in terms of the CF	remarks
interlayer	$\tau = \frac{1}{2} \dfrac{1}{|\gamma'(0)| - \frac{|\gamma''(0)|}{|\gamma'(0)|}}$	$\tau \to 0$, $\tau < l_1$
porosity	$c = 1 - \dfrac{|\gamma''(0)|}{|\gamma'(0)|^2}$	$0 \le c < 1$, PPs
spec. area surf.	$S_V = 4\dfrac{\gamma''(0)}{|\gamma'(0)|}\left(1 - \dfrac{|\gamma''(0)|}{|\gamma'(0)|^2}\right)$	for a specific L
CLD moment m_1	$m_1 = 2\tau(r) = \dfrac{|\gamma'(0)|}{|\gamma'(0)|^2 - |\gamma''(0)|}$	wall chords
CLD moment l_1	$l_1 = \dfrac{1}{|\gamma'(0)|(1-c)} = \dfrac{|\gamma'(0)|}{\gamma''(0)}$	chords of PPs
puzzle function	$\Phi_{PIM}(r,c) = \dfrac{1}{(1-c)} \cdot \dfrac{|\gamma''(r)|}{|\gamma'(r)|^2}$	$c = const.$

In cases where $0 < c < 1$, a fixed state of a *particle ensemble* is described by $\gamma(r)$ (see Fig. 7.13). After fixing the volume fraction c beforehand by another method (see also Chapter 8) $\Phi_{PIM}(r,c)$ can be examined in detail. The fitting property $\Phi_{PIM}(0, c = c_0) = 1$ can be investigated. This property is fulfilled for selected particle ensembles only.

The function $\Phi_{PIM}(r,c)$ is connected with the $\Phi_P(r)$ of single, widely separated fragments, i.e., for PPs, which are widely spread in the whole space. For $c \to 0$, $\gamma \to \gamma_P$ results. Then, $\gamma_P(r)$ represents the CF of the single (isolated) PPs of the tessellation. In agreement with the theory of Chapter 7, $\Phi_P(0+) = \gamma_P''(0)/[\gamma_P'(0)]^2 \to 1$ holds. For an infinitely diluted PIM, $m_1 \to \infty$. In a sense, $\tau = const.$ is fulfilled.

PIMs are of practical relevance (see Fig. 7.13). By substituting the PPs with (closed) pores [phase 1, $\varphi(l)$] and the intermediate phase 2 with the walls between these pores, a PIM can describe a large variety of porous materials. The pore shape is defined by the PPs of the initial tessellation. The DLm is ideal for constructing an initial model tessellation. It allows the CF of the pieces and their CLDD $\varphi(l)$ to be described analytically. One disadvantage of this approach is the fact that this model is made for non-connected pores only. Experience shows that this assumption is not very critical. A detailed comparison of a simulated PIM and a realistic sample with parameters resulting from the scattering pattern (see Fig. 7.13) is attempted in Fig. 7.14.

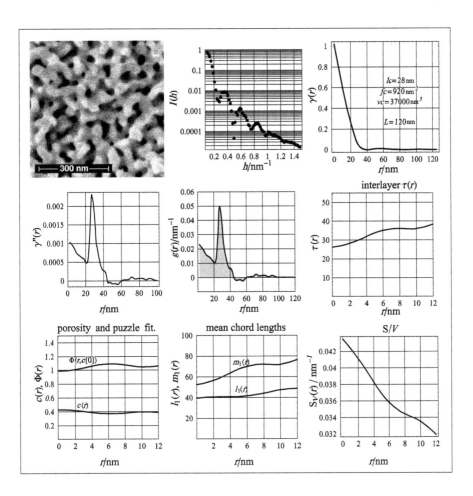

FIGURE 7.13

Analysis of a macroporous glass (MaPG) via a PIM for $L = 120$ nm. The functions $g(r)$ and $\gamma(r)$ show a spike at $r = \tau$. The estimated parameters are as follows:

Porosity $c = (40$–$50)\%$; interlayer $\tau \approx (20$–$30)$ nm; specific volume of the sample $S/V \approx 0.04$ nm^{-1}; mean chord length of the pores $l_1 \approx 40$ nm; mean chord length of the walls $m_1 = 2\tau \approx 50$ nm.

The deviations between $\Phi_{PIM}(r)$ and the horizontal line $\Phi_{PIM} \equiv 1$, where $0 < r < 12$ nm, are small (see the part *porosity and puzzle fit* of the figure). For the actual sample, the PIM is a fine approximation.

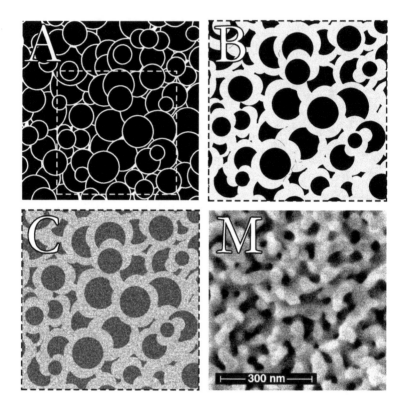

FIGURE 7.14

A,B,C: Image simulation based on a PIM with fixed parameters. M: micrograph. The simulated image is based on parameters resulting from the sample CF: $l_1 = 40$ nm, $d = 3l_1 = 120$ nm, $\tau = 25$ nm, $m_1 = 2\tau = 50$ nm and $c = 44\%$.

Micrograph (M) is close to a plane section of relatively small thickness. The difference between the const. radius of the primary spheres of the DLm tessellation and the resulting "hitting diameters" known as the Wicksell problem was considered. Thus, differently sized circle diameters fall into the plane and become the DLm grains. Based on this approximation, the plane tessellation (A) results. The grain diameters vary from zero to $d = 120$ nm. The limiting lengths 0 and d seldom occur. Clearly, some large leaves on top (large, unchanged circles as tessellation particles) survive.

All parts of the tessellation (A) drift apart. Therefore, only the pieces in the middle (indicated by a square) are analyzed further and considered. Translating these central pieces produces subfigure (B). In the plane section, a nearly constant translation parameter $\tau = 25$ nm is not achieved. Depending on the spatial direction, there exist interlayers greater than τ. However, according to the linear integration principle (Rosiwal, 1898), $m_1 \approx 50$ nm holds.

Details of the image simulation

Here are some more explanations of Fig. 7.14. The micrograph (M) and approximation (C) were considered on a length scale of about 600 nm with $L = 120$ nm. The simulation is based on the length parameters $m_1 = 2\tau$, l_1 and on the porosity c. Starting from the scattering pattern, the estimation of these parameters is explained in Fig. 7.13.

The micrograph (M) of the porous silica sample represents a nearly *two-dimensional section*. The MaPG is similar to those samples investigated by (Gille, Enke & Janowski, 2001) [79, 77, 78]. These MaPG samples were prepared based on an alkali borosilicate glass by phase separation and combined acid and alkaline extraction. The black regions are the pores and the white ones are the silica walls. The porosity of this sample is about (40–50)%.

There are disadvantages to this approach. On the one hand, the non-connected pores are not realistic. Further questions result from large primary spheres remaining in (A). Nevertheless, (C) is similar to (M); at least, the lower right corner of (C) is a fine approximation of the experiment.

Order ranges L_i that are smaller than a few nanometers are not analyzed. The silica fine structure is not included in the approach. However, the silica walls in the real sample have a fine structure [see the micrograph (M)]. But there is a fine-structure of the silica walls in the real sample; see the micrograph (M). Consequently, (B) is modified further by adapting the gray levels of (B) to those of (M). A possible way is subfigure (C), where the pores remain black, whereas the walls are modified to white/gray (similar to the micrograph).

8

Volume fraction of random two-phase samples for a fixed order range L from $\gamma(r, L)$

"A general explicit or implicit formula for determining the volume fraction (porosity) of random two-phase materials as depending on the SAS correlation function would be a miracle." The author comes to this conclusion after a long sequence of trials and experiments in this field. A similar conclusion can also be found in the textbook by Guinier and Fournét [143, p. 79]:

"This function $\gamma(r)$ contains all the information that can be obtained from the small-angle scattering experiments, but unfortunately this function does not give a direct image of the structure and is quite far from defining it. The effects of both the form of the particles and their mutual arrangement are intermixed in the single function $\gamma(r)$."

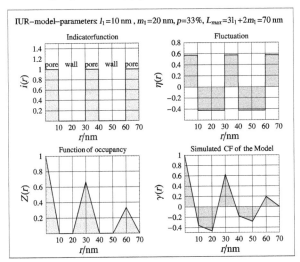

Application of a *linear simulation model* (LSM) (Gille, 2002 [82]; Enke et al., 2002 [32]) analyzing the scattering pattern of a VYCOR glass sample with $L = L_1 \approx$ (25–30) nm (see Fig. 8.5). Presupposing fixed chord lengths, it is possible to construct the theoretical behavior of $\gamma(r, L)$, the function of occupancy $Z(r, L)$ and the set covariance $C(r, c)$ as well. A 33% sample porosity results from the LSM. Sample material: PVG disks (Corning Co.,

Code 7939) with a thickness 0.4 mm were used. The PVG was prepared by phase separation of a subsequent acid leaching.

Only after making a sequence of additional assumptions can an estimation of c be carried out. A specific model that is suitable for describing this sample case must be selected. Thus, the particle volume fraction c, which is of interest here, depends on the order range L, where $c = c(L)$. Chapter 8 discusses several approaches for estimating $c(L)$ from the scattering pattern if certain additional assumptions are fulfilled. Typical examples of this are as follows:

- Approach A: The simplest case is that of a known mean chord length $l_1 = -1/\gamma_0'(0)$ of the single particle(s). Thus, for a well-defined L, the particle volume fraction c is defined in terms of the first derivative $\gamma'(0, L)$ of the sample CF in the origin and the average mean chord length of the particles, where $c = 1 - l_p(L)/l_1$. If the mean chord length l_1 is unknown beforehand, this length parameter can be estimated from the behavior of $\gamma(r, L)$ in some practical cases.

  In addition to this nearly trivial case, this chapter includes the approaches B, C, D, E, F and G. Their limits are studied and checked for selected particle ensembles. This is all based on analytic expressions for $\gamma(r, L)$. Their applicability is shown for materials samples.

- Approach B: Section 8.1 (main part of this chapter) discusses procedures for interpreting the $c(L)$ of random isotropic two-phase media. *Linear simulation models* (LSM) allow a real-space interpretation of the transformed scattering pattern (see the introductory figure of this chapter). The hollow sphere case is discussed and the typical scattering of VYCOR glass is analyzed.

- Approach C: Section 8.2 considers a stereological macropore analysis of a controlled pore glass in detail. The porosity is investigated in several ways.

- Approach D: In Section 8.3, the scattering of quasi-diluted hard particles is discussed in detail. Quasi-diluted particle ensembles (QDPEs) (see Fig. 4.1) yield straight results. Selected approaches have already been explained in Chapter 4.

- Approach E: Section 8.4 explains the fundamentals of a less known approach proposed by (Synecek, 1983) [220, 221] at the International SAS conference in Prague, which is ideal for homogeneous spherical particles (precipitates in alloys).

- Approach F: Section 8.5 considers the *Boolean model case*. The connection between c and the maximum scattering intensity is discussed.

- Approach G: Section 8.6 is about the "realistic porosities" of materials.

8.1 The linear simulation model

The linear simulation model (LSM) is an approximation of the sample CF. It describes the behavior of the three-dimensional average $\gamma(r, L)$ using a linear model, i.e., it approximates three-dimensional objects and their spatial arrangement using a one-dimensional model in the x-direction (see Fig. 1.10). The paper *"Linear simulation models for real-space interpretation of SAS experiments of random two-phase systems"* [82] explains the LSM approach.

Estimation of the functions $\gamma(x)$ and $Z(x)$ via linear probes

Two approaches for estimating c of a particle ensemble for an L are:

1. Analysis of the spatial positions of the *particle centers* (investigation of *pair distributions*) (see Chapter 4).

2. Analysis of the *particle borders* for describing the essential properties of the particle ensemble (see Chapters 2, 3 and 9).

The LSM approximation applies point 2, assuming isotropy and an indicator function $i(x)$ along a fixed straight test line of length L stretching in an arbitrary direction and intersecting the borders of phases 1 and 2. There is a great deal of geometric information in $i(x)$, i.e., depending on the length variable x along this test line, where $0 \leq x < L < x_{max}$. However, the isotropic scattering pattern $I(h)$ does not define the function $i(r_A)$, the set covariance $C(r)$ or the function of occupancy $Z(r)$ [191, 192, 193, 194]. The isotropic scattering function $I(h)$ merely defines $\gamma(r)$ and the related structure functions (see Chapter 1). Both the functions $\gamma(r)$ and $Z(r)$ [and of course the set covariance $C(r) \equiv Z(r) \cdot c$ as well] characterize the geometry of a medium the function $Z(r)$ still more precisely than $\gamma(r)$. The function $\gamma(r)$ is the normalized convolution square of the density fluctuation $\eta(r)$. However, $Z(r)$ is the normalized convolution square of the density differences.

Applying a linear probe, i.e., operating with a limited length variable x, the function $Z(x)$ can be estimated from the geometric probability $Z(x) = Pr[(0 \in phase) \wedge (x \in phase)]$. This means the following: Take a random point P_0 in *phase*. Thus, $Z(x) = Z_{phase}(x)$ is the probability that a second random point P_x, separated by a distance x from P_0, also belongs to *phase*. The function Z for *phase* involves the properties $Z(0) = 1$ and $Z(\infty) = c$, where $0 \leq Z(x) \leq 1$. For the CF, $-c/(1 - c) \leq \gamma \leq 1$ holds. The possible negative sign is a consequence of the negative parts of $\eta(x)$ in certain x-intervals (see Fig. 1.10 in Chapter 1). Eqs. (1.44) and (1.45) lead to Eq. (1.46).* For the finite, linear probe case, the connection between $Z(x)$, $\gamma(x)$ and c depends on

---

*Sometimes $Z(r)$ and c are denoted by the *variable* and *constant part* of $\gamma(r)$, respectively (Guinier & Fournét, 1955, 1966) [143, 144].

the length L of the test line, i.e., $Z(x, L)$ and $\gamma(x, L)$ must be written. The longer the test line, the more information that is collected.

Estimation of $\gamma(x)$ and $Z(x)$ in terms of $i(x)$ along the test line

For an isotropic two-phase system with volume fraction c and order range L, $Z(x)$ is defined by the (normalized) convolution square of the density differences (i equals 1 or 0) as

$$Z(x) = \frac{\int_0^L i(s) \cdot i(s+x)ds}{\int_0^L [i(s)]^2 ds} = \frac{\int_0^L i(s) \cdot i(s+x)ds}{L \cdot c}. \tag{8.1}$$

Due to the normalization of $i(x)$, the mean value $\bar{i} = 1 \cdot c + 0 \cdot (1-c) = c$ results. With $\eta(x) = i(x) - \bar{i} = i(x) - c$, an estimation of the CF is

$$\gamma(x) = \frac{\int_0^L \eta(s) \cdot \eta(s+x)ds}{\int_0^L [\eta(s)]^2 ds} = \frac{\int_0^L [i(s) - c] \cdot [i(s+x) - c]ds}{\int_0^L [i(s) - c]^2 ds}. \tag{8.2}$$

For $x = 0$, Eqs. (8.1) and (8.2) fulfill $\gamma(0) = Z(0) = 1$. Expressing the resulting product terms in the numerator of Eq. (8.2) in terms of L and c yields

$$\int_0^L c^2 ds = c^2 \cdot L, \ -\int_0^L c \cdot i(s)ds = -c^2 \cdot L, \ -\int_0^L c \cdot i(s+x)ds = -c^2 \cdot L. \tag{8.3}$$

The last integral connection in Eq. (8.3) only holds true if the particle ensemble possesses a sufficiently large L compared with the size of the single particles. With these connections [Eqs. (8.3)], Eq. (8.2) is written

$$\gamma(x) = \frac{\int_0^L i(s) \cdot i(s+x)ds + c^2 L - c^2 L - c^2 L}{\int_0^L [i(s) - c]^2 ds} = \frac{\int_0^L i(s) \cdot i(s+x)ds - c^2 L}{\int_0^L [i(s)]^2 ds - c^2 L}. \tag{8.4}$$

Because of $[i(s)]^2 = i(s)$ and with Eqs. (8.3), the integral in the denominator of Eq. (8.4) equals $c \cdot L - c^2 \cdot L$. The first numerator term in Eq. (8.4) is connected with $Z(x)$, c and L [see Eq. (8.1)]. Hence,

$$\gamma(x) = \frac{[Z(x) \cdot cL] - c^2 \cdot L}{cL - c^2 L} = \frac{Z(x) - c}{1 - c} = \frac{C(x)/c - c^2}{c(1-c)}. \tag{8.5}$$

It must be stressed again that the steps leading to Eq. (8.5) hold true for each indicator function $i(x)$, provided L is sufficiently large.

For ensembles of several particles, the connection between Z and γ, given by Eq. (8.5), holds true in a modified form. For "isotropic agglomerates" (see also the following hollow sphere case), the third term in Eq. (8.3) is written

$$\int_0^L c \cdot i(s+x)ds = c^2 \cdot f(x, L). \tag{8.6}$$

Equation (8.6) includes an *unknown function* $f(x, L)$. The term $c^2 f(x, L)$ differs from the simple result $c^2 \cdot L$ [see Eq. (8.3)]; however, it mainly depends on the behavior of the indicator function $i(x)$ in different directions for relatively small x. Hence, no final evaluation of the integral in Eq. (8.6) is possible. Case-by-case considerations are indispensable, and $f(x, L) = L$ is an approximation. Equation (8.5) is wrong for finite particle ensembles/clusters.

In order to study the limits of Eq. (8.5) and (8.6), the following can be summarized: It is necessary to analyze the order relation between the maximum size of the whole particle arrangement x_{max} and L in detail. Evidently, the upper limit of x_{max} is the sample diameter. Thus, $L \leq x_{max}$ is always fulfilled. There are the following two limiting cases:

Case 1: If L is essentially smaller than x_{max}, $L \ll x_{max}$, the parametric integral $\int_0^L i(s + x)ds = c \cdot L$ does not depend on x.

Case 2: If x_{max} is in the same order of magnitude as L, e.g., $x_{max}/2 < L < x_{max}$, no general result for the last convolution integral of Eq. (8.3) can be given. Thus, only $\int_0^L i(s + x)ds = c \cdot f(x, L)$ holds true.

Nevertheless, knowing all the details of $i(x)$ (or even knowing the density in terms of x along the test line, see Fig. 1.10) makes it possible to fix the approximations $Z(x)$ and $\gamma(x)$ from Eqs. (8.1) and (8.2) independently of one another with two separate procedures using numerical integration. This can be realized based on the patterns below (shown here without normalization for the sake of simplicity) (see [82]):

```
z[x_]:=Integrate[i[s]*i[s+x],{s,0,...singularities...,L}];
g[x_]:=Integrate[eta[s]*eta[s+x],{s,0,...singularities...,L}];
```

Instead of $i(x)$, an arbitrary positive density function can be inserted as well. Procedures for discrete convolution or discrete correlation are useful for minimizing CPU time (see [186]). For defining densities (or only indicator functions), it is a good idea to apply the function $UnitStep[...]$. Hence, $i[x] := UnitStep[x - x_1] - UnitStep[x - x_2]$ defines a constant impulse from x_1 to x_2. Such LSMs represent a simple way of finding an estimation $\gamma(x)$ for the SAS CF $\gamma(r)$. Using this approach, the curvature behavior of a sample CF $\gamma(r)$ can be compared with that of $\gamma(x)$ (see Figs. 8.1 and Fig. 8.2).

The curvature of $\gamma(r)$ and the CLDDs $\varphi(l)$ and $f(m)$

The information involved in the basic behavior of a sample CF $\gamma(r)$ can be selected via an LSM as follows: The curvature of $\gamma(r)$ is connected with the CLDDs φ and f of both phases. The behavior of $\varphi(l)$ and $f(m)$ contains stereological information about the particle ensemble. The LSM reflects the peak positions of φ and f. Restricting to one (direction) dimension yields an enormous simplification. The important assumption is the isotropy. An LSM interprets the behavior of $\gamma(r)$ based on an *inserted indicator function model* $i(x)$. Some examples given in Figs. 8.1 and 8.2 show that the approach [Eqs. (8.1) and (8.2)] can be extended to three phases. However, according to the *Babinet theorem*, different indicator functions can lead to the same CFs (see the agreement in the CFs in cases A, B and E, F of Fig. 8.1).

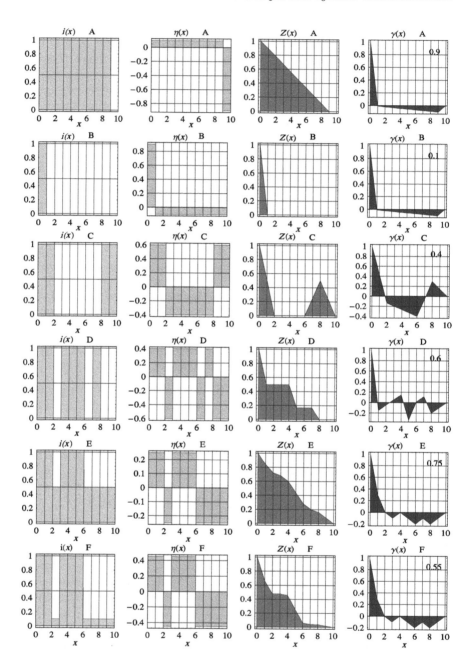

FIGURE 8.1

Several typical LSMs of linear two-phase sequences of maximum diameter $x_{max} = 10 = L$. The mean density is indicated in the CF subfigures.

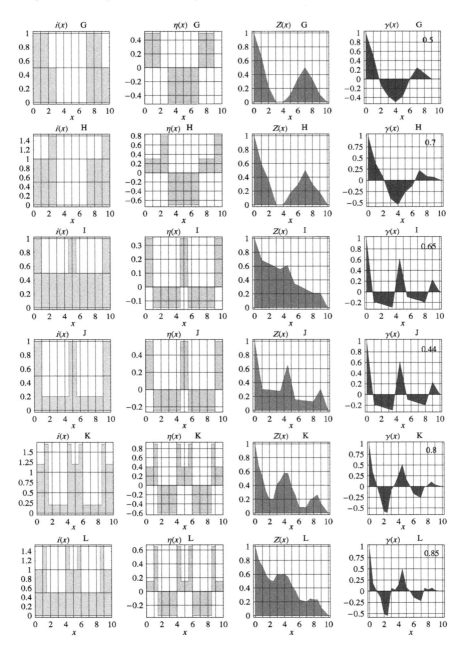

FIGURE 8.2

Selected LSMs of two-phase and three-phase ensembles. As already shown in Fig. 8.1, the singularities of $\gamma(x)$ define the typical length l_i, m_i and their sums. All the $Z(x)$ functions are different, but in cases I and J the CFs agree.

Interpretation of curvature singularities in the LSM CF

The functions $\gamma''(x)$ and $Z''(x)$ consist of a sequence of *Dirac* δ-functions.[†] These δ functions represent the limiting case of the realistic CLDDs φ and f (see examples in Figs. 8.1 and 8.2). A size distribution of the chord segments does not exist. There is no relation between a φ segment and an f segment. The function $\gamma''(r)$ can be represented by

$$l_1(1-c)\gamma''(r) = m_1 c\gamma''(r) =$$
$$[\varphi(r) + f(r) + F_{l+m+l}(r) + \ldots + \ldots] -$$
$$[2 \cdot \delta(0) + 2 \cdot F_{m+l}(r) + 2 \cdot F_{m+l+m+l}(r) + \ldots + \ldots],$$
$$(8.7)$$

where $F_\square(r)$ stands for the distribution density of the sum of the random segment lengths $m_i + l_i$ or $m_i + l_i + m_i$ or $l_i + m_i + l_i$, ... The term $2\delta(0)$ is necessary because the limiting case of chord lengths $r = 0$ is included in the functions $\gamma''(r)$, $\varphi(r)$, $f(r)$ and also in the sequence terms $F_\square(r)$. According to the terms of Eq. (8.7), $\gamma''(r)$ may come out positive [curvature to the left of $\gamma(r)$], negative [curvature to the right of $\gamma(r)$] or zero [no curvature of $\gamma(r)$]. Thus, the curvature of $\gamma(r)$ at the specific lengths $x = l_1$, $x = m_1$, $x = l_1 + m_1$, $x = l_1 + m_1 + l_1$, $x = m_1 + l_1 + m_1$ ... nearly reflects the maximum and minimum positions of $\gamma''(r)$ (Gille, 2000) [66]. Taking this into account, the behavior of $\gamma''(r)$ can be interpreted. Estimations and checks of c can be handled via Rosival's relation(s) $c = l_1/(l_1 + m_1)$ [for the phase of the l_i-chords] or $1 - c = m_1/(l_1 + m_1)$ [for the phase of the m_i-chords].

Discussion of basic configurations with LSMs

Several typical LSM cases for two-dimensional and three-phase chord segment ensembles are summarized in Figs. 8.1 and 8.2.

In agreement with the Babinet theorem, the CFs $\gamma(x)$ of the initial and inverse set agree. This is not surprising since the fluctuations of the initial and inverse set are symmetrical (see cases A and B). Furthermore, a certain constant basic level of the density (indicator function) does not influence $\gamma(x)$ (see cases D, E and I, J). However, the functions of occupancy for these configurations are different. Consequently, for two agreeing functions of $\gamma(x)$, two denotations of Z, i.e., $Z_1(x)$ and $Z_2(x)$, where $Z_1 \neq Z_2$, must be distinguished. In SAS, γ is mainly used, not Z. The equalities $I_A(h) = I_B(h)$, $I_D(h) = I_E(h)$ and $I_I(h) = I_J(h)$ are fulfilled. These examples already show that the indicator functions *cannot be figured out (or puzzled out) exclusively from the scattering pattern and correlation functions*. Furthermore, an LSM cannot detect characteristic integral parameters of the CF, like l_c, f_c and v_c. This becomes clearer when comparing the total correlation function surface areas (filled plots), which are negative in some cases. The main information

---

[†]Instead of the δ functions, more or less sharp peaks result in the case of a realistic sample.

content of the simulation lies in the curvature behavior of $\gamma(x)$, which reflects the splitting of the length intervals of the indicator function $i(x)$ [see Eq. (8.7)].

Analysis of a sphere with a central hole via an LSM

The *hole effect* has been a frequent subject of interest (see Serra, 1982) [205, exercises, p. 313, F.8.(b.)] and [206]. This simple isotropic case is fundamental for the interpretation of other three-dimensional cases. Figure 8.3 analyzes the function $\gamma''(r)$ of hollow spheres, with an outer diameter $d = const.$ and an inner diameter d_i. The sequence of segments l_i and m_i is studied for a *finite* $L = x_{max} = 8$. Here, Eq. (8.5) is too rough an approximation and $\gamma(x)$ and $Z(x)$ *are independent* functions [see Eqs. (8.1) and (8.2) and the algorithm in (Gille, 2002)] [82, p. 90].

Explanation of the hollow sphere case in Fig. 8.3: The figure shows the second derivative of the CF $\gamma''(r)$ of a sphere of diameter d with a central hole of diameter d_i (selected cases in the first row) and their interpretation based on Eqs. (8.2) and (8.7) in the second and third rows. The plots show $\gamma''(r, d, d_i)$ in several basic cases for $d = 8 = const.$ and varying d_i, where $0 < d_i \leq d$. Additionally, the first subfigure in the last line analyzes the limiting case of a layer of height $H = 1/2 = (d - d_i)/2$ (see [106]).

The LSM approach in rows 2 and 3 applies the length variable x instead of r, i.e., based on $i(x)$. Here, it is useful to compare the curvature of the approximation $\gamma(x)$ with that of the exact function $\gamma''(r)$. These curvatures agree.

The cases considered are as follows:

For $d_i = 1$, there exist two separated maximum spikes of the CLDDs φ (particle phase) and $f(m)$ (hollow phase), followed by a sharp minimum. The minimum corresponds to the negative sum part F_{l+m} for the random sum $l_i + m_i$ [see Eq. (8.7)]. The behavior with large r is based on the positive F term $F_{l+m+l}(r)$. The term $F_{m+l+m}(r)$ does not exist because there is only one hollow part that possesses m-chord segments.

The situation in the case of $d_i = 2$ is basically unchanged. Here, the superimposition of the negative sum part $< l_i + m_i >$ leads to negative values of $\gamma''(r)$ in a small r-interval. Such an interrelation always exists [see Eq. (8.7)].

In the cases $d_i = 4$ and $d_i = 5$, the curves always start with one positive peak (i.e., a superimposition of the positive originating parts φ and f) followed by the negative sum peak $F_{l+m}(r)$.

Finally, the case $d_i = 7$ and the limiting case $[d \to \infty$ and $d_i \to \infty$, but $(d - d_i)/2 = 1/2 = H]$ as well contain a clear $\varphi(l)$ peak. This sharp maximum is based on l-chords of the shell region *particle phase* of the hollow sphere.

In the case of the layer, m-chords do not exist at all. Already in the case of $d_i = 7$ the influence of these m-chords is small.

In the next section, the scattering pattern of an AlDyNi alloy is interpreted using an LSM approach.

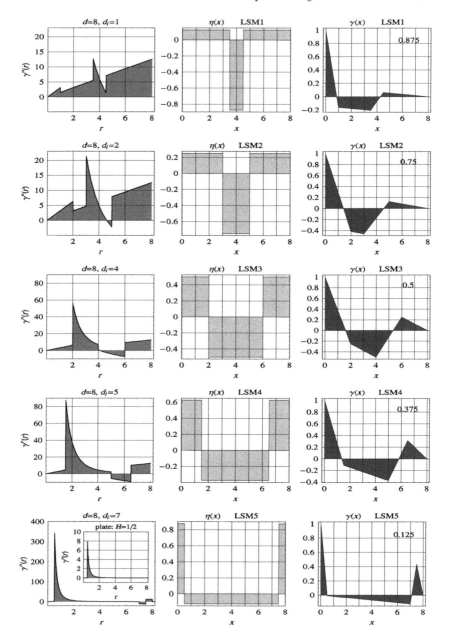

FIGURE 8.3

Exact functions $\gamma''(r)$ (left) of selected hollow spheres and the CFs $\gamma(x)$ (right) of the LSM approaches (middle). The segment lengths x cross the center of the hollow part. The smaller $d - d_i$, the more the configuration approaches the case of a lamella (see $\gamma''(r)$ of a plate, where $H = 1/2$).

8.1.1 LSM for an amorphous state of an AlDyNi alloy

The LSM approach was applied for describing the initial state of an amorphous AlDy 6at%Ni 10at% alloy.[‡]

Starting with the scattering pattern $I(h)$ to the LSM, Fig. 8.4 summarizes the details of the data evaluation. The two order ranges L_1 and L_2 can be detected from the *transformed correlation function* $\gamma_t(r)$ (Gille, 2000) [66]. For $L_2 = 40$ nm, the isotropic alloy involves two phases: a *connected* matrix phase M and a *non-connected* particle phase N. The latter is a mixture of the phases Al$_3$Dy, AlNi and amorphous material. The existence of phase N is based on *rest structures* of the melted state of the alloy. This phenomenon was discussed in detail by (Hermann, 1991) [150, p. 143, Figure 3.19] for the microheterogeneous structure of amorphous Fe-B alloys consisting of iron-rich clusters embedded in regions with an Fe$_3$B composition. The (electron) density of the matrix is greater than that of phase N. Due to the resolution limit of the experiment of $r_{min} = \pi/h_{max} = 3.5$ nm, the existence of certain boundary surface singularities in the region $r < 3$ nm cannot be detected. There is a relatively sharp transition between both phases (without any separating intermediate region). The upper resolution limit is $r_{max} = \pi/h_{min} \approx 40$ nm.

The function $\gamma''(r)$ reflects both mean chord lengths l_1 and m_1 as well as the mean values of the first sum terms of these lengths [see Eq. (8.7)]. An interpretation of $\gamma''(r)$ via an LSM yields $m_1 = 6$ nm, $l_1 = 13$ nm, $< lm > = 19$ nm± 1 nm, $< mlm > = 25$ nm± 2 nm, $< lml > = 32$ nm± 3 nm and $< mlml > = 36$ nm± 3 nm. The volume fraction of matrix phase is about $c \approx 13/(6 + 13) = 68\%$. Phase N possesses a volume fraction of $1 - c \approx 32\%$ (see Fig. 8.4). The green particles/white matrix represent a typical two-dimensional section which, of course, is an idealization of the basic state of the alloy. The same structure parameters L_2, l_1 and m_1 result regardless of the direction of the test line.

In summary, the linear simulation of the CF is based on the following parameters: the order distance $L = L_2 = 34$ nm $=< lml >$ [consider the last peak in $\gamma''(r)$], the indicator function $i(x)$ and the fluctuation $\eta(x)$

$$i(x) = \begin{cases} 00 < x < 15 : 1 \\ 15 < x < 20 : 0 \\ 20 < x < 34 : 1 \end{cases}, \quad 1 - \eta(x)) = \begin{cases} 00 < x < 15 : 0.84 \\ 15 < x < 20 : 0.16 \\ 20 < x < 34 : 0.84 \end{cases}. \quad (8.8)$$

The mean density equals $3/17$ (see Fig. 8.4). The behavior of the resulting model for $\gamma(x)$ confirms the experimental results $\gamma(r)$, $\gamma''(r)$ and $\gamma_t(r)$.

[‡]The alloy was produced by means of a melt spinning plant in an N$_2$ atmosphere at a ribbon velocity of 44 m/s. The resulting AlDyNi ribbons are amorphous. The decomposition of the AlDyNi system at an aging temperature of $T_a = 275°C$ has been investigated by several authors (see Kabisch et al., 1999) [159]. A metastable Al$_3$Dy phase and a subsequent AlNi phase are formed. It is worth mentioning that the SAS results of the 5 min aging at T_a do not show significant changes compared with the SAS results of the amorphous state.

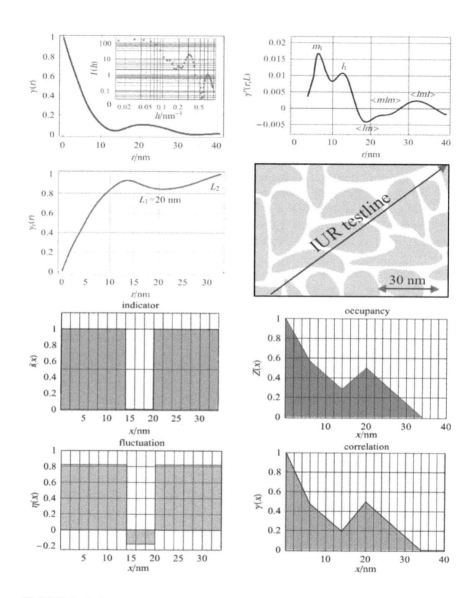

FIGURE 8.4

Scattering pattern $I(h)$ and the functions $\gamma(r)$ and $\gamma''(r)$ for $L_2 = 40$ nm. The transformed CF $\gamma_t(r)$ indicates two order ranges $L_{1,2}$. Interpretation of the structure of the AlDyNi alloy via an LSM with the following parameters: $m_1 = 6$ nm (intermediate phase), $l_1 = 13$ nm (particle phase); the sum terms are $< lm >= 19$ nm, $< mlm >= 25$ nm, $< lml >= 39$ nm, $< mlml >= 36$ nm. For the AlDy phase, $c(40\text{nm}) \approx 13/(6+13) = 68\%$ holds.

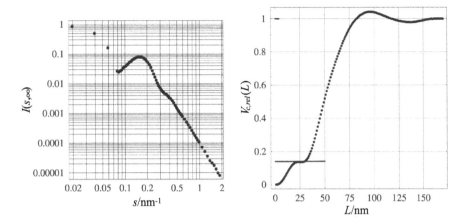

FIGURE 8.5

Scattering pattern $I(s, \infty)$ and the corresponding relative characteristic volume of a VYCOR glass sample [see Chapter 1, Eq. (1.71)].

From $v_{c,rel}(L)$, a typical value of $L = L_1 = (25...30)$ nm results. The upper resolution limit of the experiment is about $r_{max} = 40$ nm. Thus, a second range $L = L_2$, where $l_2 \approx 150$ nm, cannot be investigated (based on the actual $I(s, \infty)$).

8.1.2 LSM analysis of a VYCOR glass of 33% porosity

The order range estimation was the first step of the data evaluation for this sample. This was done using the relative characteristic volume approach (see Chapter 1 and Fig. 8.5). For the interpretation of $\gamma(r)$, an LSM with parameters $l_1 = 10$ nm and $m_1 = 20$ nm was applied (Gille, Enke & Janowski, 2002) [86]. The initial figure of this chapter shows more details. Again, the LSM is a fine approximation for the sample CF. A detailed investigation of $\gamma(r, L_1)$ and $\gamma''(r, L_1)$ even yields approximations of the complete functions φ and f. Furthermore, a comparison of the CLDD of the pores with the results of the nitrogen adsorption (NA) experiments was performed (see [86, pp. 588–589, figs. 3–4]).

For this sample, the information on the pores [CLDD $\varphi(l)$] and walls [CLDD $f(m)$] is mixed together to a great extent. The superimposition [see Eq. (8.7)] increases with increasing r. The separation of both CLDDs yields slightly modified mean chord lengths. However, these slightly modified moments yield the same porosity; i.e., c_{LSM} from the simple LSM agrees with the porosity values resulting from significantly more complicated approaches.

This LSM is a simple approach. It describes the complex geometric shape of the walls via a CLDD. A description of the wall thickness, exclusively based on only one shape variable, seems to be impossible. It is much more useful to interpret the geometry of the wall phase with a function $f(m)$ and its first moment in order to estimate the porosity of the material for a given L.[§] In fact, these are precisely the advantages of CLDDs!

8.1.3 Concluding remarks on the LSM approach

An LSM is mainly based on an *assumed one-dimensional indicator function* $i(x)$ defined in a limited x-interval, where $0 < x < L \leq x_{max}$. The model is useful for characterizing the spatial microparticle distribution. Linear intersections on a length scale from 0 to L of the particle ensemble contain the information. This works in the isotropic uniform random case.

The singularities in the behavior of $\gamma(x)$ clearly fix *all* the existing specific lengths. In the special cases A, B and C of Fig. 8.1, even the "origin" of these lengths can be found from the curvature of $\gamma(x)$. This strategy is simple. It is a special case of the application of the theory of CLDDs; however, it does not require detailed analytical calculations of the CLDDs of the particles. The latter calculations are elaborate (see Chapter 2) and it is not useful to perform this in the first step of a data evaluation. The LSM is much simpler to handle and yields an estimation of c. Three-phase systems can be considered as well.

The model case of a hollow sphere, where exact analytic expressions of $\gamma''(r)$ are available, can be interpreted in terms of the chord segments l_i and m_i as well as in terms of the random sum $l_i + m_i + l_i$. Based on the pattern $I(h)$ of an amorphous AlDyNi alloy, the moments l_1 and m_1 and the volume fraction were estimated for the basic state of a (nearly) amorphous alloy. Based on this, the model of a micrograph of the alloy for $L = L_2 = 34$ nm was derived (see Fig. 8.4).

Furthermore, an analysis was done of a VYCOR glass via an LSM. Due to the superimposition of the functions φ and f in this case, clear deviations should exist. However, the example shows that this is wrong. Even in this more complicated case, the LSM yields unexpectedly exact results.

A strategy that is very similar to an LSM will be explained in Section 8.2. Some applications follow in the subsections. In particular, the scattering patterns of porous materials are analyzed.

[§]The details of the data evaluation are based on the cylinder pore model. The diameter distribution of the cylindrical pores was estimated based on this pore shape. From this distribution, the CLDD of the pores is obtained in another way (see [79]). The function $\gamma''(r, L_2)$ involves a clear maximum at $r = 7$ nm, which indicates the pores. The walls are indicated by a shoulder at about $r = 20$ nm. Describing the shape of the pore peak and the "hidden" wall peak analytically by a smooth function makes it possible to separate parts φ and f. This allows a determination of the size distribution $V(d)$ of the random pore diameter d. Based on the sequence of three integral transformations, $V(d)$ explicitly results. In the end, it is possible to compare the pore diameter distribution resulting from NA and SAS experiments.

8.2 Analysis of porous materials via ν-chords

LSMs (see Section 8.1) use μ-chords, which is a chord type connected with the functions $\gamma''(r)$ and $g(r) = lp \cdot \gamma''(r)$. The "prefactor" r is not included.

There is another chord length strategy, which was first introduced in the textbook by (Guinier & Fournét, 1955) [pp. 13–16] [142]. This goes back to the interpretation of the function $G(r) = r \cdot \gamma''(r)$. In contrast to $-\gamma'(0) = \int_0^L \gamma''(r; L)dr$, the analysis of the function $G(r)$ yields $1 = \int_0^L G(r)dr$. Here, the relations $\gamma(0) = 1$, $\gamma(L) = 0$, $\gamma'(L) = 0$ and $\gamma''(L) = 0$ are applied. The approximation $l_1(1-c)\gamma'' \approx \varphi(l) + f(m) + R(r)$ (see Méring & Tchoubar, 1968 [182] and Section 8.1) can be written in two modified integral forms, which allows some conclusions to be made (see topics 1 and 2).

Topic 1: Integration of the function $\gamma''(r)$ yields three main parts:

$$\int_0^L \gamma''(r)dr \approx \frac{1}{l_p} \cdot \left(\int_0^L \varphi(r)dr + \int_0^L f(r)dr + \int_0^L R(r)dr \right). \qquad (8.9)$$

By performing the integrations, Eq. (8.9) leads to the identity

$$\frac{1}{l_p} \approx \frac{1}{l_p}\left(1 + 1 + \int_0^L R(r)dr\right) = \frac{1}{l_p}(1 + 1 - 1) = \frac{1}{l_p}. \qquad (8.10)$$

Based on the behavior of the functions φ and f, the integration interval in Eq. (8.9) can be split off. Hence, conclusions can be drawn for the first two integrals. The last integral remains unchanged. With $l_p = l_1(1-c) = m_1c$, the approximation

$$\int_0^L r \cdot \gamma''(r)dr \approx \frac{1}{l_p}\left(\int_0^{L_1} r\varphi(r)dr + \int_{L_1}^{L_2} rf(r)dr + \int_0^L rR(r)dr \right) \qquad (8.11)$$

results. Equation (8.11) considers the terms φ and f separately. These are written as

$$1 \approx \frac{l_1 \int_0^{L_1} r \cdot \gamma''(r)dr}{l_1(1-c)} + \frac{m_1 \int_{L_1}^{L_2} r \cdot \gamma''(r)dr}{m_1c} + \frac{1}{l_p} \cdot \int_0^L r \cdot R(r)dr. \qquad (8.12)$$

With the identity $1/l_p \int_0^L rR(r)dr = 1 - 1/[c(1-c)] = 1/l_p(l_p - (l_1 + m_1))$

$$1 \approx \frac{S_1}{1-c} + \frac{S_2}{c} + \frac{1}{l_p}\int_0^L r \cdot R(r)dr, \quad 1 \approx \frac{S_1}{1-c} + \frac{S_2}{c} + \left(1 - \frac{1}{c(1-c)}\right) \qquad (8.13)$$

results. Hence, the dimensionless integral parts $S_1 = \int_0^{L_1} r \cdot \gamma''(r)dr$ and $S_2 = \int_{L_1}^{L_2} r \cdot \gamma''(r)dr$ define c in terms of the function $r \cdot \gamma''(r)$, L_0, L_1 and L_2 as

$$S_1 = \frac{1}{1-c}, \; S_2 = \frac{1}{c} \text{ and } S_1 + S_2 = \frac{1}{c(1-c)} = \int_{L_0}^{L_2} r \cdot \gamma''(r)dr. \qquad (8.14)$$

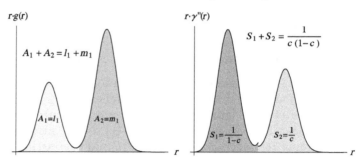

FIGURE 8.6
Interpretation of $r \cdot g(r)$ and $r \cdot \gamma''(r)$: Estimation of mean chord lengths and volume fractions.
For a random two-phase system, an idealized function $\gamma''(r, L)$ consists of two well-separated initial parts, in terms of the CLDDs $\varphi(l)$ and $f(m)$ [see Eqs. (8.14) and (8.16)].
Left: Estimation of l_1, m_1 and the sum $l_1 + m_1$ by operating with the function $r \cdot g(r) = r \cdot l_p \cdot \gamma''(r)$.
Right: Estimation of $1 - c$, c or $1/[c(1 - c)]$ by operating with the function $r \cdot \gamma''(r)$. According to Babinet's theorem, the exchanges $l_1 \leftrightarrow m_1$ and $1 - c \leftrightarrow c$ are possible.

In many practical cases, Eq. (8.14) fixes c from the first peak of $r \cdot \gamma''(r)$, where $0 < r < L_1$.

Topic 2: On the other hand, with $\int_0^L r \cdot R(r) dr = l_p - (l_1 + m_1)$, the analogous integral splitting off of $r \cdot g(r) = r \cdot l_p \gamma''(r)$ results in

$$l_p \int_0^L r\gamma''(r)dr \approx \int_0^{L_1} r \cdot \varphi(r)dr + \int_{L_1}^{L_2} r \cdot f(r)dr + \int_0^L r \cdot R(r)dr,$$
$$l_p \qquad \approx l_1 \int_0^{L_1} r \cdot \gamma''(r)dr + m_1 \int_{L_1}^{L_2} r \cdot \gamma''(r)dr + l_p - (l_1 + m_1)$$
$$l_p \qquad \approx l_1 + m_1 + l_p - (l_1 + m_1).$$
$$(8.15)$$

Consequently, l_1, m_1 and their sum follow via the corresponding *parts of the surface* of $r \cdot g(r)$. With the approximations $A_1 = \int_0^{L_1} r \cdot g(r) dr$ and $A_2 = \int_{L_1}^{L_2} r \cdot g(r) dr$ Eq. (8.15) results in

$$A_1 = l_1, \quad A_2 = m_1, \quad A_1 + A_2 = \int_0^{L_2} r \cdot g(r)dr = l_1 + m_1. \qquad (8.16)$$

Equation (8.16) is useful for an estimation of $l_1 + m_1$ from the first peaks of $r \cdot g(r)$. Altogether, Eqs. (8.14) and (8.16) involve an estimation of c and $l_1 + m_1$ (see Fig. 8.6). Practical experience shows that this approach is very precise for the first peaks of $r \cdot g(r)$ and $r \cdot \gamma''(r)$ (see a practical application of this in Fig. 8.7).

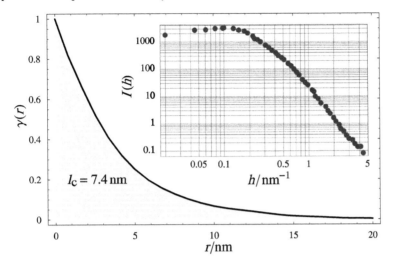

FIGURE 8.7

Scattering pattern and SAS CF of a silica aerogel for $L = 20$ nm.
The initial slope of the SAS CF results in $l_p \approx 4$ nm. Twice the filled surface area equals Porod's characteristic length of $l_c \approx 7$ nm. For an interpretation of this parameter (Guinier & Fournét, 1955) [143, pp. 13,18]. The behavior of this CF is reminiscent of an exponential function; i.e., randomly shaped and orderlessly placed particles (here pores) of varying size and shape are dominant for this order range. This is a typical gel structure (see Porod, 1951) [191].

8.2.1 Pore analysis of a silica aerogel from SAS data

Figures 8.7 and 8.8 exemplify the theory of topics 1 and 2 for a typical aerogel sample with $L = 20$ nm. Figure 8.8 exemplifies the common and different properties of the three functions γ'', $r \cdot g(r)$ and $r \cdot \gamma''$. The different normalization makes it possible to detect first CLD moments and porosity using different methods.

Via a μ-**chord length analysis**, four parameters result from the relative behavior of the function $\gamma''(r, L)$ in the first step (see the left of Fig. 8.8):

1. From the first and the second maximum, respectively, $l_1 = (5...6)$ nm and $m_1 = (15...17)$ nm result. An estimation of the porosity yields $c(L) = l_1/(l_1 + m_1) = (20...30)\%$.

2. Furthermore, integration of $\gamma''(r)$ yields $\gamma'(0, L) = -0.249$ nm^{-1} and $l_p(L) = 4.02$ nm (see also the plot of the CF). This verifies Porod's equation $1/l_p = 1/l_1 + 1/m_1$ (see Chapter 9).

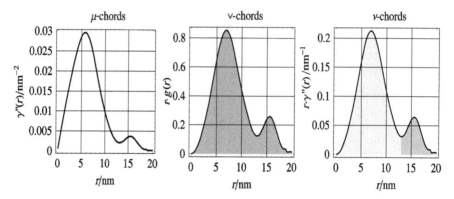

FIGURE 8.8

Comparison of the structure functions $\gamma''(r)$, $r \cdot g(r)$ and $r \cdot \gamma''(r)$. The functions result from the same scattering pattern (see Fig. 8.7).

The basic behavior is the same, but these structure functions involve different units and normalization. The structure parameters obtained are: $l_p = 4.02$ nm, $l_1 = 5.6$ nm; $m_1 = 16$ nm; $c = 27\%$. The relation $1/l_p = 1/l_1 + 1/m_1$ is correct. The relative error of the estimated volume fraction is about 15%.

The ν-**chord length analysis**, i.e., operating with the functions $r \cdot g(r)$ and $r \cdot \gamma''(r)$, leads to the same result:

1. Besides $\gamma(0) = 1$, the normalization of $g(r, L)$ is fulfilled, i.e., $\int_0^{20\text{nm}} l_p \cdot \gamma''(r)dr = 0.97 \approx 1$. Based on this, the surface parts can be interpreted.

2. The result $A_1 = \int_0^{14\text{nm}} r \cdot g(r)dr = 5.54$ nm is in agreement with the pre-estimated length $l_1 = (5...6)$ nm.

3. An analysis of the term $S_1 = \int_0^{14\text{nm}} r \cdot \gamma''(r)dr \approx 1.38$ is useful. In contrast to S_1, the integral part S_2 cannot be used. Here, the rhs influence of the sum term $< l_i + m_i >$ is too strong. Based on S_1, a porosity of $c = 1 - 1/S_1 \approx 27\%$ follows. Furthermore, based on Eq. (8.13), $1 = S_1/c + S_2/(1-c) + (1 - 1/[c(1-c)])$ or $1 = S_1/(1-c) + S_2/c + (1 - 1/[c(1-c)])$, and a porosity of 30% follows, which is a surprisingly good approximation.

Thus, the results of the approaches derived from the three functions $\gamma''(r)$, $r \cdot g(r)$ and $r \cdot \gamma''(r)$ agree to a great extend (see the three parts of Fig. 8.8). In the case of *well-shaped particles*, the application of ν-chords yields clear curves. It is important to perform the normalization check $\int_0^L r \cdot \gamma''(r)dr = 1$ before entering a detailed interpretation.

For "wild" particle shapes, e.g., in the Boolean model case, it is useful to operate with μ-chords.

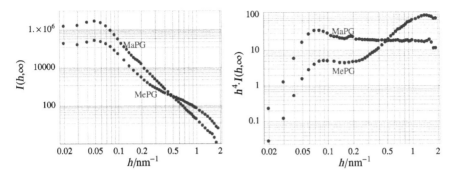

FIGURE 8.9
Relative intensities (left) and *Porod plots* (right) of the glasses (MePG and MaPG). The upper resolution limit of the experiment is about $r_{max} = (160...170)$ nm. There exist clear differences between these scattering patterns: The scattering pattern of MePG involves the information on mesopores and "hidden" macropores. This "filled system" possesses two *Porod regions* where $I(h) \sim h^{-4}$ (two different horizontal levels). One of them identifies the mesopores and the other one describes the system of the macropores of the MePG. The geometry of the MaPG only leads to one extensive *Porod region*.

8.2.2 Macropore analysis of a controlled porous glass

Here, μ-chords are applied. Two scattering patterns, i.e., those of "empty" and "filled" macropores, were analyzed in a paper by (Gille et al., 2001) [79], in which $L = 150$ nm was inserted for the data evaluation via $\gamma(r, L)$.

An initial scattering pattern (1) of a defined controlled mesoporous glass *before* and a second scattering pattern (2) *after* alkaline removal of the mesoporous colloidal silica in the macropores of the main silica framework were recorded. Consequently, in (1) a mesoporous glass (MePG) was analyzed and in (2) the immediately resulting macroporous glass (MaPG) was analyzed. In both cases, CFs and CLDDs of the macropores follow [i.e., of the filled (1) and empty (2) macropores]. There is no difference between the outer size and shape and geometrical arrangement of filled/empty macropores at all. Certainly, as a consequence of the missing filling of the macropores in the MaPG case, the scattering intensities differ by a constant factor for small h values and absolutely differ for large h values. The geometric arrangement of the macropores involves a high symmetry in the corresponding MePG and MaPG type glass.

Electron microscopy and SAS are appropriate methods for directly comparing the geometry of both glass preparation states (see Figs. 8.9, 8.10 and 8.11). Assuming that state 1 possesses two order ranges [i.e., $L_0 \approx 10$ nm and $L = (150 \pm 20)$ nm], it follows that state 2 only possesses one [i.e., $L = (150 \pm 20)$ nm].

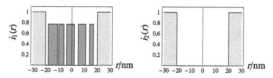

FIGURE 8.10

Weighted indicator functions i_1 and i_2 describe states 1, 2 of the glass, Fig. 8.11: Linear intersections of filled macropores (MePG state) and empty macropores (MaPG state). A contrast effect is essential: The mean density inside a "filled" macropore is smaller than that of the Si matrix, but not zero. In contrast, the density inside an empty macropore equals zero. The filled system (see left) produces an essentially smaller scattering. The density fluctuation between a filled pore and the matrix is smaller than that between an empty pore and matrix.

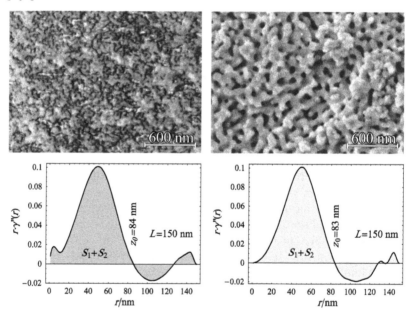

FIGURE 8.11

Analysis of the macropores of the controlled pore glass. Comparison of electron microscope and SAS results (Gille et al., 2001) [79, pp. 180–182].

Upper part: Micrographs of both glasses MePG (1) and MaPG (2) on a length scale of 600 nm.

Lower part: Comparison of the ν CLDD of the MePG (left) and the interrelated MaPG (right) for $L = 150$ nm. The differences are small. With the constant L of the macropores inserted, both curves nearly agree (see the CFs given in [79]). There is a zero at about $r = z_0 = 84$ nm. A porosity of 50% results, which is in agreement with the fact that $S_1 = S_2$ [see Eq. (8.14)].

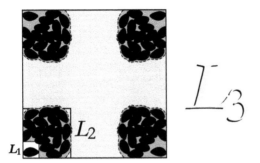

FIGURE 8.12
Three different order ranges and three different c – a plane illustration.
For $L = L_1, L_2, L_3$, where $L_1 < L_2 < L_3$, the volume fraction c depends on L
(see Fig. 1.11). The function $c(L)$ differs from sample to sample. Actually, L_1
is the largest diameter of a black primary particle, $c(L_2) \approx 0.6$ and $c(L_3) \approx$
0.3. The ensemble described by $L_3 = 3L_2 = 9L_1$ is quasi-diluted.

8.3   The volume fraction depends on the order range L

For ensembles of particles that are isotropically arranged in space, the following generally holds: Only in selected cases does c *not* depend on L (see
Fig. 8.12). For $L = L_0$, the CF of a tightly packed particle ensemble can be
written in terms of c, the single particle CF $\gamma_0(r)$ and the pair correlation
$g(s)$ of the particles as

$$\gamma(r) = \frac{1}{1-c}\left(\gamma_0(r) - c + \frac{c}{V_0} \cdot \int_{L_0}^{r+L_0} 4\pi s^2 \cdot P_{AB}(r,s,d) \cdot g(s)ds\right). \qquad (8.17)$$

The function P_{AB} is explained in Chapter 4. Those parts of Eq. (8.17) that
belong to the integral are complicated to handle. A pair correlation is available
only in special cases. However, useful conclusions can be drawn even from the
simple pre-term

$$c(r,L) = \frac{\gamma_0(r,L) - \gamma(r,L)}{1 - \gamma(r,L)}, \quad c(r) = \frac{\gamma_0(r) - \gamma(r)}{1 - \gamma(r)}. \qquad (8.18)$$

For closer particle packing, the split-off of the pre-term of Eq. (8.18) is a rough
approximation, which is more exact the smaller r is (see Fig. 8.13).
    On the one hand, $c = 1 - \gamma_0'(0)/\gamma'(0)$ results from Eq. (8.18) for $r \to 0$.
This allows c to be fixed if some information about the single particle is known
[at least $l_1 = -1/\gamma_0'(0)$]. Without more specific assumptions, additional conclusions about c would require much more effort (see Synecek's method in
Section 8.4). Furthermore, by introducing the assumption of *quasi-dilution*
(see Chapters 4 and 5), some important connections result (Gille, 2002) [87,

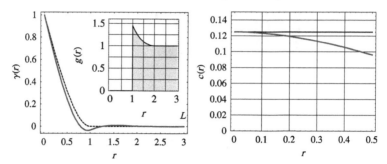

FIGURE 8.13

Analysis of the function $c(r)$ [Eq. (8.18)] based on the CF $\gamma(r, d, c)$ of a Dead Leaves model (DLm) (see left). The DLm was constructed from spherical grains of diameter $d = 1$. Close to the origin, the function $c(r)$ reaches the value $c_{DLm} = 1/8$. The horizontal line marks this value.

pp. 473–475]. If the largest particle diameter is smaller than the length of the smallest particle-to-particle chord, the integral term in Eq. (8.17) does not influence $\gamma(r) = \gamma(r, L)$. Thus, simple connections between the integral parameters of SAS and c can be derived. The restriction $r \to 0$ [see Eq. (8.18) and Fig. 8.13], is no longer necessary. Consequently, for all r in $0 \le r \le L_0$,

$$c \equiv \frac{\gamma_0(r) - \gamma(r)}{1 - \gamma(r)}; \quad c = -\frac{\gamma(L_0)}{1 - \gamma(L_0)}, \quad \gamma(L_0) \le 0 \qquad (8.19)$$

holds true. The first zero of the CF at $r = r_0$ (see Fig. 8.13) is connected with c via $\gamma(r_0) = c$. Based on Eq. (8.17), c can be estimated in terms of the CF and the maximum particle diameter L_0 [87, p. 473, equation (10)]

$$c = 1 - \frac{4\pi L_0^3/3}{\int_{L_0}^{\infty} 4\pi r^2 \gamma(r) dr}, \quad c = 1 - \frac{V_0 - \int_0^{L_0} 4\pi r^2 \gamma(r) dr}{4\pi L_0^3/3 - \int_0^{L_0} 4\pi r^2 \gamma(r) dr}. \qquad (8.20)$$

Equation (8.20) is useful for estimating relatively small c. Formally, it restricts to the case of identical particles. However, polydispersity and quasi-dilution can be combined. The mean CF of all the differently shaped particles $\overline{\gamma}_0(r)$ must be inserted instead of $\gamma_0(r)$. Summarizing all the normalized functions $\gamma_k(r)$ and $\gamma'_k(r)$ of the kth particle with volume V_k gives

$$\overline{\gamma}_0(r) = \frac{\sum V_k \gamma_k(r)}{\sum V_k}, \quad \overline{\gamma}_0(0) = 1, \quad \overline{\gamma}_0'(r) = \frac{\sum V_k \gamma'_k(r)}{\sum V_k}, \quad \overline{\gamma}_0'(0) = \frac{\sum V_k \gamma'_k(0)}{\sum V_k}. \qquad (8.21)$$

The properties of the function $P_{AB}(r, s, L_0)$ do not change. Slightly modified, Eqs. (8.17) to (8.21) still hold true. For the kth particle, $\gamma'_k(0)$ can be traced back to its volume V_k and surface area S_k via $\gamma'_k(0) = -S_k/(4V_k)$. Thus, V_k, S_k and $\gamma_k(r)$ yield all the modified terms of $\overline{\gamma}_0(r)$ instead of $\gamma_0(r)$.

For sphere diameters d_k that are equally distributed in the interval $0 < d < L_0$, the CF $\overline{\gamma}(r, L_0)$ was simulated [87, p. 474, figure 3].

8.4 The Synecek approach for ensembles of spheres

In 1983, the Czech metal physicist V. Synecek (see [220, 221]) established interesting approaches, which will be briefly explained in the following. These ideas are based on a minutely detailed analysis of the connection between the sample CF $\gamma(r)$ and the function of occupancy $Z(r)$ of an isotropic sample, i.e., $\gamma(r) = [Z(r) - c]/(1 - c)$. The idea is to operate with the identities

$$c \equiv \frac{Z(r) - \gamma(r)}{1 - \gamma(r)} \equiv 1 - \frac{Z'(r)}{\gamma'(r)} \qquad (8.22)$$

by adding additional information about the particle shape. For an isotropic ensemble of hard particles, the sample CF $\gamma(r)$ starts in the origin with $\gamma(0) = 1$ and $\gamma'(0) = -1/[l_1(1 - c)]$. Furthermore, there exists a first zero at $r = r_0$, where $\gamma(r_0) = 0$ (i.e., $Z(r_0) = c$) followed by a local minimum $r = r_{min}$, where $\gamma'(r_{min}) = 0$. The analysis restricts to the interval $0 \le r < r_0 \le r_m$. The idea introduced by Synecek was to insert a (a priori unknown, dimensionless) working function $f(r)$, which fixes a *modified function of occupancy* $Z(f(r))$. For unit spheres, i.e., for a relative length r, $Z(r)$ and $\gamma(r)$ are written as

$$Z(r, f(r)) = 1 - \frac{3}{2}f(r) + \frac{1}{2}f(r)^3, \ \gamma(r) = \frac{1 - \frac{3}{2}f(r) + \frac{1}{2}f(r)^3 - c}{1 - c}. \qquad (8.23)$$

More generally, $Z(f(r))$ must be adapted to the particle shape investigated. From this, with $c = Z(f(r_0)) = 1 - \frac{3}{2}f(r_0) + \frac{1}{2}f(r_0)^3$ and incorporating the initial conditions $f(0) = 0$, $f'(r_m) = 0$, the following differential equation for $f(r)$ in $0 \le r < r_0 \le r_m$ (Gille, 1996) [55], results

$$\gamma'(r) = \frac{\frac{3}{2}f'(r) \cdot (f(r)^2 - 1)}{1 - c} = \frac{\frac{3}{2}f'(r) \cdot (f(r)^2 - 1)}{1 - \left(\frac{1 - \frac{3}{2}f(r) + \frac{1}{2}f(r)^3 - \gamma(r)}{1 - \gamma(r)}\right)}. \qquad (8.24)$$

Equation (8.24) fixes $f(r)$ in terms of $\gamma(r)$ for all r in $0 \le r < r_0 \le r_m$ and is implemented in a *Mathematica* program. Figure 8.14 illustrates the approach for a typical scattering pattern involving an initial minimum at $h = 0$.

```
(* Determination of f[x] via the Runge-Kutta method with initial condition
    f(epsilon)=epsilon -> 0. The term g[x] denotes a spline interpolation of
    the CF and x=rm is the minimum position of g[x], g'[rm]=0 *)
epsilon = 0.01;
res     = NDSolve[g'[x] == 3(f[x]^2 - 1)(g[x] - 1)f'[x]/(f[x](f[x]^2 - 3)),
            f[epsilon] == epsilon}, f, {x, epsilon, rm},
            MaxRelativeStepSize -> 1/100, Method -> RungeKutta];
F = f /. res[[1]];   Plot[F[x], {x, epsilon, rm}]
```

The inserted DLm pair correlation function [see Eq. (4.21)] is not an approximation. Certainly, the inserted DLm restricts to $c = 1/8$. This method is ideal for analyzing precipitates in alloys. Indeed, the identities Eq. (8.22) are fulfilled (see Fig. 8.14) as shown by

$$c \equiv \frac{1 - \frac{3}{2}f(r) + \frac{1}{2}f(r)^3 - \gamma(r)}{1 - \gamma(r)} = 1 - \frac{-\frac{3}{2}f'(r) + \frac{3}{2}f(r)^2 f'(r)}{\gamma'(r)}, \ 0 \le r < r_m. \quad (8.25)$$

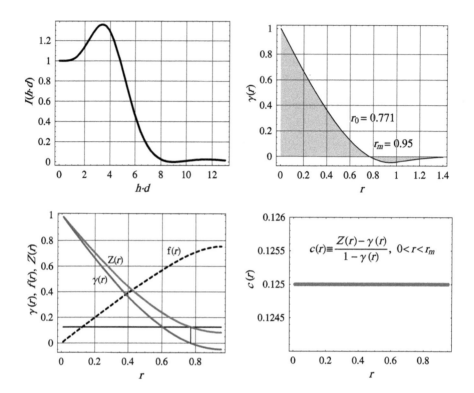

FIGURE 8.14

Verification of Synecek's method for the pair correlation function of a DLm and spherical particles (see Chapter 4).

The way of calculation from the scattering pattern $I(h)$ of the tightly packed spheres of diameter $d = 1$ to $\gamma(r)$, $f(r)$, $Z(r)$ and finally to $c(r)$ is illustrated. Actually, $r_0 = 0.771$ and $r_m = 0.95$ follow from the CF.

Finally, the function $Z(r)$ is traced back to the function $f(r)$ [see Eq. (8.23)]. Actually, $f(r_m) = 0.755$ and $f(r_0) = 0.695$ result. The horizontal line marks $c = 0.125$. Furthermore, the thin vertical line starting at $r = r_0$ marks the connection $Z(r_0) = Z(f(r_0)) = c$.

This method is very stable, as can be seen from the result $0.1248 < c < 0.12502$ (see the dark horizontal line $c(r) \equiv 0.125$). Except for the standard $\gamma'(0)$, derivatives are not required. From the estimated volume fraction c, the sphere diameter d results via $-1/[\gamma'(0) \cdot (1 - c)] = l_1 = 2d/3$.

This example is just meant to show how the method works. There exist other (much better worked out) approaches for solving this differential equation than that by Runge/Kutta.

8.5 Volume fraction investigation of Boolean models

Figuer 5.1 explains a typical case of a Boolean model (Bm) with spherical primary grains. The determination of c for samples which can be approximated by Bms has already been discussed in Chapter 5; see papers by (Hermann, 1991) [150], (Gille, 1995)[53] and (Gille, 2002) [85, p. 616, figures 5–6]. All of this is summarized in this section. Typically, the CF of a Bm is always a relatively smooth function. A direct estimation of mean chord lengths from $\gamma''(r, L)$ is a complicated task in the case of arbitrary grain shapes.

However, the determination of c for smooth grains is simple. Here, the volume fraction of the grain phase results from

$$c = \frac{\gamma''(0+)}{[\gamma'(0+)]^2}, \quad 0 \leq c < 1. \tag{8.26}$$

In fact, the infinitely diluted case $c \to 0$ is also involved in Eq. (8.26). Here, $\gamma(r) \to \gamma_0(r)$, i.e., for smooth (ellipsoidal) grains, $\gamma''(0) = 0$ and $c = 0$ result.

Denoting the typical CF of the primary grain by $\gamma_0(r)$, the Taylor series of the sample CF $\gamma(r)$ writes as

$$\gamma(r) = \frac{1-c}{c}\left[e^N \gamma_0(r) - 1\right] \approx 1 + \frac{N\gamma_0'(0)}{c}r + \frac{[N^2(\gamma_0'(0))^2 + N\gamma_0''(0)]/2}{c}r^2 + \cdots. \tag{8.27}$$

Equation (8.27) involves the two parameters c and N, which are connected by $c = 1 - \exp(-N)$. The parameter N denotes the mean *number* of germs (center points of a primary grain) which belong to a primary grain of (mean) volume V_g, i.e., $N = number/V_g$.

The length $2\bar{d}_{lm}$, which is twice the mean average chord length, is an important parameter for the estimation of c. It can be traced back to N and $\gamma_0'(0)$: Operating with the mean chord length of the grain phase $l_1 = 4Vc/S$ and describing the connected phase with $m_1 = 4V(1 - c)/S$ yields

$$2\bar{d}_{lm} = l_1 + m_1 = \frac{4V}{S}(c + 1 - c) = \frac{4V}{S} = -\frac{1}{C'(0)}. \tag{8.28}$$

Here $C'(0)$ is the first derivative of the set covariance $C(r) = 2c - 1 + (1 - c)^2 \exp[N \cdot \gamma_0(r)]$ in the origin. Consequently, $2\bar{d}_{lm}$ is written as

$$2\bar{d}_{lm} = -\frac{e^N}{N \cdot \gamma_0'(0)}, \quad 0 < N < \infty. \tag{8.29}$$

The ratio $\exp(N)/N$ involves a global minimum for $N = 1$. Thus, the minimum value of the term $2\bar{d}_{lm} = l_1 + m_1$ is $-e/\gamma_0'(0)$. For spherical primary grains of diameter d, $-1/\gamma_0'(0) = 2d/3$. Consequently, the length $2\bar{d}_{lm} = 2ed/3$ is greater than d. Since $L = d$, the sum $l_1 + m_1$ cannot be estimated from the sample CF because $\gamma(r) = 0$ for all r in the interval $L = d < r < \infty$.

See Chapter 5 for other equations that are suitable for estimating c.

8.6 About the realistic porosity of porous materials

Why do the porosity values for some materials differ significantly when estimated with different methods (e.g., c depends on the method of investigation)? Various aspects of the porosity estimation[¶] of porous materials have been summarized in the paper *About the Realistic Porosity of Porous Glasses* (Gille, et al.) [94]. This section summarizes the factors concerning the *accessibility of pores* inherent in procedures for determining c.

There exist different techniques for the preparation of mesoporous materials. One of them is the acid leaching treatment of phase-separated alkali borosilicate glasses. Furthermore, generating larger mesopores requires an additional procedure, whereby the silica gel remaining in the pore system of the glass is removed by treatment with an alkaline solution. This poses a certain disadvantage due to the fact that mesoporous glasses prepared in this way often contain residues of silica gel possessing a microporous texture. These "contaminations" can produce unspecific adsorption behavior, which leads to great differences in the results of the standard characterization techniques of *nitrogen (or argon) adsorption* (NA, AA) and *mercury intrusion* (MI). Well investigated porous glasses (VYCOR, CPG) do not contain residues of this type. Here, the results of SAS, MI, NA and scanning electron microscopy (SEM) should agree.

However, in the case of many mesoporous glasses prepared by a combined acid and alkaline leaching treatment of a phase-separated SiO_2-rich sodium borosilicate glass, the "contaminations" have a more or less strong influence on the texture properties. Different characterization techniques lead to contradictory results.

8.6.0.1 Results from MI, NA/AA and SAS

The MI, NA and AA methods have been well developed for porosity estimation and the characterization of porous materials. In NA or AA experiments, the total pore volume V_P together with the porosity c are estimated from the amount of N_2 or Ar vapor adsorbed at a high relative pressure. It is assumed that the pores are *subsequently filled with condensed nitrogen or argon* in the normal liquid state. The volume of mercury taken up by the porous material in a high-pressure test is used in MI experiments to estimate c. If portions of closed pores are present, these techniques detect the *open porosity*. The *accessibility* of the pores is an unknown parameter in most cases. Hence, correlating the *microporosity* or *mesoporosity* resulting from MI and NA/AA with those resulting from SAS *always* requires some discussion. Furthermore, for the most part, SAS does not produce any change in the sample during

---

[¶]This section mainly addresses chemists experimenting with porous materials.

preparation or measurement, whereas SEM and MI cause modifications in the sample.

However, pores and walls cannot be distinguished based on a simple scattering pattern. An SAS pattern cannot describe the "diversity" of the intermediate space between two points P_1 and P_2 [191]. The distance $\overline{P_1 P_2}$ can be any chord length that extends across many pores and walls, across a few pores and walls, across one pore (wall), or across a wall, i.e., no pore(s).

Sample preparation of the MPG: A sodium borosilicate initial glass [70 wt.% SiO_2, 23 wt.% B_2O_3, 7 wt.% Na_2O], in the shape of beads with a diameter of (0.1–0.2) mm, was annealed at 570°C for 48 h. The heat treatment of the composition in the boric acid anomaly resulted in the decomposition of the initial glass into two separated, but interconnected phases: The first one was almost pure silica. The second one was a sodium-rich borate phase with some amounts of silica dissolved in it. After the phase separation, the soluble sodium-rich borate phase was removed by leaching with 3 M hydrochloric acid at 90°C for 3 h. After this, the colloidal silica deposits formed in the mesopores during the acid leaching treatment were removed by treatment with 0.5 M sodium hydroxide solution at room temperature for 2 h. This prepared mesoporous glass (PMPG) was washed three times with distilled water and air dried.

SAS experiment (see Fig. 8.15): SEM investigation of the mesoporous glass sample was carried out on a Phillips ESEM XL 30 FEG microscope using micrographs with different magnification. The order range was estimated.

A slit collimated Kratky camera using nickel-filtered CuK_α radiation was applied. The pattern $I(h)$ was recorded in the h-interval $h_{min} = 0.05$ nm$^{-1}<$ $h < h_{max} = 6.5$ nm^{-1}. After background subtraction, the data was analyzed in the region of 0.05 nm$^{-1}< h < 1.55$ nm^{-1}. The functions $\gamma(r)$, $\gamma'(r)$ and $\gamma''(r)$ were determined for $L \approx 60$ nm, i.e., $\gamma(r) \equiv 0$ if $L < r$ (see Chapter 1). From $I(h)$, a normalized Porod plot $P_1(h)$ of the two-phase sample results (see Chapter 1). The functions $\gamma(r)$ and $\gamma''(r)$ were extrapolated for $r \to 0$ and plotted in the interval 2 nm $< r < (50...60)$ nm. The maximum and minimum positions of the curvature of the CF nearly reflect the first moments (and their sums) of the CLDDs, φ and f. The first moments agree. The behavior of γ was ascertained by use of an LSM with parameters $l_1 = m_1 = 16$ nm and L, i.e., $c = 50\%$ porosity (see p. 185, figure 5) [94]. There is no indication of a significantly smaller porosity.

For the inserted L, $l_p \approx 8$ nm follows. The function $\gamma(r)$ possesses a maximum curvature between $r = 10$ nm and $r = 20$ nm. Furthermore, $\gamma(r) \approx 0$ if 50 nm$< r < 60$ nm and $\gamma(r) \equiv 0$ if $L < r$. There is a point of inflection at $r_{inf} \approx 28$ nm [see the zero in $\gamma''(r)$] and another maximum of curvature to the right in the interval 30 nm$< r <45$ nm at $r = \overline{d}_{lm}$ marking the mean average chord length $(l_1 + m_1)/2 = \overline{d}_{lm}$ (Gille, 2002) [85, p. 613], where $2 \cdot \overline{d}_{lm} = (33 \pm 1.2)$ nm. Thus, $c = 0.5 \pm 0.06$ follows from the relation $c(1 - c) = 1/(2\overline{d}_{lm} \cdot |\gamma'(0, L)|)$ [85, 84].

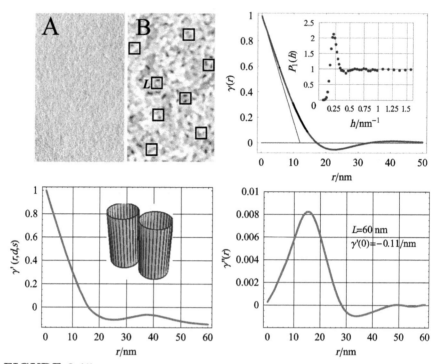

FIGURE 8.15

From SEM micrographs and the scattering pattern of the PMPG to the SAS structure functions for $L = 60$ nm.

With a lower magnification, the material looks relatively homogeneous (see the left SEM). The right SEM involves additionally inserted squares of edge length L. Each of the squares stretches over two pores and two walls. Smaller squares, e.g., $L < 30$ nm, lead to a loss of information. Larger squares do not contain more information about pores and walls. In this regard the whole image of the micrograph is useful, but much too large.

Detected parameters: $\gamma'(0) = -0.12$ nm^{-1}; hence, $l_p = 8.3$ nm $= l_1(1-c) = m_1 c$. There is a local minimum/maximum of the CF at $r = 28$ nm and $r = 40$ nm, respectively. This behavior can be explained by the model of two parallel cylinders; see (Gille, 2001) [76, p. 186, figure 3] and (Gille, 2005) [99, p. 62, figure 7]. The basic model $\gamma(r, s, d) = Z(r, s, d)$ of two parallel cylinders is explained in Section 4.3 (see Fig. 4.6). The parameters $d = 16$ nm and $s = 36$ nm yield the plot of the modified model function $\gamma^*(r, s, d) = [Z(r, s, d) - c_0]/(1 - c_0)$ with $c_0 = 0.155$. This function possesses a behavior similar to the experiment $\gamma(r)$. However, this two-cylinder model possesses the theoretical full-order range $L \to \infty$. There is the sub-order range $L = s + d = (36 + 16)$ nm $= 52$ nm ≈ 60 nm. In this approximation, the value c_0 is twice the cylinder volume over the whole volume available for both cylinders, $c_0 \approx (\pi d^2/4)/s^2$ [76, p. 190].

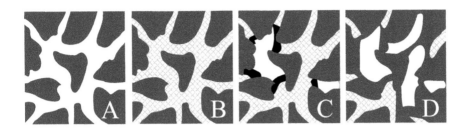

FIGURE 8.16

Cases A–D explain what realistic porosities mean as follows:

A: Porous material, $c = 0.3$; all pores can be reached via a suited adsorbent.

B: Interconnected pores, completely filled with adsorbent, $c = 0.3$.

C: Pores are not completely connected. Here, only part of the whole pore volume can be reached via the adsorbent; thus $c < 0.3$ results from techniques that measure a *degree of filling*.

D: Some parts of the pore system are (temporarily) not accessible to the adsorbent. Thus, the porosity estimated by *adsorbent using procedures* is (essentially) smaller than c.

Mesoporous and macroporous glasses, which are created by the preparation process with phase-separated alkali borosilicate glasses, can still possess a considerable degree of microporosity. These micropores are "generated" by residues of silica gel remaining in the mesopore or macropore system of the porous glass after the combined acid and alkaline leaching treatment of the phase-separated initial glass. These residues are located as small flakes, "plugs" or layers inside the pore channels of the glass.

1. The real porosity of a porous material is connected with the ratio of open and closed pores and their accessibility.

2. Furthermore, the porosity depends on L, which can be directly taken into account via SAS.

Closed pores (D) are not accessible to nitrogen or mercury in the standard characterization techniques or to reactant molecules in porous catalysts. There is what is called the permanent and temporary non-accessibility of the pores. Permanent non-accessibility results from the microstructure of the material itself. The temporary non-accessibility is generated during the characterization process or through deactivation phenomena (i.e., coking) in heterogeneous catalysis. Such processes or phenomena can be described with the percolation theory.

Hence, in many home-made mesoporous glasses, residues of silica gel (resulting from the preparation process) lead to differences in the results of the standard characterization techniques NA and MI. These contradicting results can be explained by going back to first geometrical principles and combining SAS and SEM.

Aside from this strategy, $\gamma''(r)$ already contains all the stereological information about the sample porosity. The function $\gamma''(r)$ possesses the one and only positive and nearly symmetrical main peak at $r = l_1 \approx m_1 \approx 16$ nm. Furthermore, $\gamma''(r)$ contains a minimum at nearly $r = l_1 + m_1 \approx 34$ nm. Hence, a 50% porosity results. A numerical check for the inserted order range is $\int_0^L \gamma''(r, L)dr = -\gamma'(0, L) = 0.12$ nm$^{-1}= 1/l_p$. This result nearly agrees with the two-cylinder approximation.

The porosity value obtained from the MI agrees with the results from SEM and SAS, but the value $c = 0.19$ determined by NA does not reflect the realistic texture of this mesoporous glass. Assuming long cylindrical pores of diameter d, a diameter distribution density $V(d)$ with first moment $\bar{d} = (15 \pm 1)$ nm follows [94, p. 184, figure 4]. The behavior of the functions $\gamma''(r)$ and $V(d)$ is similar. This indicates the existence of relatively long stretched cylindrical pores.

8.6.0.2 Interpretation of estimated volume fractions (porosities)

Generally, the leaching with acidic solution removes the alkali-rich borate phase of phase-separated alkali borosilicate glasses, which leaves a porous glass structure. However, the soluble phase often contains certain amounts of dissolved silica as a result of a secondary phase separation. The quantity depends on the initial glass composition, temperature and time of the phase separation process. Due to the very low solubility of silica in acids, it remains inside the formed pores after the leaching procedure. The finely dispersed silica gel masks the pore structure of the resulting porous glasses. It can be easily removed by treatment with an alkaline solution. This process leads to an enlargement of the pore diameter and of the specific pore volume of the corresponding porous glass.

Controlling the conditions of the alkaline treatment is useful for protecting the porous glass skeleton for dissolution. Overly mild conditions (short-time treatment, small concentration of the alkaline solution) lead to incomplete removal of the silica gel. The residues can cause differences in the value of the porosity determined by various techniques.

As discussed in Section 8.15, NA and MI yield completely different porosities. The result of about $c = 0.5$ from the MI agrees with the results obtained with SAS and SEM. The value of $c = 0.19$ determined by NA does not reflect the realistic texture of the glass.

Furthermore, the porosity depends on the parameter L (see Gille, 2002) [85, pp. 613–616, figures 1–4] as shown by

$$c = c(L) = \frac{l_1(L)}{l_1(L) + m_1(L)}. \tag{8.30}$$

Additional porosity parameters, such as the *pore connectivity number* and *pore accessibility*, would have to be introduced. A radial distribution of the porosity and the connected parameters can occur. For PVG, the porosity is homogeneous (see Fig. 8.15). Most porous glasses involve a complex, highly networked, three-dimensional, disordered microstructure, i.e., most pores are interconnected [Fig. 8.15(A,B)]. In addition, *blind pores* exist. There is always a period of temporary non-accessibility if porous glasses are characterized by *low temperature NA* or used as a catalyst support.

For any material, the selection of a certain order range L (a typical size of the analyzed section of the whole sample volume) is as important as the consideration of *permanent or temporary non-accessibility*. With porous materials possessing a complex microstructure, the application of only one characterization technique (NA, AA, MI, SAS, SEM) can lead to a completely misleading picture of the realistic porosity. The results of several characterization techniques have to be analyzed simultaneously.

9

Interrelations between the moments of the chord length distributions of random two-phase systems

Statistically isotropic arrangements of hard, homogeneous particles embedded in a homogeneous matrix (constant density) can be analyzed by two (naturally different) chord length distributions. Their moments are interrelated and connected with the sample CF. Integral transformations, which directly define the moments in terms of the scattering pattern, are investigated.

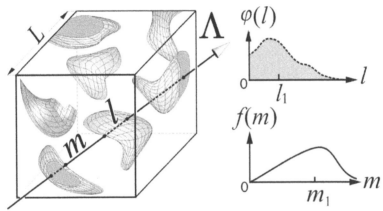

If a straight line Λ intersects particles, random chord lengths result. Based on such a measurement along a linear probe, second-order characteristic parameters that are closely connected with the scattering pattern result. The length segments along Λ inside and outside the (convex or non-convex) particles (i.e., l and m) vary. The distribution laws of these random variables are described by the distribution densities $\varphi(l)$ and $f(m)$ defining the moments l_1, l_2, l_3 ... and m_1, m_2, m_3 ..., respectively.

It is known that the first moments l_1 and m_1 are interrelated with Porod's length parameter l_p [191]. Furthermore, there exist connections between higher CLD moments and typical scattering and experimental parameters applied in other fields. For example, CLDs are intimately related to the transport properties of porous media. Here, a parameter $l_e \equiv l_p$ is referred to as the *effective mean free path length* due to the molecule-solid collisions of the transport phenomena [169].

9.1 Single particle case and particle ensembles

Chord length distribution densities (CLDDs) (see the analytic expressions given in Chapter 2) are also described and applied in stochastic geometry and statistical physics [3, 205]. The fundamental mathematical results obtained in this field can also be applied in materials science [225, 150], crystallography [16, 7], astronomy and astrophysics [4], as well as in the acoustic design of auditoria and archeology.

By utilizing suitable physical apparatuses, CLDDs can be obtained from elastic scattering experiments and used for interpretation [165, 107]. CLDDs are applied in the single particle case [217, 175] as well as for particle ensembles [156, 66]. One of the first fundamental results obtained as early as 1898 is Rosiwal's linear integration principle [197].

Let the density correlation function of the two-phase sample $\gamma(r) = \gamma(r, L)$ (see [16, p. 220]) be the working function for the following: The term $\gamma(r, L)$ describes a fixed order range L of the sample material and the particles are defined by an indicator function (see the introductory part of Sukiasian's paper [217]). The CLDD of a single homogeneous convex particle denoted by $\varphi(r)$ is $\varphi(r) = \gamma_0''(r)/|\gamma_0'(0)|$. For a single homogeneous particle (largest particle diameter L_0, volume V_0 and whole surface area S_0), the first moment l_1 of $\varphi(l)$, $l_1 = \bar{l} = 1/|\gamma_0'(0)| \equiv 4V_0/S_0$ results. This relation includes the non-convex single particle case [175]. For ensembles of quasi-diluted homogeneous particles (the largest particle diameter L_0 is greater than the smallest chord length between any two particles), the scattering pattern defines γ_0'' and likewise the function $\varphi(r)$ in the interval $0 \leq r < L_0$.

The extension of these quite plain connections to the tightly packed particle case [particle volume fraction c, where $0 \leq c < 1$, $\gamma_0(r, L_0) \to \gamma(r, L)$] is not trivial. Chapters 5–8 include applications for several special cases. The analytic results of $\gamma(r, L)$ for Boolean models (Bms) were discussed in Chapter 5. Special rules were established for ensembles of puzzle fragment particles resulting from a Dead Leaves model (DLm) and which fit together up to a volume fraction $c \to 1$ (see Chapters 7–8). A sequence of approaches for estimating c based on $\gamma(r, L)$ and geometric parameters are explained in Chapters 4, 5 and 8. After handling such a sequence of special models, the central question arises whether there is a general connection between γ, φ, f and c for random two-phase particle ensembles. As early as in 1951, the great Porod made observations on such general formulas [191, 192]. He recognized that the function $\gamma''(r)$ is of importance.

To determine whether there is a connection, the deciding step is an analysis of the function $g(r) = l_p \cdot \gamma''(r) + 2\delta(r)$, where $0 \leq r < L_0 < L$. To do this, a more general length parameter $l_p = 1/|\gamma'(0)|$ is included instead of l_1 and $\gamma''(r)$ involves all the information in the length interval $0 \leq r < L_0 < L$. In the very first step, an analysis of the 0th moment of $g(r)$ gives $l_p \equiv 1/|\gamma'(0)|$.

In addition, the chord segments of random length m outside the particle phase (outside the single particles) can also be analyzed. The random variable m possesses the distribution density $f(m)$. The first moment is $\overline{m} = m_1$. The connection $1/|\gamma'(0)| = l_p = 4V \cdot c(1 - c)/S$ holds true [165, 48, 80] (see also Eqs. (1.42), (1.43) and Fig. 1.9). Here, S denotes the total surface area of the particle phase (of all particles) possessing the whole volume cV. The connection between the first moments l_1, m_1 and l_p is

$$|\gamma'(0)| \equiv \frac{1}{l_p} = \frac{1}{l_1} + \frac{1}{m_1}. \tag{9.1}$$

Equation (9.1) involves the limiting case of a diluted particle ensemble. If the ensemble of unchangeable particles becomes more diluted (limiting case $c \to 0$, $m_1 \to \infty$), $l_p = 1/|\gamma_0'(0)| \to l_1$ results.

Equation (9.1) is a useful approach for a wide class of statistically isotropic systems of arbitrary microgeometry (see [169, p. 6474] and [165, p. 778]). Equation (9.1) involves a sequence of applications in various fields [224]. This includes astronomy, where linear probe statistics is applied, based upon linear probing of complex patterns and measurement of large and superlarge scale structures of the universe [8]. Other examples are studies of matter distribution in the universe in terms of Gaussian fields or occultation experiments [4].

Equation (9.1) has been verified analytically for the hard particle "Dead leaves" model (DLm) [100] (see Chapter 6). In this model, hard spheres of constant diameter d are arranged in space via a special, easily manageable pair correlation function resulting in the order range $L = 3d$.

For this pair correlation, analytic expressions of the function $\gamma''_{DLm}(r)$ (interval splitting, three r-intervals), $f_{DLm}(m)$ and the corresponding characteristic function $q_{DLm}(t)$ are obtained (see Section 4.4 and Fig. 4.8). The derivative of $q_{DLm}(t)$ in the origin is $q'_{DLm}(0) = 14d \cdot i/3$. This obeys

$$\int_0^{L=3d} \gamma''_{DLm}(r)dr = |\gamma'_{DLm}(0)| = \frac{12}{7d} = \frac{1}{l_{pDLm}}. \tag{9.2}$$

In point of fact, Eq. (9.2) satisfies Eq. (9.1), where $l_1 = 2d/3$, $m_1 = 14d/3$ and $c = 1/8$.

Combining the Babinet theorem with Rosiwal's linear integration principle [85], the solutions $c = l_1/(l_1 + m_1)$ for phase 1 and $1 - c = m_1/(l_1 + m_1)$ for phase 2 give the quadratic equation

$$c \cdot (1 - c) \cdot (l_1 + m_1)/l_p = 1, \quad 0 < l_p \le l_1, m_1. \tag{9.3}$$

The parameters l_1, m_1 and l_p fix c and $1 - c$.

This introductory part only refers to *first moments*, i.e., l_1, m_1 and $l_p = M_1$. In the following section, interrelations between higher CLD moments of statistically isotropic ensembles of hard, homogeneous particles embedded in a homogeneous matrix will be analyzed. Obviously, Eqs. (9.1) and (9.3) are the simplest of a set of more general connections. It will turn out to be the simplest case of a more general theory, which interrelates *all* CLD moments.

9.2 Interrelations between CLD moments of random particle ensembles

The moments M_n of $g(r)$ expressed as

$$M_n = \int_0^L r^n \cdot g(r)dr = \int_0^L r^n \cdot \left[l_p\gamma''(r) + 2\delta(r)\right] dr, \qquad (9.4)$$

where $n = 0, 1, 2, \ldots$, are connected with the scattering pattern of the particle ensemble as well as with the CLDDs φ and f. Figure 1.9, Chapters 2–3, Fig. 5.1, Chapter 5, the starting figure of this chapter and Fig. 9.5 illustrate the meaning of the functions $\varphi(l)$ and $f(m)$. The term $2\delta(r)$ in Eq. (9.4) is of minor importance for the following. It ensures that $g(r)$ is the continuous part of $l_p\gamma''(r)$. Via the function $\gamma(r)$, all the M_n are connected with the *characteristic integral parameters** l_c, f_c and v_c [205, 150]. Equation (9.4) defines the M_n in the most general case. The first terms are $M_0 = 1$ and $M_1 = l_p = 1/ \mid \gamma'(0) \mid$. The next subsections explain that the moments $M_{2,3,4,...}$ are interrelated with the characteristic integral parameters $M_2 = l_p \cdot l_c$, $M_3 = 3l_pf_c/\pi$, $M_4 = 3l_pv_c/\pi$ and the scattering pattern Eq. (9.10).

Obviously, for single particles considered in Chapter 2, $M_2 = l_2$, $M_3 = l_3$ and $M_4 = l_4$. For a plane geometric figure [surface area $S = S_0 = f_c$ and (whole) perimeter u], the special cases $M_2 = x_2$, $M_3 = x_3$ and $M_4 = x_4$ result. In the following, more general interrelations are derived and summarized. In this context, the case $c \to 0$ is a special case of the interval $0 \le c < 1$ considered.

1. Moments of $g(r)$ in the single particle case

An analysis[†] of the first four moments $M_{1,2,3,4}$ of $g(r) = l_p \cdot \gamma''(r)$ for an order range L yields surprisingly compact expressions. The resulting M_i only involve *characteristic integral parameters* of the CF, l_p, l_c, f_c and v_c. These useful parameters were introduced by the great Austrian physicist G. Porod (1951) [191]. He used the German term *"Charakteristik"* instead of the English *"correlation function"* as it is known today.

The M_i can be specialized for the single particle case (see Table 9.1). The M_i fix the parameters l_p, l_c, f_c and v_c. In the plane single particle case, the terms $x_{1,2,3,4}$ define the surface area $S_0 = S = f_c$ and perimeter u of

---

[*]$l_c = 2\int_0^L \gamma(r)dr$, $f_c = \int_0^L 2\pi r\gamma(r)dr$, $v_c = \int_0^L 4\pi r^2\gamma(r)dr$.

[†]In summary, this analysis consists in applying partial integration by taking into account trivial limiting conditions like $\gamma(L) = 0$, $\gamma'(L) = 0$ and $\lim_{L\to\infty}[L^2\gamma'(L)] = 0$. For example, with M_4, a term $T(r) = \int_0^r (r - x)^2/2 \cdot \gamma(x)dx$ can be introduced. It follows $T'(r) = \int_0^r (r - x) \cdot \gamma(x)dx$, ..., $T^{(5)} = \gamma''(r)$. The integral $M_4 = \int_0^L r^4 \cdot l_p \cdot T^{(5)}(r)dr$ is easy to handle.

the particle. In the spatial case, the particle volume $V_0 = V = v_c$ and surface area S result from $l_{1,2,3,4}$. The last line of Table 9.1 involves the

TABLE 9.1
Moments of the functions $g(r)$, $B(x)$ and $A(r)$ (see Chapter 2) as depending on characteristic particle parameters in the two- and three-dimensional case.

$g(r)$-moment	moment x_i (plane case)	moment l_i (spatial case)
$M_1 = l_p$	$x_1 = \pi S_0/u$	$l_1 = 4V_0/S_0$
$M_2 = l_p \cdot l_c$	$x_2 = \pi S_0/u \cdot \int_0^{L_0} 2\beta(x)dx$	$l_2 = 4V_0/S_0 \cdot \int_0^{L_0} 2\gamma(r)dr$
$M_3 = 3l_p f_c/\pi$	$x_3 = 3S_0^2/u$	$l_3 = 12V_0/(\pi S_0) \cdot \int_0^{L_0} 2\pi r\gamma(r)dr$
$M_4 = 3v_c l_p/\pi$	$x_4 = 3v_c S_0/u$	$l_4 = 12V_0^2/(\pi S_0)$

denotation v_c in the plane case $x_4 = 3v_c S/u$. Here, v_c simply denotes the integral $v_c = \int_0^{L_0} 4\pi x^2 \cdot \beta(x)dx$ as formally applied to $\beta(x)$. This is not a particle volume. The calculations below lead to the results in Table 9.1.

2. Derivation of the moments of the function $g(r)$

First, $M_0 \equiv 1$ and $M_1 \equiv l_p$. The second moment M_2 results from the definitions $F(r) = \int_0^r \gamma(r)dr$, $F''(r) = \gamma'(r)$ and $F^{(3)}(r) = \gamma''(r)$, and taking into account $F(0) \to 0$, $F(L) \to l_c/2$, $F'(L) \to \gamma(L)$, $F''(L) \to \gamma'(L)$, $\gamma(L) \to 0$,

$$M_2 = \int_0^L r^2 \cdot l_p F^{(3)}(r)dr = l_p \cdot [l_c + L^2 \cdot \gamma'(L)]. \tag{9.5}$$

Without introducing any restrictions for L, the term $L^2 \cdot \gamma'(L)$ can represent an indeterminate expression $[\infty \cdot 0]$. The analysis of the limit yields $\lim\limits_{L \to \infty} \frac{L \cdot \gamma'(L)}{1/L} =$
$\lim\limits_{L \to \infty} [\gamma'(L) - \frac{L \cdot \gamma''(L)}{1/L^2}] = 0$. Consequently, Eq. (9.5) results in $M_2 = l_p \cdot l_c$.

Based on the working definitions $P(r) = \int_0^r (r - x) \cdot \gamma(x)dx$, $P'(r) = \int_0^r \gamma(x)dx = l_c/2$, $P''(r) = \gamma(r)$, $P^{(3)}(r) = \gamma'(x)$ and $P^{(4)}(r) = \gamma''(r)$,

$$M_3 = \int_0^L r^3 \cdot l_p P^{(4)}(r)dr$$
$$= l_p \cdot [6P(0) - 6P(L) + 6L \cdot P'(L) - 3L^2 P''(L) + L^3 P^{(3)}(L)]$$
$$= l_p \cdot [3L \cdot l_c - 6P(L)] \tag{9.6}$$

is obtained. Hence, Eq. (9.6) gives $M_3 = l_p \cdot (3L \cdot l_c - 6 \cdot \int_0^L (L - x) \cdot \gamma(x)dx) = l_p \cdot [3L \cdot l_c - 6 \cdot \int_0^L L \cdot \gamma(x)dx + 6 \cdot \int_0^L x \cdot \gamma(x)dx] = l_p \cdot [3L \cdot l_c - 3L \cdot 2 \int_0^L \gamma(x)dx + 6 \cdot \int_0^L x \cdot \gamma(x)dx]$. Consequently, $M_3 = l_p [6 \int_0^L x \cdot \gamma(x)dx]$ follows. Since the last integral term equals $f_c/(2\pi)$, $M_3 = l_p \cdot 6 \cdot f_c/(2\pi) = 3l_p f_c/\pi$ results. Operating with a working function $T(r)$ leads to the the moment M_4:

$T(r) = \int_0^r (r-x)^2 \gamma(x)/2dx$. Then, $T'(r) = \int_0^r (r-x) \cdot \gamma(x)dx$, $T''(r) = \int_0^r \gamma(x)dx$, $T'''(r) = \gamma(r)$ and $T^{(4)}(r) = \gamma'(r)$, $T^{(5)}(r) = \gamma''(r)$ follow. Now, the integration $M_4 = \int_0^L r^4 l_p T^{(5)}(r)dr$ yields $M_4 = l_p \cdot (-24T[0] + 24T[L] - 24LT'[L] + 12L^2T''[L] - 4L^3T^{(3)}[L] + L^4T^{(4)}[L])$. By use of the limiting conditions $\{T(0) \to 0, T^{(3)}(L) \to 0, L^4 \cdot T^{(4)}(L) \to 0\}$, $M_4 = l_p \cdot [24T(L) - 24LT'(L) + 12L^2T''(L)]$ results. The substitution of the corresponding integrals yields

$$M_4 = l_p \cdot \int_0^L [12(L-x)^2 - 24L(L-x) + 12L^2] \cdot \gamma(x)dx. \qquad (9.7)$$

Equation (9.7) gives the term M_4 as

$$M_4 = l_p \cdot \int_0^L [12L^2 - 24xL + 12x^2 - 24L^2 + 24Lx + 12L^2] \cdot \gamma(x)dx. \qquad (9.8)$$

Simplification of Eq. (9.8) and insertion of v_c leads to $M_4 = l_p \cdot \int_0^L 12x^2 \cdot \gamma(x)dx = l_p 12 \int_0^L x^2 \cdot \gamma(x)dx = l_p \cdot 12v_c/(4\pi) = 3v_c l_p/\pi$. By doing this, the fourth moment of the function $g(r)$ is given in terms of the characteristic parameters l_p and v_c.[‡]

Thus, single characteristics of the particle can be detected depending on the four independent parameters $M_{1,2,3,4}$. Clearly, for tightly packed particle arrangements, the $g(r)$ moments remain as independent parameters. Their interpretation depends on the particle (volume and shape), on the pair correlation of the particles and on the volume fraction c. All these parameters depend on L. Figures 9.1 and 9.2 describe this approach with special micropowders. Powder samples for scattering experiments can be prepared very simply. In contrast, sample preparation procedures for electron microscopes require more effort and time and sometimes even modify the sample.

3. Chord length analysis of a ceramic micropowder for $L_1 = 18$ nm

From the viewpoint of a scattering experiment, the chord lengths inside and outside the particle are intermixed. Clear and simple exceptional cases exist. The following example shows that, in special cases, *single particle parameters* can be detected directly from $\gamma''(r)$. The information content of the function $g(r, L) \sim \gamma''(r, L)$ is demonstrated in Fig. 9.1 for a realistic two-phase sample with $L_1 = 18$ nm. The mean chord lengths l_1 and m_1 clearly differ [see Eq. (9.1)]. This leads to the following clearly arranged situation: The function γ'' starts with two nearly completely separated positive peaks. The first

---

[‡]By extending the approach, still higher moments can be analyzed. Certainly, specific denotations for the sequence of similar typical integrals do not exist (any more) in the field of SAS. In stochastic geometry these moments are interpreted in connection with another function. This function is denoted by "The Linear Erosion" (see [205] and Chapter 6).

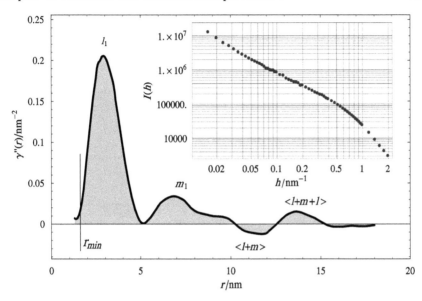

FIGURE 9.1

Chord length analysis of a micropowder via the function $\gamma''(r)$. The resolution limit r_{min} is marked.

The primary particles (4–6) nm are investigated with $L_1 = 18$ nm via the pattern $I(h)$ of an Al_2O_3 powder used for ceramic applications. The spatial arrangement of these particles is characterized by $l_p = 2.1$ nm, $l_1 = (2.5–3)$ nm, $m_1 = (7–8)$ nm, $< l + m >= 11$ nm and $< l + m + l >= 14$ nm. The volume fraction of the primary particles is about $c(L_1) = 30\%$. Further existing secondary particles (in other order ranges L_i) are not inspected. The insert shows the experimental data with the limiting values $h_{max} = 1.9$ nm^{-1}, $h_{min} = 0.011$ nm^{-1}, $r_{min} = 1.6$ nm and $r_{max} = 260$ nm. The sample was prepared for analysis by spraying the powder on a 0.01 mm carrier foil.

corresponds to particle chords l_i, but the second (weaker peak) can be traced back to the particle-to-particle chords m_i. Even the distribution laws of the *chord length sum-terms* [66, 70, 82] (i.e., the sequences $< l_i + m_i >$ and $< l_i + m_i + l_i >$), can be clearly inspected.

The volume fraction results from Rosiwal's relation $c = l_1/(l_1 + m_1)$ (see Chapter 8). Actually, from the first peak, the moments of the CLDD $\varphi(r)$, $l_1 = 3.3$ nm, $l_2 = 10.6$ nm^2 and $l_3 = 36$ nm^3 follow. This leads to $l_c = l_2/l_1 = 3.2$ nm, $f_c = \pi l_3/(3l_1) = 11.3$ nm^2 and $v_c = \pi l_4/(3l_1) = 40$ nm^3 for the (single) primary particles. These values cannot be found directly from the sample CF $\gamma(r)$, where the separation between the particles and intermediate spaces is not so obvious. By interpreting γ'' or g, the functions φ and f can be determined. Hence, a nearly complete separation of both CLDDs is achieved.

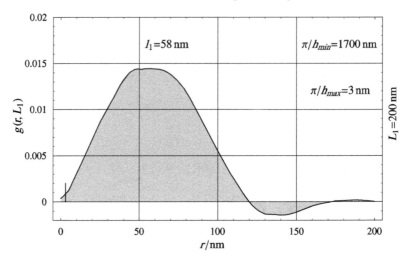

FIGURE 9.2

The function $g(r, L_1)$ of a CuPc micropowder analyzed in Figs. 1.20 and 1.25. Based on the SAS CF, the (initial) analysis yielded the set of characteristic parameters $l_c = 56$ nm, $f_c = 3\,000$ nm^2 and $v_c = 71\,000$ nm^3. The moments M_i belonging to this function $g(r)$ are $M_1 = 55$ nm, $M_2 = 3\,200$ nm^2, $M_3 = 200\,000$ nm^3 and $M_4 = 4 \cdot 10^6$ nm^4. The errors were estimated at about 4% for M_1 and 16% for M_4.

4. Moments and characteristic parameters from the scattering pattern of a copper (I) cyanide/platinum micropowder

Figure 9.2 shows another example of a $g(r, L_1)$ function. It corresponds to the CF already analyzed in Figs. 1.20 and 1.25. If $L_1 = 200$ nm is assumed, $g(0) \to 0$ is obtained. This indicates that very small particles do not exist. This means that L has been fixed correctly. A line marks the lower resolution limit $r = r_{min} = 3$ nm. This L excludes a simultaneous inspection of particles possessing very different diameters.

The minimum at $r_2 = 140$ nm is a consequence of particle/particle interferences. A typical distance between two particle centers is denoted by $r = r_2$. This minimum is not directly interrelated with the chord segments m_i in phase 2. The moment m_1 is much bigger than r_2. The restriction to L_1 excludes information about intermediate chord lengths in the order range of about $L = L_3 = 1\,000$ nm.

The relations between the M_i and the characteristic SAS parameters (see Table 9.1) can be verified. This includes $M_1 = 1/|\gamma'(0)|$ (compare with Fig. 1.20). The two-phase approximation is fulfilled for $L_1 = 200$ nm. This remains true for bigger L, but appears questionable for smaller L.

In connection with Fig. 9.2, it must be emphasized that the normalization of $g(r)$, $\int_0^{r=L_1} g(r, L_1)dr = 1$, depends on $\gamma'(0, L_1)$ *and* on the upper integration limit $r = L_1$. With this strategy, all the parameters and functions estimated depend on L_1. This point can be irritating if an experimenter is using a computer program that does not require an input length $L = L_1$. The user of the computer program should at least know the programming lines that define L.

There are many examples in materials sciences, which demonstrate the strategy of CLDs. The general analysis of CLDs is summarized in a compact paper by Ruland and Burger (2001) [7]. Ruland and Smarsly (2005) [200] gave an overview of the SAS of cylinder arrays. Chord simulations of special particle shapes for applications in polymer chemistry were performed by D. Fanter (1977) [34]. His thesis is entitled *"Modelluntersuchungen zur Auswertung von Roentgenstreukurven mittels eines Fourierverfahrens."*

A wide range of polyethylene samples can be investigated effectively with SAS. For this job, Stribeck and Ruland (1978) introduced the *Interface Distribution Function* (IDF) (see [215]), which is closely connected to $g(r)$ (see also [66]).

There are numerous papers for many types of samples. One example is the morphology of *isomorphous substituted hectorites* (Gille, Koschel & Schwieger, 2002) [83]. The q-parameter is useful for describing and distinguishing *connected and non-connected pores* in microporous and mesoporous materials. In general, this parameter was introduced for a characterization of porous glasses via SAS (Gille, Reichl, Kabisch, Enke & Janowski, 2002) [84]. Essentially, the pore size distribution and the CLD of porous VYCOR glass (PVG) were determined from the SAS data (Gille, Enke & Janowski, 2002) [86] (see also the considerations regarding the *realistic porosity* of porous glasses) (Gille, Enke, Janowski & Hahn, 2003) [94].

Scattering experiments are frequently applied in the (large) field of carbon samples. Activated and non–activated carbon samples (see Braun & Gille (2003) [92]) were investigated applying linear simulation models (LSM) (see Chapter 8).

There are many applications in the field of metal physics. For example, *structure changes and precipitation kinetics in melt spun and aged Al-Li-Cu alloys* were investigated via CLDs (see Dutkievicz et al. (2001) [29]). Such investigations are of importance in the aircraft industry for reaching an optimum strength/mass ratio of the aluminum alloys applied. Scattering patterns of Al-2Li-5Mg-0.1Zr alloys were correlated with measurements of the electrical resistivity of this material (Truong et al., 2002) [227].

Several papers compare the results obtained via SAS with other characterization techniques for two-phase porous silica. For example, Gille, Enke & Janowski (2001) studied order distances in porous glasses via the approach of the *Transformed CF of SAS* [77]. Mesopores inside controlled pore glasses were characterized (Gille, Enke & Janoswki, 2001) [78]. Other papers deal with the *"Structure and Texture Analysis of Colloidal Silica in Porous Glasses"* or with the *"Stereological Macropore Analysis of a Controlled Pore Glass"* (see Enke et al. (2001) [31] and also Gille, Enke and Janowski

(2001) [79]). A summary of SAS models for the analysis of porous materials can be found in a paper (Gille, 2003) [90].

Actual issues regarding application in the field of catalysis have been studied in several papers, such as in *"Application of SAS for the identification of small amounts of platinum supported on porous silica"* (see Gille, Enke, Janowski & Hahn (2003) [93]. Additionally, the influence of support geometry was investigated in *"Platinum dispersion analysis depending on the pore geometry of the support"* (Gille, Enke, Janowski & Hahn, 2004) [97]. These papers apply the theory presented in the following subsections.

5. Scattering pattern $I(h)$ and higher moments of $g(r)$

The moments M_n, where $4 \le n$, are components of a Taylor series of $I(h, L)$.§
From

$$I(h, L) = \frac{4\pi}{v_c \cdot l_p} \int_0^L g(r) \cdot \frac{2 - 2\cos(hr) - hr\sin(hr)}{h^4} \, dr, \qquad (9.9)$$

$$I(h, L) = \frac{4\pi}{l_p \cdot v_c} \left(\frac{M_4}{12} - \frac{h^2 M_6}{180} + \frac{h^4 M_8}{6720} - \frac{h^6 M_{10}}{453600} + \ldots - \ldots \right) \approx 1 - \frac{R_g^2}{3} h^2 + \cdots \quad (9.10)$$

results. Equations (9.9) and (9.10) hold true for isotropic samples and homogeneous particles. Equation (9.10) connects the moments $M_{4,6,8,\ldots}$ with the scattering pattern for a specific L. It is applied in Fig. 9.3 for a Boolean model with $c \approx 0.7$.

For (very) diluted particle ensembles, the quadratic term in Eq. (9.10) can be related to a *radius of gyration*, where $R_g = \sqrt{\pi M_6 / (15 \cdot l_p v_c)}$; see Table 1.1 and Fig. 1.1. However, R_g is not a useful parameter for tightly packed particle ensembles (see dashed line in Fig. 9.3).

More intrinsic properties of the M_n terms can be studied by tracing back Eq. (9.4) to the CLDDs φ and f. For a given particle ensemble, there exist interrelations between the functions g, φ and f. Consequently, the moments of $g(r)$ can be traced back to those of φ and f, l_1, $l_2 \ldots$ and m_1, $m_2 \ldots$ A generalization of Eq. (9.1) for higher moments is achieved with this concept.

In many practical cases (random porous materials, nuclear filters, ceramic micropowders, adsorption coal products), the shape of a pore (particle) does not depend on that of the adjacent pore (particle) [225, 77, 31, 79]. It follows then that the chord lengths along the arbitrary test line Λ (see the initial figure of this chapter) are independent. Therefore, a description via a *renewal process of independent events* [23] is nearly a perfect approximation of reality. This *non-correlation approximation* has frequently been applied for modeling scattering patterns [182, 48, 100, 107, 165]. This approach is the starting point for the formulas in the next section.

---

§The limit $\lim_{h \to 0} [2 - 2\cos(hr) - hr\sin(hr)]/h^4 = r^4/12$ is significant for verification of the normalization $I(0) = 1$.

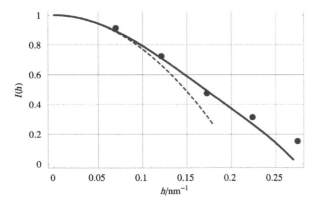

FIGURE 9.3

Four selected points in the initial part of a scattering pattern $I(h, L = 25 \text{ nm})$ of a Boolean model (see Fig. 9.4) are approximated using Eq. (9.10) (solid line).

The values of the parameters l_p, v_c, M_4, M_6 and M_8 are $l_p = 8.25$ nm, $v_c = 3\,520$ nm^3, $M_4 = 27\,700$ nm^4, $M_6 = 9.5 \cdot 10^6$ nm^6 and $M_8 = 3.7 \cdot 10^9$ nm^8. The connection $1 = 4\pi M_4/(12 l_p v_c)$ holds true and is not an approximation (see Table 9.1).

The behavior of the additionally inserted dashed line results from the *constant term* plus the *quadratic term* $R_g = R_g(M_6, l_p, v_c) = 8.3$ nm in Eq. (9.10). The *model of the radius of gyration R_g is not useful for the interpretation of such a scattering pattern.*

9.2.1   Connection between the three functions g, φ and f

Length independent segments l and m are connected via a sequence of convolution operations (symbol $*$), where $g(r) = l_p \cdot \gamma''(r) + 2\delta(r) = [\varphi(r) + f(r) - 2 \cdot \varphi(r) * f(r)] * [\delta(r) + \varphi(r) * f(r) + \varphi^{2*}(r) * f^{2*}(r) + \varphi^{3*}(r) * f^{3*}(r) + \ldots + \ldots]$. After introducing the characteristic functions $p(t)$, $q(t)$ (see [23]), $p(t) = \int_0^{L_0} \varphi(r) \exp(itr) dr$, $q(t) = \int_0^L f(r) \exp(itr) dr$ and the function $Q(t) = \int_0^L g(r) \exp(itr) dr$, from the convolution theorem results

$$Q(t) = \frac{p(t) + q(t) - 2p(t)q(t)}{1 - p(t)q(t)}, \quad Q(t) - 2 = \frac{p(t) + q(t) - 2}{1 - p(t)q(t)}. \tag{9.11}$$

The normalization of the functions $\varphi(r)$, $f(r)$ and g involves $p(0) = q(0) = Q(0) = 1$. Equation (9.11)¶ has been applied to determine the function $f(r)$ if both $\varphi(r)$ and $g(r)$ are known [100]. In addition, knowledge of $f(r)$ and

---

¶Equation (9.11) has been known for a long time; however, from time to time, it is rediscovered in different fields.

$\varphi(r)$ for selected special cases allows the functions $g(r)$ and $I(h)$ [165, 107] to be determined in general and for specific models.

Based on these actual connections, $M_{1,2,3}$ can be fixed without introducing special function types for g, φ or f. The moments in question will be derived in terms of derivatives of characteristic functions in the origin.

9.2.2 The moments M_i, l_i, m_i in terms of $Q(t)$, $p(t)$, $q(t)$

The functions Q, p and q involve information about the moments of g, φ and f. The nth moments l_n and m_n of the random variables l and m (for the first moments $\bar{l} = l_1$ and $\bar{m} = m_1$ is also common) result as depending on the derivatives $p^{(n)}(0)$ and $q^{(n)}(0)$: $l_n = \int_0^{L_0} r^n \varphi(r) dr = \lim\limits_{t \to 0+} \frac{p^{(n)}(t)}{i^n}$ and $m_n = \int_0^{L} r^n f(r) dr = \lim\limits_{t \to 0+} \frac{q^{(n)}(t)}{i^n}$. Thus, for $n = 1, 2, 3, \ldots$, $\{\, p'(0) = i \cdot l_1,\ q'(0) = i \cdot m_1\,\}$, $\{p''(0) = -l_2,\ q''(0) = -m_2\,\}$, $\{\, p^{(3)}(0) = -i \cdot l_3,\ q^{(3)}(0) = -i \cdot m_3\,\}$ ... is an effective approach for determining the moments l_n and m_n.

Similarly, the $g(r)$ moments can be obtained as depending on the derivatives of $Q(t)$ in the origin. In summary, if the nth moment M_n exists, this term is then defined by Eq. (9.4) and follows from the corresponding derivative $M_0 = Q(0)$ or $M_1 = Q'(0)/i$ or $M_2 = -Q''(0)$ or $M_3 = -Q^{(3)}(0)/i$, $\ldots$

The derivatives of Eq. (9.11) with respect to t, $Q^{(n)}(t)$ and $t \to 0$, lead to indeterminate expressions such as $[0/0]$, where $n = 0, 1, 2\ldots$ By use of the Bernoulli-L'Hospital rule,

$$M_0 = Q(0) = \lim_{t \to 0} \frac{p(t) + q(t) - 2 \cdot p(t) \cdot q(t)}{1 - p(t) \cdot q(t)} = \lim_{t \to 0} \frac{p'(t) + q'(t)}{p'(t) + q'(t)} = \frac{p'(0) + q'(0)}{p'(0) + q'(0)} = 1 \tag{9.12}$$

and

$$M_1 = \frac{Q'(0)}{i} = \frac{1}{i} \cdot \lim_{t \to 0} \frac{\partial}{\partial t}\left(\frac{p(t) + q(t) - 2 \cdot p(t) \cdot q(t)}{1 - p(t) \cdot q(t)} \right) \tag{9.13}$$

$$= \frac{1}{i} \cdot \lim_{t \to 0} \frac{p'(t) \cdot q'(t)}{p'(t) + q'(t)} = \frac{il_1 \cdot im_1}{il_1 + im_1} = \frac{l_1 \cdot m_1}{l_1 + m_1} = l_p$$

result. Equation (9.12) is trivial. Equation (9.13) is in agreement with Eq. (9.1). The next subsections investigate M_2 and M_3.

9.2.3 Analysis of the second moment $M_2 = -Q''(0)$

From the term

$$Q''(0) = \lim_{t \to 0} Q''(t) = \lim_{t \to 0} \frac{\partial^2}{\partial t^2}\left(\frac{p(t) + q(t) - 2}{1 - p(t)q(t)} \right), \tag{9.14}$$

the second moment is determined in Section 9.2.3.1. Equation (9.14) gives

$$\frac{Q''(0)}{Q'(0)^2} = \frac{p''(0)}{p'(0)^2} + \frac{q''(0)}{q'(0)^2} - 2. \tag{9.15}$$

The transformation of Eq. (9.15) into real space followed by splitting a factor $l_p{}^2 \equiv M_1{}^2$ yields

$$M_2 = l_p{}^2 \cdot \left(\frac{l_2}{l_1{}^2} + \frac{m_2}{m_1{}^2} - 2 \right) = l_p{}^2 \cdot \left(\frac{l_2 - l_1{}^2}{l_1{}^2} + \frac{m_2 - m_1{}^2}{m_1{}^2} \right). \tag{9.16}$$

Furthermore, by using central moments [the central moment $\mu_{2m} = m_2 - m_1{}^2$ of the distribution density $f(m)$ and $\mu_{2l} = l_2 - l_1{}^2$ of $\varphi(l)$], a more compact representation follows from Eq. (9.16):

$$M_2 = l_p{}^2 \cdot \left(\frac{\mu_{2l}}{l_1{}^2} + \frac{\mu_{2m}}{m_1{}^2} \right) = (1 - c)^2 \cdot \mu_{2l} + c^2 \cdot \mu_{2m}. \tag{9.17}$$

Accordingly, the second moment of $g(r)$ equals the sum of the (second) central moments averaged by the squares of the volume fraction of both phases.

A certain check of Eqs. (9.16) and (9.17) is possible by considering the limiting case of an infinitely diluted particle ensemble. Regardless of the particle size, $m_{1,2} \to \infty$ holds for the moments. Obviously, $M_2 = l_2$ and $M_1 = l_p = l_1$ are fulfilled for the single particle. Hence, it can be concluded that for $c \to 0$, the ratio $m_2/m_1{}^2$ possesses the property $m_2/m_1{}^2 \to 2$. This is a fundamental property of the CLDD of the connected phase. It is fulfilled in the well-investigated cases of Bms (see Chapter 5) and is also discussed in the context of a DLm case (see Section 9.2.3.1).

Otherwise, the limiting case of fitting together *puzzle fragment particles* (see Chapter 7) (e.g., those originating from a random tessellation, see Chapter 6) involves $c = 1$ as a possible case. Evidently, here $f(m) \to \delta(m)$ and $m_{1,2} \to 0$ require *special particle shapes*. Certainly, more intrinsic assumptions about the function $\varphi(l)$ must be fulfilled. First, Eq. (9.17) must remain true. In addition, a special equation for the investigation of c based on the analysis of M_2 is described (Sections 9.2.3.1 and 9.2.3.2).

9.2.3.1 Diluted and tightly packed particle ensembles

The results obtained in Section 9.2.2 involve the limiting case of an infinitely diluted particle ensemble. For a fixed particle size $l_1 > 0$, the relation $m_1 \to \infty$ is required in order to fulfill $c \to 0$. Obviously, $M_1 \to l_1$, $M_2 \to l_2$ and $M_3 \to l_3$ result. In order to fulfill this (in addition to the triviality $m_1/m_1 \to 1$), the additional relationships $m_2/m_1{}^2 \to 2$, $m_3/m_1{}^3 \to 6$ can be deduced. Such an additional set of relationships between moments that are not typical of a distribution density $f(x)$ in mathematical statistics must be fulfilled here.

Example 1: In order to show that these relationships are fulfilled, hard spheres of constant diameter d, arranged in space with a special pair correlation resulting from a DLm [80] will be considered. For this, $c = 1/8$ represents a sufficiently small volume fraction (see the function $\gamma''(r, d)$ of a DLm) [100]. Based on the characteristic function $p(t, d)$ of the CLDD of a single sphere and γ'' and by use of Eq. (9.11), the moments $m_{1,2,3}$, $m_1 = 14d/3$, $m_2 = 42.2823 \cdot d$

and $m_3 = 574.554 \cdot d$ result from $q(t) = [Q(t) - p(t)]/[1 - 2p(t) + p(r)Q(t)]$. Inserting these terms (of course, the terms for c are somewhat greater than 0) into the moment ratios in question gives $m_2/m_1^2 = 1.9415\ldots \approx 2$ and $m_3/m_1^3 = 5.6534\ldots \approx 6$. This confirms the $f(m)$ moment relation.

Example 2: The additional relationships are fulfilled for Boolean models [210, 156, 3] (see Chapter 5). Phase 2 (the anti-grain phase) is described by the CLDD $f(m) = 1/m_1 \cdot \exp(-m/m_1)$, where $0 \le m < \infty$. This CLDD possesses the moments $m_0 = 1$, m_1, $m_2 = 2m_1^2$ and $m_3 = 6 \cdot m_1^3$.

In contrast, a general analysis of a *"completely filled case"* (e.g., $m_1 \equiv 0$) or the *"maximally filled case"* (e.g., $m_1 \to m_{1min}$) requires more extensive investigations. In order to do this, a detailed shape analysis of the particles is indispensable. A situation $m_1 \equiv 0$ can be reached only in selected special cases, such as for the *puzzle-interlayer particles* model [107] (see the PIm in Chapter 7). For simply inserted elementary particle shapes, ellipsoids, cones, cylinders and tetrahedrons, a maximum limiting volume fraction c_{max} for the isotropic particle arrangement (corresponding to a certain lower limit of the moment m_1, m_{1min}) exists. Such c_{max} values are surprisingly small. A forced increase of c_{max} (i.e., the wish to fix c_{max} at a relatively large value) leads to ill-posed problems. For more detailed investigations, very special assumptions about the particle shape are indispensable [225].

9.2.3.2 The volume fraction and the first two moments of $f(m)$

The combination of Eq. (9.13) and Eq. (9.17) involves information about c. Evidently, c cannot be derived only from information about the size and shape of the non-connected particle phase. However, c is defined by the size and shape of the connected phase, which is reflected in the behavior of $f(m)$. In this regard, f essentially contains more information about the particle ensemble than φ does.

Elimination of the parameter l_1 from $c = l_1/(l_1 + m_1)$ and $l_p = l_1 m_1/(l_1 + m_1)$ yields $c = l_p/m_1; (m_1 \neq 0)$. Additionally, taking into account Eq. (9.16) [i.e., $M_2 = l_p^2 \cdot (l_2/l_1^2 + m_2/m_1^2 - 2)$], elimination of the parameters l_p and l_1 yields an intermediate relation

$$-2cl_2 + c^2(l_2 - 2m_1^2 + m_2) = -l_2 + M_2. \tag{9.18}$$

Equation (9.18) again requires the limiting assumption $m_1 \neq 0$. This is not surprising since it reflects the trivial fact that arbitrarily shaped particles do not fit together (see the considerations at the end of Section 9.2.3.1). A non-trivial connected phase requires $c < 1$. Eq. (9.18) gives

$$M_2 = (1-c)^2 l_2 + c^2(m_2 - 2m_1^2) = (1-c)^2 \cdot l_2 + c^2 \cdot m_2 - 2l_p^2. \tag{9.19}$$

As expected for $c \to 0$, Eq. (9.19) leads to $M_2 \to l_2$. The limiting case $c = 1$ can be excluded. The particle volume fraction c is defined by Eq. (9.19) as depending on the moment parameters involved. This can be verified for a hard particle DLm [100] [see Section 4.4, Eq. (4.27), Fig. 4.8] where Eq. (9.19) is written

$$1.04347277615000702 \cdot d^2 = (1 - c)^2 d^2 / 2 + c^2 \cdot 42.282257673600 \cdot d^2. \quad (9.20)$$

The positive solution $c_1 = 0.125$ is the correct volume fraction with this model. Actually, the second (negative) solution c_2 of Eq. (9.20) is meaningless.

It is advantageous to apply Eq. (9.19) for estimating very small c values from second moment CLD data (e.g., from the occultation data of cosmic objects in astrophysics). Here, M_2 is a bit smaller than l_2 and the term c^2 can be neglected. Thus, according to the moment relations for small c, both factors of the product $c^2 \cdot (m_2 - 2m_1^2)$ disappear. Consequently, Eq. (9.19) gives an approximation of $c = 1 - \sqrt{M_2/l_2} \approx (1 - M_2/l_2)/2$ as $c \to 0$. In a certain way, this is the "extrapolation" of $c \equiv 1 - M_1/l_1$.

Analysis of the third moment $M_3 = -Q^{(3)}(0)/i$

The term M_3 can be traced back to $Q^{(3)}(0)$. By analogy with M_2 and M_3 follows an analysis of the moment equation $Q^{(3)}(0) = \lim_{t \to 0} Q^{(3)}(t)$. An analytic expression for M_3 is

$$M_3 = l_p{}^3 \cdot \left(6 - \frac{6l_2}{l_1{}^2} + \frac{l_3}{l_1{}^3} - \frac{3l_2{}^2}{2l_1{}^3 m_1} + \frac{l_3}{l_1{}^2 m_1} - \frac{6m_2}{m_1{}^2} + \frac{3l_2 m_2}{l_1{}^2 m_1{}^2} - \frac{3m_2{}^2}{2l_1 m_1{}^3} + \frac{m_3}{m_1{}^3} + \frac{m_3}{l_1 m_1{}^2} \right).$$
$$(9.21)$$

The large number of terms in Eq. (9.21) complicates the general discussion. The additional introduction of third central moments of the distribution densities $f(m)$ and $\varphi(l)$ [expressed by $\mu_{3m} = m_3 - 3m_1 m_2 + 2m_1{}^3$ and $\mu_{3l} = l_3 - 3l_1 l_2 + 2l_1{}^3$] leads to a more symmetric representation:

$$M_3 = M_1{}^2 \cdot \left(\frac{\mu_{3l}}{l_1{}^2} + \frac{\mu_{3m}}{m_1{}^2} - \frac{l_1 + m_1}{2} \right) - \frac{3M_1{}^3 \cdot (m_1 \mu_{2l} - l_1 \mu_{2m})^2}{2l_1{}^3 m_1{}^3}. \quad (9.22)$$

In the limiting case $c \to 0$, $m_3/m_1{}^3 \to 6$ can be concluded from the analysis of Eq. (9.22). More details are included in Section 9.2.3.3, in which the question of obtaining a recurrence formula is also discussed.

9.2.3.3   General relations between all moments

Based on Eqs. (9.12), (9.13), (9.17) and (9.22), several properties of the moments can be derived. The moments M_n, where $n = 0, 1, 2$, fulfill a general law of representation $M_n = (1 - c)^n \cdot l_n + c^n \cdot m_n - n! \cdot M_1{}^n$. However, the extrapolation to $n = 3, 4, 5 \ldots$ is wrong.

Nevertheless, for the nth derivatives of M_n with respect to the moments l_n and m_n, the connections

$$\frac{\partial M_n}{\partial l_n} = \frac{m_n{}^2}{(l_n + m_n)^2} = (1 - c)^2 \quad and \quad \frac{\partial M_n}{\partial m_n} = \frac{l_n{}^2}{(l_n + m_n)^2} = c^2, \ n = 1, 2, 3, \ldots$$
$$(9.23)$$

hold. Furthermore, if the distances between the particles increase continuously (i.e., for $m_1 \to \infty$ and $c \to 0$), the limiting relation $\lim (c^n \cdot m_n) = n! \cdot l_1{}^n$, $n = 1, 2, 3, \ldots$ holds true.

An elaborate representation of the third moment M_3 which is nearly analogous to Eq. (9.17) is

$$M_3 = (1 - c)^3 \cdot l_3 + c^3 \cdot m_3 - 6M_1{}^3 +$$

$$M_1{}^3 \cdot \left[3\left(2 - \frac{l_2}{l_1{}^2} \right)\left(2 - \frac{m_2}{m_1{}^2} \right) + \frac{l_3 - 3l_2{}^2/(2l_1)}{m_1 l_1{}^2} + \frac{m_3 - 3m_2{}^2/(2m_1)}{l_1 m_1{}^2} \right].$$

$$(9.24)$$

Compared with Eq. (9.17), Eq. (9.24) involves an extensive additional term (enclosed in brackets) which is multiplied by the factor $M_1{}^3$. In the limiting case of an infinitely diluted particle ensemble [$l_n = const.$, but $c \to 0$ and $m_1 \to \infty$], the terms inside the brackets disappear. As expected, $M_3 = l_3$ results.

Figure 9.4 shows an application in materials science. Parameter estimations from the scattering pattern of a platinum carbon catalyst sample are explained and illustrated. Following are details of parameter estimations from a scattering pattern $I(h)$, $0.07 \text{ nm}^{-1} < h < 2.5 \text{ nm}^{-1}$: Starting with the micrograph (transmission electron microscope) of the silica sample, an order range $L = 25$ nm was estimated and inserted into the data evaluation. The size distribution density of the (overlapping) primary grains, expressed as the function $f(d)$, is equally distributed on the interval $20 \text{ nm} < d < 24 \text{ nm}$. Hence, the mean grain diameter is about 22 nm. The mean volume of the primary grains equals about $5\,600 \text{ nm}^3$.

The volume fraction of the grain phase results from $\Phi(r) = \gamma''(r)/[\gamma'(r)]^2$, $c = \Phi(0+) \approx 0.73$ (see Eq. (8.26), Fig. 5.2 and [53]). The subfigure to $I(h)$ shows the extrapolation of the function $\Phi(r)$ to the interval $0 < r < 2$ nm. From the SAS CF $\gamma(r, L = 25 \text{ nm})$, the parameters $l_p = 8 \text{ nm}$, $l_c = 13 \text{ nm}$, $f_c = 205 \text{ nm}^2$, $v_c = 7\,000 \text{ nm}^3$ result. A computer simulation based on these parameters is similar to the micrograph.

9.3   CLD concept and data evaluation: Some conclusions

CLDs contain information about the second-order characteristics of statistically isotropic two-phase systems. Equations (9.16) and (9.17) define the moments M_2 and M_3 of the (intersect distribution) function $g(r)$ in terms of the (central) moments of the functions φ and f.

Statistically isotropic ensembles of hard, homogeneous particles (fixed volume fraction c) embedded in a homogeneous matrix can be analyzed by the following two chord length distribution densities: the distribution density $\varphi(l)$

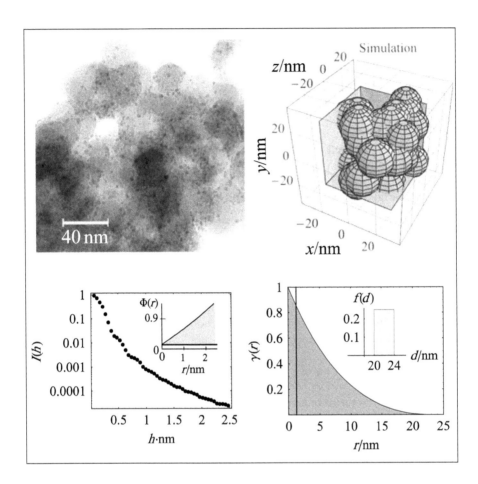

FIGURE 9.4

Analysis of a carbon catalyst sample made up of 10% platinum via a Boolean model of $\approx 70\%$ volume fraction.

A weak size distribution of the grains follows from the micrograph. Based on Eq. 5.2, these grains were approximated by spheres of varying diameter (mean diameter $d_0 \approx 22$ nm).

Significant moments of the CLDDs of the grain phase and the matrix phase are $l_1 = 30$ nm, $l_2 = 1\,400$ nm^2, $m_1 = 11$ nm and $m_2 \approx 250$nm^2, respectively. From this, $c = l_1/(l_1 + m_1) = l_1{}^2 l_c/(m_1 l_2) = 0.73$ results.

A narrow grain size distribution density $f(d)$ belongs to the model applied. Nevertheless, the weakest and most disputed point in the data evaluation of this scattering pattern is the approximation of the primary grain shape by the sphere model.

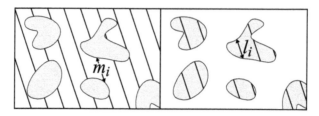

FIGURE 9.5

Chord lengths (m_i) outside the connected phase (left) exhibit much more entropy than the chords (l_i) inside (see right). In a way, the outside chords m_i are similar to particle pair correlation distances. This fact illustrates the huge information content contained in a pair correlation function [see Eq. (4.1), Section 4.4, Fig. 4.9 and Fig. 4.26].

In summary, the world outside cannot be realized via chords l_i inside a particle. By restricting to l_i only, no information about the volume fraction or particle-to-particle distances can be derived.

of the random chord length l of the non-connected (particle) phase 1 and the distribution density $f(m)$ of the random chord length m of the connected (matrix) phase 2. An unexpected but basic conclusion is explained in Fig. 9.5 (see also Fig. 4.26).

The moments l_i and m_i of φ and f, respectively, are related to the moments M_i of the function $g(r) = \gamma''(r)/|\gamma'(0)| + 2\delta(r)$. The formula for the second moment of $g(r)$ in terms of the central moments μ_{2l} and μ_{2m} of $\varphi(l)$ and $f(m)$ is simply $M_2 = (1-c)^2 \cdot \mu_{2l} + c^2 \cdot \mu_{2m}$. This is an extension of Eq. (9.1), i.e., $1/M_1 = 1/l_p = |\gamma'(0)| = 1/l_1 + 1/m_1$, which has been known for a long time (Porod, 1951) [191]. The moments in question are closely connected with the isotropic scattering pattern of the particle ensemble.

The first four terms of the Taylor series of the characteristic function $Q(t)$ of g are analyzed in the next subsection [see Eq. (9.25) and Fig. 9.6]. The parameter set of the moments analyzed is not independent of other SAS parameters. Maximum numbers of independent parameters as depending on c are discussed at the end of Chapter 9.

9.3.1   Taylor series of $Q(t)$ in terms of the moments M_n of the function $g(r)$

Based on the *non-correlation assumption*, characteristic functions have been applied for investigating the connections between the nth moment M_n of $g(r)$ with the first n moments of the functions φ and f [see the connections Eq. (9.23)]. Moreover, there is a general connection between the $g(r)$ moments and the characteristic parameters of scattering experiments, which is that the density autocorrelation function $\gamma(r, L)$ defined by $I(h)$ fixes the

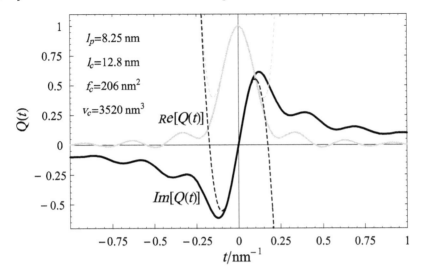

FIGURE 9.6
The characteristic function $Q(t)$ (see two solid lines) for the Bm investigated in Fig. 9.4 is compared with the series Eq. 9.25 (see two dashed lines).
Both real and imaginary parts of $Q(t)$ are plotted. In the interval $-0.1 < t \cdot \text{nm} < 0.1$, the approximation is perfect, i.e., here the deviations from the exact characteristic function $Q(t)$ can be neglected.

characteristic function $Q(t)$ in

$$Q(t) = 1 + il_p \cdot t - \frac{l_c l_p}{2} \cdot t^2 - i\frac{l_p f_c}{2\pi} \cdot t^3 + \frac{l_p v_c}{8\pi} \cdot t^4 \pm \dots \qquad (9.25)$$

(see the example given in Fig. 9.6). Apart from the *non-correlation approximation*, the terms of this series reflect the connection between $Q^{(n)}(0)$ and the characteristic parameters. Equation (9.25) also holds true in the one- and two-dimensional case and for plane sections of the three-dimensional case (e.g., for the faces of the test cube in the initial figure of this chapter).

The right-hand side of the final Eqs. (9.17) and (9.22) involve very different parameters. The central moments allow simplified expressions.

The advantage of these equations is not limited to the CLD approach with regard to scattering methods (e.g., it can be used in crystallography); in addition, it can be applied in occultation experiments in the field of astronomy [Eq. (9.19) in Section 9.2.3.2].

9.3.2 Sampling theorem, the number of independent SAS parameters, CLD moments and volume fraction

The results can be interpreted in the context of the *sampling theorem of the information theory* for scattering experiments. Each scattering curve $I(h)$ is recorded on a limited h interval $h_{min} < h < h_{max}$. The better the experimental equipment is, the more the lower limit approaches zero ($h_{min} \to 0$). For h_{max}, an exact upper limit cannot be given. The particle model fixes the upper limit up to $h_{max} \approx (1.5\text{--}2)$ nm^{-1}. According to the sampling theorem, the number of independent parameters N that can be obtained at all depends on the maximum diameter of the largest particle L_0 (Damaschun et al., 1971) [26]. Based on the *sampling point distance* $\Delta h = \pi/L_0$, an experimental curve in a typical case (say $L_0 = 10$ nm for SAS experiments) defines $N \leq (h_{max} - h_{min})/\Delta h = L_0 \cdot (h_{max} - h_{min})/\pi$ parameters.

For a case $L_0 = 10$ nm and $(h_{max} - h_{min}) = 2$ nm^{-1}, about $N < 8$ *independent parameters* can be detected. This statement, which clearly applies to single particles, also holds true for a whole ensemble of tightly packed particles with the order range L, where $L_0 < L$. Regardless of the experimental conditions or the actual problem, a limited sequence of parameters can be obtained.

A systematic order for the sequence of parameters should exist. Which conclusions can be drawn with respect to the CLD moments? As shown, the volume fraction, characteristic parameters and CLD moments are not independent parameters. These parameters are (partly) interrelated. A parameter space of N independent parameters does not increase with a formal addition of "new" parameters. For example, c, l_p, l_c, f_c, v_c and the second moments l_2, m_2 (or the central moments μ_{l2} and μ_{m2}) represent *an independent parameter space*. Obviously, the first moments l_1, m_1 are not independent parameters since they depend on c and l_p. On the other hand, the $3 + 3 = 6$ moments $l_{1,2,3}$ and $m_{1,2,3}$ of the distribution densities $\varphi(l)$ and $f(m)$ would represent six independent parameters. Furthermore, other parameters like the volume fraction c, the radius of gyration R_g (Guinier & Fournét, 1955) [143] and the moments $M_{1,2,3}$ depend on the moments l_n and m_n.

In the limiting case of an infinitely diluted particle collection (i.e., for single particles), three of these parameters namely $m_{1,2,3}$, do not exist at all. The remaining three parameters $l_{1,2,3}$ describe the single particle. Consequently, for $c \to 0$ *additional connections* must appear between the moments of the CLDD of the connected phase. Hence, the existence of additional limiting relations like $m_2/m_1^2 \to 2$, which are really surprising at first, is not a weak point of the approach presented; it is actually in agreement with the sampling theorem.

10

Exercises on problems of particle characterization: examples

This chapter deals with some of questions that frequently arise in practice. These problems are arranged in four subsections. The solutions are explained briefly. Corresponding plots are reduced to the bare minimum. In order to understand the solutions, it is useful to study the corresponding chapters.

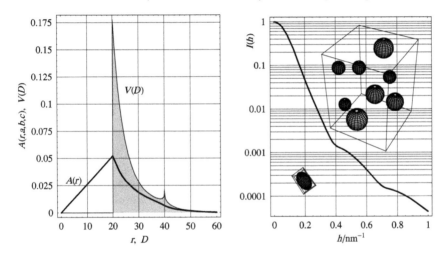

Illustration of the scattering equivalence between differently sized spheres and ellipsoids of unique size (see Section 10.2.6). Both particle ensembles are supposed to be infinitely diluted.

An ellipsoid (IUR orientation) of semiaxes $a = 10$ nm, $b = 20$ nm and $c = 30$ nm possesses the CLDD $A(r)$ (left: bold solid line). This CLDD corresponds to the scattering pattern $I(h)$ (right). The same curve $I(h)$ results from an ensemble of differently sized spheres with random diameters D if these diameters possess the distribution density $V(D)$, as shown by the filled plot (left). The first moment of $V(D)$ is $D_1 = 27$ nm. Clearly, both systems involve the same order range $L = 60$ nm, i.e., the maximum diameter $2c = 60$ nm of the ellipsoid coincides with the diameter of the largest sphere. This model case had already been considered by the great Porod [191] in 1950 (see the exercise in Section 10.2.6).

10.1   The phase difference in a point of observation P

This addition involves useful details concerning elementary waves that interfere at a point of observation P (see the initial figure of Chapter 1).

Let $\Delta\varphi(P)$ be the phase difference between two waves at a point of observation P. In SAS, it is assumed that all distances between the sample and
P are much bigger than all the distances r between the scattering centers
inside the interradiated volume V_i of the sample. According to the derivation made in Chapter 1, Eq. (1.6) is mainly based on the approximation
$\Delta\varphi(P) \approx (2\pi/\lambda_0) \cdot [\mathbf{r_A} \cdot (\mathbf{s_0} - \mathbf{s})] = -(2\pi/\lambda_0)\,\mathbf{r_A} \cdot \mathbf{h}$. What is the reason for
this simple interference condition?

The basic task is to illustrate these questions by use of a simple figure, which
involves the x, y, z sample coordinate system with origin $\mathbf{0}$, the unit vectors
$\mathbf{s_0}$ and $\mathbf{s}$, a sample point A (position vector $\mathbf{r_A}$) and the point of observation
P (position vector $\mathbf{r_0}$).

Is it possible that both paths for reaching point P (i.e., the direct $path_1$
and the second $path_2$ which involves point A) agree and consequently require
the same time for a wave propagation?

Answer and conclusions

The vector stretching from sample point A to the point of observation P is
denoted by $\mathbf{r_p}$. At time $t = 0$, the plane wave of phase velocity v_p starts in
the origin $\mathbf{0}$ and propagates through the sample in direction y. The wave has
not yet reached point A or point P at $t = 0$. Two path lengths, which are
denoted as *path length*$_1$ for the direct path and *path length*$_2$ for the longer,
indirect path, are of importance for detecting $\Delta\varphi(P) = (2\pi/\lambda_0)[$*path length*$_2-$
path length$_1]$ in P (see Fig. 10.1).

Operating with the approximation $\lambda_0 = const.$ (inside and outside the
sample) results in *path length*$_2 = \mathbf{r_A} \cdot \mathbf{s_0} + r_p$. Furthermore, *path length*$_1 = r_0$.
The exact phase difference is $\Delta\varphi(P) = 2\pi/(\lambda_0)[\mathbf{r_A} \cdot \mathbf{s_0} + r_p - r_0]$. By use of
the cosine theorem and operating with the scalar product $\mathbf{r_A} \cdot \mathbf{s}$, the length
difference $r_p - r_0$ can be approximated by $r_p \approx r_0 - \mathbf{r_A} \cdot \mathbf{s}$ [see Eqs. (10.1) and
(10.2)]:

$$r_P{}^2 = r_0{}^2 + r_A{}^2 - 2r_0 r_A \cos(\alpha), \tag{10.1}$$

$$\frac{r_p}{r_0} = \sqrt{1 + \frac{r_A{}^2}{r_0{}^2} - \frac{2r_A}{r_0}\cos(\alpha)} \approx 1 - \frac{r_A}{r_0}\cos(\alpha). \tag{10.2}$$

Equation (10.2) results in $r_p \approx r_0 - r_A \cdot \cos(\alpha) = r_0 - \mathbf{r_A} \cdot \mathbf{s}$.

Hence, in a first, but very excellent approximation, $\Delta\varphi(P)$ can be entirely
expressed by the vectors $\mathbf{s_0}$, $\mathbf{s}$ and by the lengths r_A and λ_0. This result
essentially simplifies the theory of data evaluation in SAS. The important
assumption of this approximation is the order relation $r_A \ll r_0$, which is

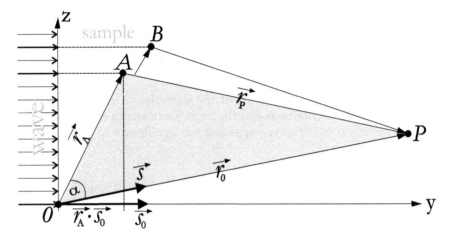

FIGURE 10.1
Analysis of the path difference $\Delta\varphi$ at a point of observation P for the scattering of a plane wave at two or more points.
The wave of wavelength λ_0 propagates in the y direction. Points A, B, ... belong to the irradiated volume of the thin sample. It is placed perpendicular to the y, z-plane, i.e., parallel to the x, z-plane. The vectors $\mathbf{r}_P$ and $\mathbf{r}_0$ are nearly parallel under the experimental conditions. The thickness of the sample is very small compared with all those distances stretching between the point of observation P to points A, B, ... This is the reason for the resulting interference condition $\Delta\varphi(P) \approx -(2\pi/\lambda_0)\,\mathbf{r_A} \cdot \mathbf{h}$, where $\mathbf{h} = \mathbf{s} - \mathbf{s_0}$.

excellently fulfilled in typical experimental cases. Typically, $r_A \approx (3...3\,000)$ nm and the distances r_0 and r_P are in an order of magnitude ranging from centimeters to meters. There exist physical devices in the field of neutron scattering which operate with $r_0 > 20$ m.

Conclusions: The time difference $\Delta t(P) = t_2 - t_1$ of the propagation of the wave along $path_2$ and $path_1$ results from $\Delta t(P) = [\mathbf{r_A} \cdot (\mathbf{s_0} - \mathbf{s})]/v_p = \Delta\varphi(P)\lambda_0/(2\pi v_p)$. There exist many positions of point P where $t_2 = t_1$, i.e., $\Delta\varphi(P) = 0$. Such positions of P are fixed by the condition $\mathbf{h} \cdot \mathbf{r_A} = 0$. In this case, the scattering vector $\mathbf{h}$ is perpendicular to the position vector $\mathbf{r_A}$.

These results do not restrict to the one sample point A, but rather hold true for any irradiated sample point, especially for a point B (see Fig. 10.1 and the first figure of Chapter 1).

In many cases, it is advantageous to characterize a scattering pattern $I(\theta)$ by $I(h \cdot R) = I((4\pi/\lambda_0)\sin(\theta/2) \cdot R)$. Here, the length R is a typical length parameter of the particle ensemble. By doing this, the argument value $v = hR$ of $I(v)$ is dimensionless. However, there are exceptional cases where such a generalized representation of the scattering pattern is not possible.

10.2   Scattering pattern, CF and CLDD of single particles

This section deals with questions about the scattering patterns of single particles, particle size distributions and the mean chord length of a particle. The exercises include considerations regarding the significance of the correlation function.

10.2.1   Determination of particle size distributions for a fixed known particle shape

Interrelating the SAS CF of single particles $\gamma_0(r)$ with those of diluted particle ensembles $\gamma(r)$, the particle size distribution density of a random size parameter can be detected. Particle size distributions of diluted ensembles of equally shaped particles are simple to handle. How can such a size distribution be determined?

Let $V(x)$ be the distribution density of a random size variable x, describing the polydispersity of a certain particle type. How can a random sphere diameter $x = D$ and the random height of a lamella $x = H$ be investigated? How can the normalized functions $V(D)$ and $V(H)$ be determined in terms of the sample CF $\gamma(r)$? How can a term $\left(r^3 \cdot \gamma''(r)\right)'$ be traced back directly to the scattering intensity $I(h)$?

Answer and conclusions

Let $\gamma_0(r, D)$ be the CF of the spheres. Operating in an r-interval $0 \leq r < L$, the sample CF results by averaging $\gamma_0(r, D)$ with $V(D)$ as follows:

$$\gamma(r) = \frac{\int_{D=r}^{L} \frac{\pi}{6} D^3 \cdot \gamma_0(r, D) \cdot V(D) dD}{\int_0^L \frac{\pi}{6} D^3 \cdot V(D) dD}, \quad \overline{V} = \frac{\pi}{6} \cdot M_3 = \frac{\pi}{6} \cdot \int_0^L D^3 \cdot V(D) dD.$$

(10.3)

Aside from the many existing special types of (correctly normalized) distribution densities $V(D)$, the explicit result for $0 \leq D \leq L$ is

$$V(D) = \frac{-2\overline{V}}{\pi} \cdot \left(\frac{\gamma''(D)}{D}\right)', \quad \overline{V} = \frac{\pi}{2 \cdot \gamma'''(0)}.$$

(10.4)

A similar approach is possible for a lamella of random thickness H described by $V(H)$. Let $\bar{l}$ be the mean average chord length of all lamellas $\bar{l} = -1/\gamma'(0)$. It is simpler to average over the CLDD of the lamellas, i.e., over the term $2H^2/r^3$, rather than over the CF. This gives the pair of equations

$$\bar{l} \cdot \gamma(r) = \int_{H=0}^{H=r} \frac{2H^2}{r^3} \cdot V(H) dH, \quad V(H) = \frac{(H^3 \cdot \gamma''(H))'}{2 \cdot |\gamma'(0)| \cdot H^2}.$$

(10.5)

The solutions $V(D)$ and $V(H)$ fulfill $\int_0^L V(x)dx = 1$.

Clearly, both these results [see Eqs. (10.4) and (10.5)] are formulated for zero volume fraction, i.e., $c \equiv 0$. Nevertheless, in some special cases, $V(D)$ and $V(H)$ are useful for tightly packed particle ensembles as well. In this case, these functions are composed of a sum of distribution densities (possessing negative parts) and are referred to as an *equivalent diameter distribution (density)* and *interface distribution (density) function*. Surprisingly, the term $(r^3 \cdot \gamma''(r))'$ can be easily traced back to the pattern $I(h)$ with

$$\frac{(r^3 \cdot \gamma''(r))'}{r^2} = \frac{3r^2\gamma''(r) + r^3\gamma'''(r)}{r^2} \approx -\int_0^\infty h^4 I(h) \cdot \cos(hr)dr. \quad (10.6)$$

Starting from $\gamma(r)$ [see Eq. (1.33)], Eq. (10.6) can be proved by a line from *Mathematica*, which yields the result $-h^4 \cdot i[h] \cdot \cos[hr]$.

```
Simplify[(1/r^2)*D[r^3*D[h^2*i[h]*(Sin[h*r]/(h*r)), {r, 2}], {r, 1}]]
```

10.2.2   About the P_1 plot of a scattering pattern

What are the basic properties of a P_1 plot? How can this be determined for a given scattering pattern $I(h)$ [where $I(0) = 1$] of a single (not necessarily convex) particle of volume V_0 and surface area S_0? Limiting cases of plane sets of surfaces in space (which are not of interest here) require other considerations and strategies. How is a P_1 plot calculated for a single sphere of radius R?

Answer and conclusions

Such a plot is a special dimensionless Porod plot, expressed as $P_1(h) \sim h^4 I(h)$, which applies the normalization strategy $P_1(\infty) = 1$. For a single particle of volume V_0, $0 < V_0$, $\int_0^\infty h^2 I(h)dh = 2\pi^2/V_0$. Operating with a mean chord length of $l_1 = 4V_0/S_0$ gives $P_1(h) = \pi l_1 \cdot h^4 I(h) / (4 \int_0^\infty h^2 I(h)dh) = V_0^2 h^4 \cdot I(h)/(2\pi S_0)$, i.e., a single sphere of radius R results in

$$P_1(h) = \frac{V_0^2}{2\pi S_0} \cdot h^4 \cdot \left[3 \cdot \frac{\sin(hR) - hR\cos(hR)}{h^3 R^3}\right]^2. \quad (10.7)$$

A check of the units of Eq. (10.7) yields $\left[\frac{m^6}{m^2} \cdot \frac{1}{m^4} = 1\right]$. Such plots are useful for analyzing the scattering behavior of relatively big h. A double linear representation $P_1(h)$, where $0 < h < h_{max}$, is the optimum. Particle volume V_0 and particle surface area S_0 follow from $V_0 = 2\pi^2 / \int_0^\infty h^2 I(h)dh$ and $S_0 = V_0^2/(2\pi^2) \cdot \lim [h^4 \cdot I(h)]$, where $I(0) = 1$.

In the case of a sphere, $\lim [h^4 \cdot I(h)] = 2\pi S_0/V_0^2 = 9/(2R^4)$. The following *Mathematica* lines produce plots $h^4 \cdot I(h)$ and $P_1(h)$. The curves oscillate around the asymptotes $9/2$ and 1.

```
(* Porod plots for a sphere of radius R *)
i[h_, R_] := (3*((Sin[h*R] - h*R*Cos[h*R])/(h^3*R^3)))^2;
P1[h_, R_] := (((4/3)*Pi*R^3)^2/(2*Pi*4*Pi*R^2))*h^4*i[h, R];
Plot[{h^4*i[h,1],1,9/2,P1[h,1]},{h,0,11},AxesLabel->{"h","h^4*I(h), P1(h)"}]
```

These kinds of plots, which sometimes exhibit a high degree of oscillation, fix V_0 and S_0. The plots are smoother and the amplitude of the oscillations is damped for polydisperse or randomly shaped particle ensembles. In this case, the mean values $\overline{V_0}$ and $\overline{S_0}$ are obtained.

10.2.3   Comparing single particle correlation functions

The SAS CF, set covariance and all the other structure functions can be determined from linear, plane and spatial sets, i.e., in $\mathbb{R}^1$, $\mathbb{R}^2$ and $\mathbb{R}^3$. Sometimes it is not useful to compare functions defined in different dimensions or spaces.

In which cases is a comparison of two SAS CFs not useful? There are publications in which the CF of a single sphere is compared with that of a circle or a rod. Are such comparisons always a good idea?

Answer and conclusions

No, not at all. It is not a good idea. If two structure functions are defined in the same space, only those two should be compared with each other. Especially in the classical case, the functions $\alpha(r, L) = 1 - r/L$, where $0 \leq r \leq L$ and $\gamma(r, d) = 1 - (3/2) \cdot (r/d) + (1/2) \cdot (r^3/d^3)$, where $0 \leq r \leq d$, cannot be directly compared. This becomes clear if an attempt is made to determine the characteristic volume v_c "inherent in these functions." The correct result $v_c = \pi d^3/6$ characterizes the sphere case. On the other hand, the volume term $\int_0^L 4\pi r^2 \cdot (1 - r/L) dr = \pi L^3/3$ does not have a practical meaning in this connection.

Now consider the SAS CF of a homogeneous cylinder of height H and diameter D. There exist (at least) two limiting cases A and B.

Case A: $H = const. > 0$, but $D \to \infty$ yields the CF $\gamma_{la}(r, H)$ of a layer with thickness H in $\mathbb{R}^3$. This function is given by the two terms in the r-intervals, $\gamma_{la}(r, H) = 1 - r/(2H)$, where $0 \leq r < H$; and $\gamma_{la}(r, H) = H/(2r)$, where $H < r < \infty$. Hence, $\gamma'(0) = -1/l_1 = -1/(2H)$. In this case, for the characteristic volume $v_c \to \infty$ holds. Case A is simple and quite clear. Based on $\gamma_{la}(r, H)$, the CLDD $A_{la}(r) = l_1 \cdot \gamma''(r) = 0$ results. It follows that $A_{la}(r) = 0$, if $0 < r < H$; and $A_{la}(r) = 2H^2/r^3$, if $H < r < \infty$. The first moment of $A_{la}(r)$ is $l_1 = 2H$ (see Fig. 10.2).

Case B: $H = const. > 0$, but $D \to 0$ yields a line of length H. However, the CF of a line in $\mathbb{R}^1$ completely differs from that of a thin cylinder in $\mathbb{R}^3$. Both limiting cases are characterized by different measures. A "jump" exists.

Care must be taken if two CFs which result from different measures are plotted together in the same coordinate system.

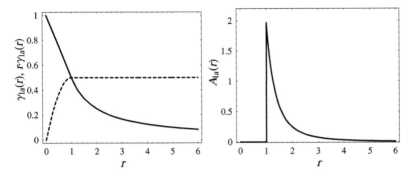

FIGURE 10.2
Analysis of the CF and CLDD of a layer of height $H = 1$.
Left: The CF $\gamma_{la}(r)$ (solid line) and the function $r \cdot \gamma_{la}(r)$ are useful for detecting H (dashed line).
Right: The CLDD $A_{la}(r)$ possesses a finite jump size of $2/H$ at the abscissa $r = H$. The case of a general flat particle was investigated in detail (Glatter, 1982) [129].

10.2.4 Scattering pattern of a hemisphere of radius R

The (isotropic) scattering pattern of a single particle can be traced back to its CLDD (IUR chords) [see Eq. (2.29)]. It is useful to apply this exercise to the hemisphere case and check the invariant of the resulting scattering pattern for $R = 1/2$ and $R = 3/2$.

Answer and conclusions

The CLDD of the hemisphere is formulated by Eq. (2.47) restricting to the case $R = 1$. Hence, a transformation like Eq. (2.32) is required. These steps can be formulated via some *Mathematica* lines. On the one hand, the first moment of the CLDD $A_\mu(r, 1/2)$ results via numerical integration. On the other hand, the *Mathematica* term l1hemis[R]=l_1 = $4V/S$ is involved. Hence, $l_1 = 4/9$ for $R = 1/2$.

```
(* hemisphere, R=1 *)
Aunitrad[r_]:= Which[0<=r<=1,(4*Pi*r^4+r*Sqrt[4-r^2]*(2+3*r^2)-8*ArcSin[r/2])/
(6*Pi*r^3), 1<=r<=2,(r*Sqrt[4-r^2]*(2+3*r^2)+8*ArcCos[r/2])/(6*Pi*r^3),True,0];
Plot[Aunitrad[r], {r, 0.001, 2.2}]

(* definition of the CLDD of a hemisphere with arbitrary radius R *)
Amy[r_, R_] := Module[{x}, x = r/R; Aunitrad[x]/R];
(* R=1/2 *)  Plot[Amy[r, 1/2], {r, 0.01, 1.1}];
moment0 = NIntegrate[ Amy[r, 1/2], {r, 0, 5}]
moment1 = NIntegrate[r*Amy[r, 1/2], {r, 0, 5}];
l1hemis[R_] := 4*(2/3)*Pi*(R^3/(2*Pi*R^2 + Pi*R^2));
{l1hemis[1/2], moment1}
```

The scattering pattern $I(h) =$ ihemi[h,R] of the hemispheres in question can be formulated in the following way:

```
(* pattern i[h,R] of the hemisphere of radius R in terms of its CLDD *)
ihemi[h_, R_] := ((4*Pi)/((((2*Pi)/3)*R^3)*11hemis[R]))*
NIntegrate[Amy[r, R]*((2-2*Cos[h*r]-h*r*Sin[h*r])/h^4), {r,0,R,2*R}];
Plot[Log[10, N[ihemi[h, 1/2]]],{h, 0.01, 20}];
```

A logarithmic plot, where $0.01 < h < 20$, shows the details. Taking into account the terms in front of the integral leads to $I(0) = 1$. The invariant of the scattering pattern can be checked as follows:

```
(*check of the invariant for 2 hemispheres of diameter R=1/2 and R=3/2;
  inserting the option PrecisionGoal->2 *)
{R = 1/2, NIntegrate[h^2*ihemi[h, R], {h, 0., 2000},
   PrecisionGoal -> 2]*(2/3)*Pi*R^3, N[2*Pi^2]}
{R = 3/2, NIntegrate[h^2*ihemi[h, R], {h, 0., 1500},
   PrecisionGoal -> 2]*(2/3)*Pi*R^3, N[2*Pi^2]}
```

The invariant does not depend on R, which is the way it should be. For the normalization used, $invariant = v_c \cdot \int_0^\infty h^2 I(h) dh = 2\pi^2$ holds. For higher precisions, bigger upper integration limits can be inserted. This is important for smaller R, $R \to 0$. Of course, a higher precision (e.g., *PrecisionGoal* $\to$ 10) of the result requires a bigger working precision, say *WorkingPrecision* $\to$ 50 or higher.

10.2.5   The "butterfly cylinder" and its scattering pattern

This exercise concerns an interesting geometric figure called the "butterfly cylinder" (BC) (see Fig. 10.3). The fundamentals explained in Chapter 1 (two touching spheres) and Chapter 3 (right cylindrical particles) are applied. The analytic expression of the SAS CF $\beta_0(r)$ of a unit butterfly (BF), (i.e., of a plane figure) is known (Ciccariello, 2009) [21]. This is the starting point for the following exercise.

How is the SAS CF $\gamma_{BC}(r, H)$ of a BC determined, i.e., the CF of a right cylinder of height H of the right section (RS) of a "unit butterfly"? What are the properties of the distance distribution density $p_{BC}(r, H)$ and the scattering pattern $I_{BC}(h, H)$? What is the influence of the parameter H on these functions?

Answer and conclusions

IUR orientation of the BCs is assumed. The formulation of the function $\beta_0(r) = \beta_{BF}(r)$ of the BF of surface area $S_0 = 1$ requires r-interval splitting. It is useful to perform a normalization check of the function $2\pi r \beta_0(r)/S_0$. By use of the *Mathematica* function Which[], this can be written as follows:

```
(* CF betaBF[r] of a butterfly: triangle sides = {1,1,Sqrt[2]}.
   normalization check and two plots *)
betaBF[r_]:=Which[Inequality[0, LessEqual, r, Less, 1/Sqrt[2]],
           1-((4 + 2*Sqrt[2])*r)/Pi + ((3*Pi + 8)*r^2)/(4*Pi),
           Inequality[1/Sqrt[2], LessEqual, r, Less, 1],
           -1-((4+2*Sqrt[2])*r)/Pi-((5*Pi-8)*r^2)/(4*Pi) +
           (Sqrt[72]*Sqrt[r^2 - 1/2])/Pi +
           (4*(1 + r^2)*ArcSin[1/(r*Sqrt[2])])/Pi,
           Inequality[1, LessEqual, r, Less, Sqrt[2]],
           ((4 - 2*Sqrt[2])*r)/Pi - r^2/4,
           Inequality[Sqrt[2], LessEqual, r, Less, 2],
           -((2 + Sqrt[2]*Sqrt[r^2 - 2])/Pi) + (4*r)/Pi +
           (r^2*(Pi - 4*(1 + ArcSin[Sqrt[2]/r])))/(4*Pi),
           True, 0];

NIntegrate[2*Pi*r*betaBF[r], {r,0,1/Sqrt[2],1,Sqrt[2],2},
PrecisionGoal->50, WorkingPrecision->100]

Plot[{2Pi*r*(betaBF[r]/1),betaBF[r]},{r,0,2.2},PlotRange->All]
```

The three-dimensional CF $\gamma_0(r, H)$ results in terms of $\beta_0(r)$ via the transformation (see [21])

$$\gamma_0(r, H) = \frac{1}{r} \int_{u(r,H)}^{r} \left[\frac{x}{\sqrt{r^2 - x^2}} - \frac{x}{H} \right] \cdot \beta_0(x)dx, \quad \gamma_0(0, H) = 1. \qquad (10.8)$$

The function $u(r, H)$ is written $u(r, H) = \sqrt{r^2 - H^2}$ if $H \leq r < \infty$; else $u(r, H) = 0$. The parameter H sensitively influences the CF of the cylinder. Equation (10.8) defines the distance distribution density function $p_{BC}(r, H) = 4\pi r^2 \gamma_{BC}(r, H)/V_0$, $V_0 = V_{BC} = a^2 H$. Equation (10.8) fixes the scattering pattern of BCs for arbitrary heights $0 < H < \infty$ (see Fig. 10.4). The invariant integral of the scattering pattern [see Eqs. (1.37) and (1.38)] does not depend on H, which is a sensitive check of this approach.

Conclusion: Based on Eq. (10.8), isotropic ensembles of parallel right cylinders of constant H can be analyzed. In the limiting case $H \to \infty$, the second integral term $\left[-x/H \right]$ disappears and the integral simplifies via a two-step partial integration with respect to the variable r. Finally, twice differentiating with respect to r leads to Eq. (3.2).

There is a certain similarity between the case of a BC and two touching spheres (see Chapter 4). The curves of the BCs are the smoother, the longer the cylinders. For detecting the size parameters of a BC, the method described in Chapter 3 cannot be applied. There is no oval RS and there are surface singularities.

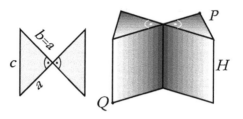

FIGURE 10.3

The triangular "wings" of the butterfly considered have the side lengths $a = 1$, $b = 1$ and $c = \sqrt{2}$. The height H of the three-dimensional "butterfly cylinder" varies, $0 < H < \infty$. In this exercise, the interval $0 < H \leq 2a$ is investigated (see Fig. 10.4) i.e., the triangular double cylinder remains relatively flat. The largest particle diameter L of this model is $L = \overline{PQ}$.

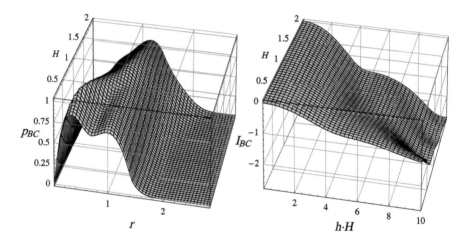

FIGURE 10.4

Scattering properties of "butterfly cylinders" of height H, where $0 < H < 2a$. Distance distribution density $p_{BC} = p_{BC}(r, H)$ (see left); logarithmic plot of the scattering pattern $I_{BC} = I_{BC}(r, H)$, where $0 < h \cdot H < 10$, normalization $I_{BC}(0, H) = 1$ (see right). The order range of the figure is $L = \sqrt{4a^2 + H^2}$. The bigger H is, the more structured the function $p_{BC}(r, H)$ is, i.e., for $a < H$ the curve is relatively smooth and possesses only one maximum near the side length $r = a = 1$ of the butterfly triangle. The bigger H is, the bigger the largest particle diameter L and the dynamic of the scattering pattern are. More intrinsic properties about the body cannot be derived by restricting exclusively to relatively small scattering vectors like $0 < hH < 1$ in this case. This phenomenon holds true for most particle shapes (compare with Fig. 1.1).

10.2.6   Scattering equivalence of (widely separated) particles

There are exceptional cases where one and the same scattering pattern of two really different particle ensembles exists. An example can be easily constructed for *infinitely diluted isotropic particle ensembles*. Here, the scattering of equally sized ellipsoids and differently sized spheres is considered. How can a scattering equivalence between *equally sized ellipsoids of IUR orientation* and *differently sized spheres* be obtained?

Answer and conclusions

Let an ensemble of equally sized ellipsoids of semiaxes $a = 10$ nm, $b = 20$ nm and $c = 30$ nm be given. Equation (2.43) describes the CLDD $A(r) = A(r, a, b, c)$. Obviously, $L = 2c = 60$ nm. The function $A(r)$ fixes the corresponding scattering pattern $I(h) = I(h, a, b, c)$ [see Eq. (2.29)].

On the other hand, the function $A(r)$ can be constructed by summing up CLDDs of differently sized spheres of diameter D, where $2a \leq D \leq 2c$. Let D be a random variable with distribution density $V(D)$. As the CLDD of a single sphere is written $2r/D^2$, where $0 \leq r \leq D$ and the surface area of a sphere is $\pi \cdot D^2$, the function $V(D)$ is implicitly fixed by

$$A(r) = \frac{\int_{D=r}^{2c} \pi D^2 \cdot \frac{2r}{D^2} \cdot V(D)dD}{\int_0^{2c} \pi D^2 \cdot V(D)dD}, \quad \overline{S} = \int_0^{2c} \pi D^2 \cdot V(D)dD, \quad \overline{S} = 2\pi/A'(0).$$

$$(10.9)$$

By use of the limit $A'(0)$, this equation is correctly normalized. In the limiting case $r \to 0$, $A(r)/r \to A'(0) = 2\pi/\overline{S}$ results. Hence, $\overline{S} = 2\pi/A'(0)$. The explicit solution of Eq. (10.9) is written (by use of the variable D instead of r)

$$V(D) = -\frac{\overline{S}}{2\pi} \cdot \left(\frac{A(D)}{D}\right)', \quad 2b \leq D \leq 2c.   (10.10)$$

Equation (10.10) takes into account $A'(r)/r \equiv 0$ if $0 \leq r \leq 2b$. The latter holds true since the CLDD of the ellipsoid is a linear function in the first r-interval. For more details about the functions $A(r)$, $V(D)$ and $I(h)$, see the figure on the first page of this chapter.

It can be concluded that exceptional cases really do exist, where scattering patterns of two different particle ensembles agree. A correct interpretation of the scattering pattern $I(h)$ requires information from other experimental techniques.

10.2.7   About the significance of the SAS correlation function

How significantly is a certain particle shape reflected in the SAS CF? Let $\gamma_1(r)$ and $\gamma_2(r)$ be the SAS CFs of two different single particles in the isotropic case. It is assumed that both the functions are available only in a certain r-interval near the origin, where $0 \leq r < r_m$. Is it possible then at least approximately to detect an approximation of the particle shape *based exclusively on this limited information?*

Answer and conclusions

There does not exist a one-to-one connection between the particle shape and the SAS CF [115]. Based on the information involved in the behavior of the CF in a limited r-interval, no conclusions should be drawn.

In order to explain this, it is useful to study an example. Let $\gamma_1(r)$ be the CF of an infinitely long circular cylinder of diameter $d = 2$ and $\gamma_2(r, d)$ the CF of a sphere of diameter $d = 3$. Using *Mathematica* notation, analytic expressions for the CFs of the different particle shapes can be formulated for $0 < d < \infty$ as follows:

```
(* CF g1[r, d] of the circular rod of diameter d *)
g1[r_, d_] := Which[0<r<d, 1-(r*Hypergeometric2F1[-(1/2),3/2,2,r^2/d^2])/d -
        r^3*Hypergeometric2F1[1/2, 3/2, 3, r^2/d^2])/(4*d^3),
        d<r<Infinity, 1-Hypergeometric2F1[-(1/2), 3/2, 2, d^2/r^2] -
        d^2*Hypergeometric2F1[1/2, 3/2, 3, d^2/r^2])/(4*r^2), True, 0];
(* CF g2[r,d] of the sphere *)
g2[r_,d_]:=Which[0<=r<= d, 1-(3/2)*(r/d)+(1/2)*(r^3/d^3),True,0];
```

Hence, the rod CF in the first interval can be formulated in terms of hypergeometric type functions [98] as

$$\gamma_1(r) = \gamma_{cyl}(r,d) = 1 - \frac{{}_2F_1\left(\frac{1}{2}, \frac{3}{2}; 3; \frac{r^2}{d^2}\right)r^3}{4d^3} - \frac{{}_2F_1\left(-\frac{1}{2}, \frac{3}{2}; 2; \frac{r^2}{d^2}\right)r}{d}, \quad 0 \leq r < d. \tag{10.11}$$

The second derivative of the CF in this r-interval involves the elliptic integrals K and E in

$$\gamma_1''(r) = \gamma_{cyl}''(r) = \frac{4[(2d^2 + r^2)K(r^2/d^2) - 2(d^2 + r^2)E(r^2/d^2)]}{3\pi dr^3}, \quad 0 \leq r < d. \tag{10.12}$$

Additionally taking the single sphere CF and its second derivative (see the left of Fig. 10.5), it is useful to plot both the CFs simultaneously, where $0 \leq r \leq 6$. Obviously, for $0 < r < 3$ there is not a significant difference, although the particle shapes differ completely.

Remark: The first terms of the Taylor series at $r = 0$ of the cylinder CF are $\gamma_1(r, d) = 1 - r/d + r^3/(8d^3) + r^5/(64d^5) + 5r^7/(1024d^7) + O[r]^9$. It is not difficult to find other curves and examples which show better agreement.

It can be concluded that the analysis of an SAS CF should – with much greater effort – of course include the interval $0 < r < L$, which reflects distances over the whole particle. More intrinsic details of the particle shape are reflected when interpreting the second derivative of the CF (see right of Fig. 10.5). However, as the actual case demonstrates, even the difference $\gamma_1''(r) - \gamma_2''(r)$ is still extremely small.

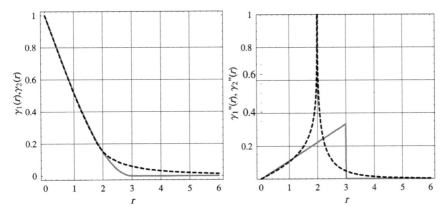

FIGURE 10.5

Comparison of a sphere with an infinitely long cylinder: sphere of diameter $d = 3$ (solid line) and infinitely long cylinder of diameter $d = 2$ (dashed line). Left: The CFs $\gamma_1(r)$ and $\gamma_2(r)$; right: The functions $\gamma_1''(r)$ and $\gamma_2''(r)$. For the definition of each of the two functions, two essential r-intervals must be distinguished. In both cases the mean chord length equals $l_1 = 2$.

10.2.8   The mean chord length of an elliptic needle

Each three-dimensional particle involves a mean chord length l_1. An infinitely long circular cylinder possesses a finite l_1, i.e., l_1 is limited as in $l_1 < Konstant$. It equals the cylinder diameter $d = l_1$. What about other cases of long figures?

The elliptic needle is a limiting case of an ellipsoid. Let such an ellipsoid of semiaxes a, b ($a \le b$) and c ($b \le c < \infty$) be given. An elliptic needle results if the length c gets bigger and bigger relative to a and b, i.e., $c \to \infty$ for $a = const.$ and $b = const.$ Additionally, it is useful to compare this needle case with the case of a long parallelepiped of edges $2a$, $2b$, $2c$ under the condition $2c \to \infty$.

Answer and conclusions

The behavior of the mean chord length l_1 as depending on a, b and c will be investigated by a simple plot. For both, the elliptic needle and the long stretched parallelepiped, the connection $l_1(a, b, c) = 4V(a, b, c)/S(a, b, c)$ holds. Inserting the volume $V = (4/3)\pi abc$ and the surface area of the ellipsoid, the connection is described by the following:

```
(* surface area of the ellipsoid in terms of the semi axes *)
S[a_, b_, c_] := 2*a^2*Pi + 2*b*Sqrt[-a^2 + c^2]*Pi*
EllipticE[ArcSin[Sqrt[-a^2+c^2]/c],((-a^2+b^2)*c^2)/(b^2*(-a^2+c^2))]+
  (2*a^2*b*Pi*
EllipticF[ArcSin[Sqrt[-a^2+c^2]/c],((-a^2+b^2)*c^2)/(b^2*(-a^2+c^2))])/
  Sqrt[-a^2 + c^2];
```

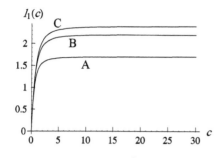
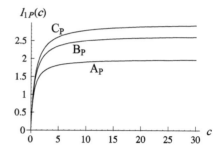

FIGURE 10.6

The mean chord length l_1 as depending on the length c for three ellipsoids (left) and for three parallelepipeds (right).

Left: Three cases with semiaxes $(a = b = 1;$ curve A$)$, $(a = 1, b = 2;$ curve B$)$ and $(a = 1, b = 3;$ curve C$)$ are considered. The bigger the length parameters, the bigger the mean chord lengths. For $c \to \infty$, all curves tend to a limiting value of the needle $l_{1n}(a, b)$.

Right: Investigation of parallelepipeds with half-edges denoted by a, b, c. The edge lengths are $2a$, $2b$ and $2c$.

The (mean) chord length $l_{1P}(c)$ of a parallelepiped (see curves A$_P$, B$_P$ and C$_P$) is somewhat bigger than that of the corresponding ellipsoid.

In this program, the function EllipticF[] denotes the elliptic integral of the first kind and EllipticE[] denotes the elliptic integral of the second kind. It must be emphasized that the definitions of both these functions in *Mathematica* do not agree with the traditional ones (for details see [234]).

```
(* mean chord length(s) l1abc[a,b,c] of the ellipsoid(s) for IUR chords
    in terms of a, b and c *)
l1abc[a_,b_,c_]:=4*(4/3)Pi a*b*c/S[a,b,c];
```

In the limiting case $c \to \infty$, the mean chord length of the elliptic needle is $l_1 \to l_{1n}(a, b, \infty) = 2a/\left(3E\left[1 - a^2/b^2\right]\right)$ (see left of Fig. 10.6). This is analogous to the case of a parallelepiped with edges $2a$, $2b$, $2c$, which results in the limit $l_{1P}(a, b, \infty) = 4ab/(a + b)$ (see right of Fig. 10.6).

Conclusion: In many cases, mean chord lengths of infinitely long figures are limited. This simplifies the analysis of volume fractions via the Rosiwal ratio $l_1/(l_1 + m_1)$ for many long stretched particles. The results for the elliptic needle and long stretched parallelepiped are similar. The existing limits characterize the infinitely long geometric figures. The scattering behavior of infinitely long cylinders is simple to describe (see Chapter 3). Evidently, the particle shape cannot be reconstructed based on the parameters V, S and l_1. Nevertheless, the parameter l_1 includes a certain specific information about the particle.

10.3 Structure functions parameters of special models

The exercises in this section analyze model parameters and their practical application.

10.3.1 The first zero of the SAS CF

Based on Chapter 4, an interpretation of the first zero point of the SAS CF of isotropic two-phase particle ensembles can be performed. Equation (4.27) has many applications and allows interesting conclusions to be drawn. In particular, it describes the first zero point $r = r_1$ of the SAS CF for an isotropic sample.

In the case of a Boolean model (Bm) (see Chapter 5), the CF does not possess negative values. There *does not exist a zero point* $r = r_1 < L$. The position $r = L$ is the only existing zero point.

In contrast to this, as many examples of *ensembles of hard particles* show, $\gamma(r)$ can possess one or more zero points r_i. This phenomenon can be traced back to a sequence of SAS structure parameters and structure functions: volume fraction c, the single particle CF $\gamma_0(r)$, the pair correlation function $g(r)$ and the function $P_{AB}(r, l)$ (see Chapter 4). How can the first zero $r = r_1$ of the sample CF $\gamma(r)$ be interpreted in detail? A possible way to handle this matter is to consider the well-known pair correlation $g(r) = g_{DLm}(r)$ of the Dead Leaves model (DLm) [see Eq. (4.21)]. How can the background of the zero $r = r_1$ be interpreted for a quasi-diluted particle ensemble?

Answer and conclusions

The sample CF $\gamma(r)$ of a DLm of spherical grains of diameter d is fixed by the program, which follows after Eq. (4.30) (see the considerations regarding the DLm in Chapter 4). The normalized scattering pattern case $d = 1$ results from

```
(* Test of the normalized scattering pattern i[h,d] *)
denomi    = NIntegrate[4*Pi*r^2*gamma[r, 1], {r, 0, 1, 3}]
i[h_,d_]:= NIntegrate[4Pi*r^2*gamma[r,1]*(Sin[h*r]/(h*r)),{r,0,1,3}]/denomi;
LogPlot[i[h, 1], {h, 0.01, 20}, AxesLabel -> {"h", "I(h)"}]
```

The resulting plot yields the first zero point at $r_1 = 0.771622$ (see Fig. 10.7). How can this position be interpreted? The length r_1 is connected to the particle-to-particle pair correlation function $g(r)$ of particles of a (mean) CF $\gamma_0(r)$. In the zero case $\gamma(r_1) = 0$, it follows from Eq. (4.27) that

$$0 \equiv \gamma_0(r_1) - c + \frac{c}{(\pi d^3/6)} \cdot \int_0^{l=3d} 4\pi l^2 g(l) \cdot P_{AB}(r_1, l) dl = \gamma_0(r_1) - c + \frac{c \cdot T(r_1)}{(\pi d^3/6)}.$$
$$(10.13)$$

The term $P_{AB}(r, l, d)$ in the integrand describes the following: For two single particles A and B, whose centers are separated by a distance l, where $L_0 \leq l$, the function $P_{AB}(r, l, L_0)$ is the geometric probability that a point X_B, which is placed at a fixed distance r from a point X_A, lies in particle B. From Eq. (10.13), c explicitly results in terms of L, r_1, γ_0, V_0, g and P_{AB}, as shown by

$$c = \frac{\gamma_0(r_1)}{1 - \frac{1}{V_0} \cdot \int_0^{l=L} 4\pi l^2 g(l) \cdot P_{AB}(r_1, l)dl}. \tag{10.14}$$

In order to illustrate this equivalence, the right side of Eq. (10.14) is considered for all r in $0 \leq r \leq r_1 < L = 3d$. For the DLm considered, the theoretically expected value $c = 0.125$ results (see Fig. 10.7).

Special cases: For $c \to 0$, $\gamma(r) = \gamma_0(r)$ results. It follows then that $r_1 = L_0$ (the largest diameter of the largest particle) is the only zero of the SAS CF.

This theory also holds true for quasi-diluted particle ensembles (i.e., if the smallest chord length between any two particles l_m is bigger than the maximum diameter of the largest particle). Actually, $L_0 = d$ and $g(l) = 0$ if $0 \leq l < L_0 + l_m$. Taking into account $0 < r_1 \leq L_0 + l_m$, the integral term $T(r)$ involved in Eqs. (10.13) and (10.14), where

$$T(r) = \int_{L_0+l_m}^{\infty} 4\pi l^2 \cdot g(l) \cdot P_{AB}(r_1, l)dl, \tag{10.15}$$

disappears at $r = r_1$, where $T(r_1) = 0$. Consequently, $\gamma(r_1) = (\gamma_0(r_1)-c)/(1-c)$ holds (see Fig. 10.8). Operating with the zero of $\gamma(r)$ at $r = r_1$, $\gamma_0(r_1) = c$ follows. Hence, Eq. (10.14) is fulfilled, i.e., $c \equiv \gamma_0(r_1)/(1 - 0/V_0) = \gamma_0(r_1)$. Conclusion: The interpretation of the first zero point of $\gamma(r)$ is a simple task for quasi-diluted particle ensembles. However, a potentially existing anisotropy of the particle ensemble can lead to $\gamma_0(r_1) \neq c$.

10.3.2 Different models, different scattering patterns

It is interesting to consider the scattering of hard, tightly packed spheres for $c = 1/8$ in detail. The *hard particle Dead Leaves model* (HPDLm) [see Eq. (4.29)] operates with the special pair correlation function $g(r)$ [see Eq. (4.21)]. Completely independent of this, there is also the Percus-Yevick (PY) approximation (see Chapter 4, [188, 232, 233]). It should be interesting to compare the scattering patterns $I_{DLm}(h)$ and $I_{PY}(h)$ of both these models. Is the normalization strategy $I(0) = 1$ useful for this project?

Answer and conclusions

The answer is yes. The strategy $I(0) = 1$ is not wrong; however, for the actual comparison, it is a better idea to normalize the scattering behavior to a certain characteristic volume v_c. A simple normalization $I(0) = 1$ does not allow a comparison of $I_{DLm}(h)$ and $I_{PY}(h)$ for big h. The scattering patterns in

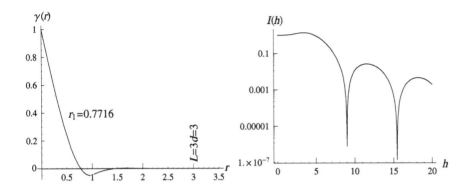

FIGURE 10.7
SAS CF and scattering pattern of a DLm with spherical grains of diameter $d = 1$.
The order range of the model is $L = 3d$. The existence of one or more zero points is typical for ensembles of hard particles.
The scattering pattern results from the CF [see Eq. (4.27) and right.

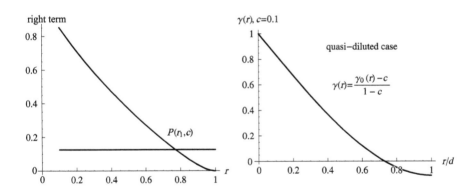

FIGURE 10.8
Investigation of the terms of Eq. (10.13).
Left: Both terms agree at the zero $r = r_1$ (hitting point P). The volume fraction c can be detected operating with Eq. (10.14).
Right: In the quasi-diluted case of spheres of diameter d with $c = 0.1$, $r_1 = 0.728 \cdot d$ results. The behavior of the CF for bigger r is not involved in the model of quasi-dilution.

question should agree better the larger h is. In addition, they should coincide with the single sphere pattern $I_S(h)$, i.e., $I_{DLm}(h) = const.I_{PY}(h) = I_S(h)$.

The comparison will be performed for $d = 2 \cdot R = const.$ by applying numerical integration. The corresponding scattering pattern follows immediately from the CF of the HPDLm [see Fig. 4.8 and Eq. (4.29)]. The following *Mathematica* lines fix the scattering pattern $I_{PY}(h)$ [163]. This function is traced back to that of a single sphere $I0[h, R]$. Two functions $I0[h, R]$ and $S[h, R, c]$ are defined. The final definition is $S(h, R, c) = 1/(1 + 24c \cdot G(A)/A)$. In the end, $I_{PY}(h)$ follows from these terms. For $R = 1/2$ and $c = 1/8$, this curve possesses a maximum at $h \approx 2.5/d$. A test value is $IPY[6, 1/2, 1/8] = 0.356923\ldots$.

```
(* single sphere and PY-approximation *)
I0[h_, R_] := (3*((Sin[h*R] - h*R*Cos[h*R])/(h*R)^3))^2;
S[h_,R_,c_]:=(G=((1+2*c)^2*(-2*h*R*Cos[2*h*R]+Sin[2*h*R]))/(4*(1-c)^4*h^2*R^2)-
    (3*(1 + c/2)^2*c*(-2 + (2-4*h^2*R^2)*Cos[2*h*R] + 4*h*R*Sin[2*h*R]))/
    (4*(1 - c)^4*h^3*R^3) + (1/(64*(1 - c)^4*h^5*R^5))*(c*(1 + 2*c)^2*
    (-16*h^4*R^4*Cos[2*h*R] + 4*(6 + (-6 + 12*h^2*R^2)*Cos[2*h*R] +
    (-12*h*R + 8*h^3*R^3)*Sin[2*h*R])));
    1/(1 + 24*c*(G/(2*h*R))));
IPY[h_, R_, c_] := S[h,R,c]*((1+2*c)^2/(1-c)^4)*I0[h, R];
Plot[IPY[h, 1/2, 1/8],{h, 0, 10}]
N[IPY[6, 1/2, 1/8]] (*0.356923...*)
```

From the scattering pattern of the PY model, the corresponding CF $\gamma_{PY}(r)$ follows in a second step. This CF yields a characteristic volume $v_c = \int_0^L 4\pi r^2 \cdot \gamma_{PY}(r)dr$, where $v_c = 0.224d^3$. This value is used for the normalization (see Fig. 10.9). For comparing both cases, two h-intervals are used, where $0 \leq h \cdot d \leq 10$ and $10 \leq h \cdot d \leq 40$. The three curves already nearly agree for $15 \leq h \cdot d$. It can be concluded that scattering patterns of different models can agree by restricting to certain h-intervals. Inversely, this also holds true for the Fourier-transformed intensity, i.e., the SAS CFs. With small r, the deviations between the CFs $\gamma_{DLm}(r)$ and $\gamma_{PY}(r)$ can be neglected. However, there exist huge deviations for relatively big r.

10.3.3   Boolean model contra hard single particles and quasi-diluted particle ensembles

The set referred to as the Boolean model (Bm) is a special isotropic random two-phase system of volume fraction c. Generally, for such a two-phase system, $m_1 \cdot c = (1 - c) \cdot l_1 = l_p = -1/\gamma'(0)$ holds [Eq. (1.42)]. Here, l_1 and m_1 denote the mean chord lengths of the particle phase 1 and the matrix phase 2, respectively. Let $\gamma_0(r)$ be the SAS CF of the grains (i.e., of the grain particles) of a Bm, Fig. 5.1. The sample CF, i.e., the CF of the whole Bm set, is denoted by $\gamma(r)$.

Does this mean that the volume fraction c, sample CF $\gamma(r)$ and grain CF $\gamma_0(r)$ are connected via $1 - c = \gamma_0'(0)/\gamma'(0)$?

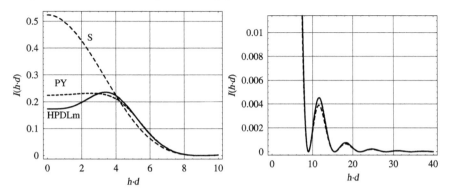

FIGURE 10.9

Comparison of the scattering curves of a single sphere (S), an HPDLm and the PY approximation.

Clearly, three different models lead to three different scattering patterns. The approximations are the dashed lines. The results are clearly different for relatively small h values, but agree for $h \to \infty$. The three curves are normalized with respect to the characteristic volume v_c, which results from the PY approximation, i.e., $I_{PY}(0) = v_c$.

Answer and conclusions

No, not at all. There are significant differences between the approaches used in Chapters 4 and 5 (see also Chapter 8). The mean chord length l_1 of the particle phase (phase 1) cannot be directly derived from the grain CF $\gamma_0(r)$, i.e., $l_1 = -1/\gamma_0'(0)$ is wrong. At first the parameter l_1 in the fundamental equation $(1 - c) \cdot l_1 = l_p = -1/\gamma'(0)$ is unknown for the Bm. Intense considerations are required to express l_1 in terms of the functions given (see Chapter 5). However, there exists the trivial limiting case $c \to 0$, where $\gamma_0 = \gamma$ and $l_1 = -1/\gamma'(0) = -1/\gamma_0'(0)$. In this case, there is no overlapping of any grains, i.e., the grains are single particles.

10.3.4 A special relation for detecting the c of a Bm

Let M_2 be the second moment of the function $g(r) = l_p \cdot \gamma''(r)$ and $l_p = -1/\gamma'(0)$. In Chapter 9 (i.e., for the general random two-phase system), it is shown that the CLDD moments $l_{1,2}$ and $m_{1,2}$ are interrelated, where $M_2 = l_p{}^2 (l_2/l_1{}^2 + m_2/m_1{}^2 - 2)$. Some special cases can be picked out from this: How can this connection for a Bm and a single particle be simplified? It is useful to work with the connection between M_2, l_p and l_c (see Table 9.1).

Answer and conclusions

For a Bm, $m_2/m_1{}^2 - 2 \equiv 0$ (see Chapter 5). Hence, $M_2 = l_p{}^2 \cdot l_2/l_1{}^2$ results. With $M_2 = l_p \cdot l_c$ (see Table 9.1) and $l_p = (1-c) \cdot l_1$, $1-c = l_c \cdot l_1/l_2$ follows, where $0 \le c < 1$. In the limiting case $c \to 0$, i.e., for a single particle, $l_2 = l_1 \cdot l_c$ results. The particle(s) do not need to be convex.

10.3.5   Properties of the SAS CF of a Bm

For a Bm with a certain volume fraction c, the general connection between the SAS CF $\gamma_0(r)$ of the grain particle and the CF of the sample $\gamma(r)$ is known (see Fig. 5.2). These considerations have shown that for smooth grains, i.e., $\gamma_0''(0) = 0$, c is defined via $c = \gamma''(0)/[\gamma'(0)]^2$. The case of non-smooth grains is more general than that of smooth grains.

Is there an extension of this connection for grains possessing the property $0 \le \gamma_0''(0) < \infty$? How can the volume fraction for a Bm be fixed in terms of the sample CF $\gamma(r)$ and the grain CF $\gamma_0(r)$ by means of a simple equation?

Answer and conclusions

Deleting the restriction of smooth grains and starting from $\gamma(r) = (1-c) \cdot \left(e^{N\gamma_0(r)} - 1\right)/c$ with $N = -\ln(1-c)$, the sample CF is written $\gamma(r) = (1-c) \cdot \left(e^{-\ln(1-c) \cdot \gamma_0(r)} - 1\right)/c$. From this, $\gamma_0(r)$ explicitly results in $\gamma_0(r) = \ln\left(\frac{1-c}{\gamma(r) \cdot c - c + 1}\right)/\ln(1-c)$. The Taylor series of this expression in the origin gives

$$\gamma_0(r) = 1 - \frac{c\gamma'(0)}{\ln(1-c)} \cdot r + \frac{c[c\gamma'(0)^2 - \gamma''(0)]}{2\ln(1-c)} \cdot r^2 - \ldots + \ldots, \qquad (10.16)$$

which results in the terms $\gamma_0'(0) = -\frac{c\gamma'[0]}{\ln[1-c]}$ and $\gamma_0''(0) = \frac{c(c\gamma'(0)^2 - \gamma''(0))}{\ln[1-c]}$. After eliminating the logarithmic term,

$$c = \frac{\gamma''(0)}{\gamma'(0)^2} - \frac{1}{\gamma'(0)} \cdot \frac{\gamma_0''(0)}{\gamma_0'(0)} = \frac{\gamma''(0)}{\gamma'(0)^2} - U(0), \quad U(r) = \frac{1}{\gamma'(r)} \cdot \frac{\gamma_0''(r)}{\gamma_0'(r)} \qquad (10.17)$$

is obtained. Two special cases can be introduced: Smooth grains are characterized by $\gamma_0''(0) = 0$. Furthermore, the limiting case of very small grains independent of $\gamma_0''(0)$, $\gamma_0'(0) \to -\infty$ can be introduced.

Conclusion: The functions γ_0 and γ are closely interrelated for a Bm. For a given sample CF $\gamma(r)$, the volume fraction c can be obtained via Eq. (10.17). Since the term $U(0)$ is non-negative, the order relation $0 \le c \le \gamma''(0)/\gamma'(0)^2$ is fulfilled.

There is a far-reaching analogy to the puzzle fitting function $\Phi(r)$ for smooth and non-smooth grains (see Chapter 7). Hemispherical grains were considered in detail [see Eq. (7.3) and Fig. 7.1].

10.3.6 DLm from Poisson polyhedral grains

The step from the grains of the DLm to the resulting SAS structure functions of the puzzle cells (PCs) of that model is an interesting subject of investigation. The SAS CF of a Poisson polyhedron of mean chord length l_1 is given by the simple exponential function $\gamma_{Poisson}(r) = \exp[-r/l_1]$, where $0 < r < \infty$. The use of this function as a grain function allows some simple calculations. The case can be discussed analogously to that of spherical grains (see Chapter 6).

Based on a specific grain CF $\gamma_G(r)$, the following functions describing the puzzle cells (PCs) are defined (see Chapter 6):
1. The SAS CF $\gamma_{PC}(r)$ of a prototype PC
2. The function $U_{PC}(r)$
3. The linear erosion $P_{PC}(r)$ of a prototype PC
4. The CLDD $f_{PC}(r)$ of a prototype PC

Which of these 4 functions can be fully detected from SAS experiments, i.e., in terms of scattering patterns?

Answer and conclusions

In general, only the SAS CF and $U_{PC}(r)$ can be detected from the scattering pattern. Nevertheless, $P_{PC}(r)$ and $f_{PC}(r)$ are useful for model calculations. Furthermore, $f_{PC}(r) \approx U_{PC}(r) = l_p \cdot \gamma''_{PC}(r)$, i.e., the behavior of both these functions is very similar (see Chapter 6).

For the functions considered, Eqs. (10.18) to (10.21) result (see the illustration in Fig. 10.10):

$$\gamma_{PC}(r) = \frac{\gamma_G(r)}{2 - \gamma_G(r)} = \frac{2 \cdot \exp(r/l_1)}{[2\exp(r/l_1) - 1]^2 \cdot l_1}, \quad \gamma'_{PC}(0) = -\frac{2}{l_1}. \quad (10.18)$$

$$U_{PC}(r) = \frac{\gamma''_G(r)}{|\gamma'_G(0)|} = \frac{\exp(r/l_1)(1 + 2\exp(r/l_1))}{[2\exp(r/l_1) - 1]^3 \cdot l_1}, \quad \int_0^\infty r \cdot U_{PC}(r)dr = \frac{l_1}{2}. \quad (10.19)$$

$$P_{PC}(r) = \frac{\gamma_G(r)}{1 - \gamma_G'(0) \cdot r} = \frac{\exp[-r/l_1]}{1 + r/l_1}. \quad (10.20)$$

$$f_{PC}(r) = \frac{P''_{PC}(r)}{|P'_{PC}(0)|} = \frac{\exp[-r/l_1] \cdot (5l_1{}^2 + 4l_1 r + r^2)}{2(l_1 + r)^3}, \quad \int_0^\infty r \cdot f_{PC}(r)dr = \frac{l_1}{2}. \quad (10.21)$$

The distribution densities $U_{PC}(r)$ and $f_{PC}(r)$ fulfill the normalizations $\int_0^\infty U_{PC}(r)dr = 1$ and $\int_0^\infty f_{PC}(r)dr = 1$. The scattering pattern of the typical PC prototype results in terms of $\gamma_{PC}(r)$ or in terms of $U_{PC}(r)$ [see Eqs. (1.33) and (1.34)]. An upper integration limit $L \to \infty$ must be inserted for this integration. The model discussed does not involve a finite order range.

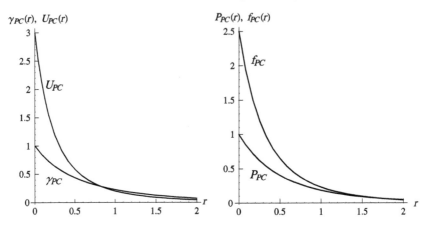

FIGURE 10.10
A plot of the four structure functions describing the PCs that result for a DLm, where the grain is a Poisson polyhedron of mean chord length $l_1 = 1$. On the one hand, $\gamma_{PC}(0) = 1$ and $U_{PC}(0) = 3$ holds. On the other hand, $P_{PC}(0) = 1$ and $f_{PC}(0) = 2.5$ holds. The corresponding functions are similar and possess similar normalization properties; however, they do not agree.

10.4   Moments of $g(r)$, integral parameters and c

There exist many useful relations between the moments of chord length distributions in random two-phase systems and integral parameters like l_c, f_c, v_c and the scattering intensity $I(h)$ (see Chapters 8 and 9). Equations like Eq. (9.10) and all the connections given in Table 9.1 are of practical relevance. The following exercises make this clear.

10.4.1   Properties of the moment M_2 of $g(r)$

A basic property of the moment M_2 of the function $g(r) = l_p \cdot \gamma''(r)$ is $M_2 = l_p \cdot l_c$. How can this connection be specialized for a single particle $\gamma(r) \to \gamma_0(r)$ (e.g., for a sphere)? Which integral parameters are interrelated with the radius of gyration R_g? What are the connections of these moments to $I(h)$?

Answer and conclusions

In the limiting case $c \to 0$ (i.e., for a single particle), $l_2 = M_2$ and $l_1 = l_p$. The relation

$$\frac{M_2}{l_p \cdot l_c} \to \frac{l_2}{l_1 \cdot l_c} = 1 \qquad (10.22)$$

fixes l_2 of a single particle in terms of l_1 and l_c, where $l_2 = l_1 \cdot l_c$. This is of general importance. A single sphere of diameter d results in the following three terms:

$$l_2 = \int_0^d \left(r^2 \cdot l_p \cdot \frac{3r}{d^3}\right) dr = \frac{9d^2}{8}, \; l_p = \frac{3d}{2}, \; l_c = 2\int_0^d \left(1 - \frac{3r}{2d} + \frac{r^3}{2d^3}\right) dr = \frac{3d}{4}. \quad (10.23)$$

Thus, Eq. (10.22) is fulfilled. This means that the information content of the moment l_2 (and the other higher moments as well) is restricted. Adding a new moment M_{i+1} to i known moments M_i does not necessarily add an independent structure parameter. The parameters in question are interrelated.

In a similar way, this can also be concluded from the well-known presentation of the radius of gyration R_g of a particle (see first pages of Chapter 1 and details in Fig. 1.1):

$$R_g = \sqrt{\frac{\int_0^{L_0} 4\pi r^4 \gamma_0(r) dr}{2 \cdot \int_0^{L_0} 4\pi r^2 \gamma_0(r) dr}} = \sqrt{\frac{\int_0^{L_0} 4\pi r^4 \gamma_0(r) dr}{2 \cdot V_0}}. \quad (10.24)$$

By considering the quadratic term $\sim (-)R_g{}^2 \cdot h^2/3$ [see also Eqs. (9.9) and (9.10)], Eq. (10.24) follows starting from the Tayler series of the scattering intensity $I(h) = 1 - c_2 h^2 + c_4 h^4 + \dots$ in the origin [see Eq. (1.33)]. In fact, the SAS CF $\gamma_0(r)$ of the particle fixes R_g of the particle.

10.4.2 Tests of properties of M_2 special model parameters

The *hard particle Dead Leaves model* (HPDLm) with spherical grains is an approach that has been well investigated (see Chapter 9). Operating with the analytic expressions known for this approach, interesting tests can be performed for tightly packed particle ensembles. In particular, an exact verification of the relation $M_2 = l_p \cdot l_c$ can be entered. Is this relation exactly fulfilled? The problem should be considered numerically for the sake of simplicity (6 digit precision).

Answer and conclusions

Yes, the relation is fulfilled. Based on the analytic expressions known for the SAS CF $\gamma(r)$ of an HPDLm (see Chapter 4), the three parameters $l_p = -1/\gamma'(0) = 0.583417 \cdot d$,

$$M_2 = \int_0^{L=3d} r^2 \cdot l_p \gamma''(r) dr = 0.36297 \cdot d^2 \text{ and } l_c = 2\int_0^{L=3d} \gamma(r) dr = 0.622144 \cdot d$$
$$(10.25)$$

result. The relation in question is fulfilled, i.e., $0.36297 = 0.583417 \cdot 0.622144$. Furthermore, the integrals Eq. (10.25) can be handled analytically; however, this requires a greater effort. The HPDLm is not an approximation like the Percus-Yevick approach [188] [see Eq. (4.39) and Fig. 4.14]. Certainly, for spheres, circles and line segments, the HPDLm restricts to $c = 1/8$ in $\mathbb{R}^3$, $c = 1/4$ in $\mathbb{R}^2$ and $c = 1/2$ in $\mathbb{R}^1$.

10.4.3   Volume fraction and integral parameters

Chapter 9 shows that the volume fraction is fixed by several equations involving CLD moments and Porod's length parameter l_p. According to these considerations, does c always seem to be defined by a *quadratic equation* involving at least l_p, which then yields two solutions for c and $1 - c$?

How can a simple equation fixing c in terms of l_1, l_2, m_1, m_2 and l_c be written?

Answer and conclusions

Based on the connections derived in Chapter 9,

$$M_2 = l_p{}^2 \cdot \left(\frac{l_2}{l_1{}^2} + \frac{m_2}{m_1{}^2} - 2 \right), \; M_2 = l_p \cdot l_c, \; m_1 = (1 - c) \cdot l_1 = l_p. \quad (10.26)$$

Therefore the relation

$$l_c = l_p \cdot \left(\frac{l_2}{l_1{}^2} + \frac{m_2}{m_1{}^2} - 2 \right) \quad (10.27)$$

follows. By substituting l_p by m_1 and c or by l_1 and $1 - c$, the equations

$$c = \frac{l_c}{m_1 \cdot \left(\frac{l_2}{l_1{}^2} + \frac{m_2}{m_1{}^2} - 2 \right)}, \; 1 - c = \frac{l_c}{l_1 \cdot \left(\frac{l_2}{l_1{}^2} + \frac{m_2}{m_1{}^2} - 2 \right)} \quad (10.28)$$

result. Equations (10.26) to (10.28) can be verified for a hard particle DLm (see Fig. 10.9). For spherical grains of diameter d, the parameters $l_p = 7d/12$, $l_c = 0.622144 \cdot d$, $l_1 = 2d/3$, $l_2 = d^2/2$, $m_1 = 14d/3$ and $m_2 = 42.2823 \cdot d^2$ follow. Inserting these parameters on the right side of Eqs. (10.28) yields the expected value $c = 0.125$ (for this type of model). Equations (10.28) define c and $1 - c$ in a simple way and are the most suitable for describing the two special cases below. The parameter $l_c = 2 \int_0^L \gamma(r) dr$ plays an essential role in detecting volume fractions, involving information about the whole particle ensemble.

Special case 1: The first equation in Eqs. (10.28) collapses to $c \to 0$ for $m_1 \to \infty$ (this is the case of an infinitely diluted particle ensemble). Here, the particle-to-particle chord lengths with the moments $m_{1,2,3...}$ get bigger and bigger. A discussion of this case based on the introduction of smaller and smaller particle sizes $l_1 \to 0$, $l_2 \to 0$ and $l_c \to 0$ is more complicated.

Special case 2: The second equation of Eqs. (10.28) collapses to $c \to 1$ for $l_1 \to \infty$ (this is the limiting case if the particles grow more and more and the particle-to-particle intermediate spaces get smaller and smaller until the whole sample is *one and the same huge particle*). Hence, the simple relations in Eqs. (10.26) to (10.28) substitute a quadratic equation for the variable c.

In certain special cases, there exist interrelations between the moments of the same phase. For example, for a Boolean model (Bm), $m_2/m_1{}^2 = 2$. In this case, the corresponding term in Eq. (10.26) and all the following equations disappear, which reduces the number of independent parameters. The parameters l_2 and l_1 only describe the particle phase 1. The volume fraction c cannot be detected exclusively from particle parameters $l_{1,2,3,...}$. A relation $c = f(l_1, l_2, l_3, ...)$ is impossible (see Fig. 9.5 and related considerations).

10.4.4   Application of Eq. (9.16) for a ceramic micropowder

For the ceramic micropowder already analyzed in Fig. 9.2, the parameters $M_1 = l_p = 55$ nm, $M_2 = 3\,200$ nm^2, $l_c = 56$ nm and $l_1 = 58$ nm were detected from $\gamma(r)$, $\gamma''(r)$.

Is it possible to detect more parameters from this data? How can an approximation for the moment m_2 in terms of these parameters be determined even if parameter l_2 is unknown?

Answer and conclusions

The parameter m_2, defined by $m_2 = \int_0^L r^2 \cdot f(m)dm$, denotes the second moment of the CLDD $f(m)$ of the space outside the particles of the powder, i.e., the "*whole air volume*" between the particles. This region is characterized by a mean chord length m_1. This length can be estimated via $c \cdot l_p = m_1 = 1\,100$ nm, where c can be traced back to l_1 and l_p via $c = 1 - l_p/l_1 = 1 - 55/58 \approx 5\%$. The fundamental Eq. (10.28) fixes m_2 in terms of the other parameters. However, the moment l_2 is unknown. Furthermore, m_2 depends on l_2 through a linear connection [see Eq. (10.29)]:

$$m_2 = m_1 \cdot \left[\frac{l_c}{c} + \left(2 - \frac{l_2}{l_1{}^2} \right) \cdot m_1 \right]. \tag{10.29}$$

Operating with the plausible assumption $0 < l_2 < 2000$ nm^2 (i.e., the particles are smaller than the particle-to-particle intermediate space), $m_2 = (3.3 \cdot 10^6 \pm 0.6 \cdot 10^6)$ nm^2 follows. This interval can be derived from a simple plot with a variable parameter $l_2 = x$, where $0 < x < 2\,000$ nm^2, which nearly results in a horizontal line. Doing this gives an estimation of m_2. Such an approach is useful for practice. Experience shows that the sample preparation technique can really modify m_2, even when operating at room temperature.

Conclusion: The moments of the CLDDs $\varphi(l)$ and $f(m)$ are interrelated. In the actual case, the problem consists in the fact that the scattering pattern was detected assuming an order range at about $L_1 = 200$ nm. Nevertheless, it is possible to estimate the parameters c and m_1 and from these an estimation of m_2. It can be useful to vary one (or more) unknown parameters in order to obtain an estimation for a parameter of interest. Experience shows that a relative deviation of 5% to 15% for second order moments like M_2, l_2 and m_2 is a realistic specification. In the actual case, $M_2 = l_c \cdot l_p$ is nearly fulfilled.

References

[1] Aharonyan NG and Ohanyan VK. Chord length distributions for polygons. *J. Contemp. Math. Anal.*, 40/4:43–56, 2005.

[2] Ambartzumian RV. *Combinatorial Integral Geometry with Applications to Mathematical Stereology.* John Wiley, Chichester, 1982.

[3] Ambartzumian RV, Stoyan D, and Mecke J. *Stochastische Geometrie.* Akademie-Verlag, Berlin, 1995.

[4] Bardeen JM, Bond JR, Kaiser N, and Szalay AS. The statistics of peaks of Gaussian random fields. *Astrophysical Journal*, 304:15–61, 1986.

[5] Borak TB. A method for computing random chord length distributions in geometrical objects. *Radiat. Res.*, 137:346–351, 1994.

[6] Brunner-Popela J and Glatter O. Small-angle scattering of interacting particles: Basic principles of a global evaluation technique. *J. Applied Cryst.*, 30:431–442, 1997.

[7] Burger C and Ruland W. Analysis of chord length distributions. *Acta Cryst A*, 57:482–491, 2001.

[8] Buryak OE, Doroshkevich AG, and Fong R. Measurement of large- and superlarge-scale structures of the universe. *Astrophysical Journal*, 434:24–36, 1994.

[9] Cabo AJ and Baddeley AJ. Line transects, covariance functions and set convergence. *Adv. Appl. Prob.(SFSA)*, 27:585–605, 1995.

[10] Cahn R, Haasen WP, and Kramer EJ. *Phase Transformations in Materials, Experimental Techniques for Studying Decomposition Kinetics.* Material Science and Technology vol. 5, section 4.3. pp. 229-338. VCH Verlag, Weinheim, Germany, 1991.

[11] Ciccariello S and Benedetti A. Singularities of phase boundaries and values of the second-order derivative of the correlation function at the origin. *Physical Review*, B26:6384–6389, 1982.

[12] Ciccariello S. Some parameterizations useful for determining the specific surfaces and the angularities of amorphous samples. *J. Appl. Phys.*, 56:162–168, 1984.

[13] Ciccariello S. Small-angle techniques for the asymptotic analysis of X-ray diffraction peaks. *J. Appl. Cryst.*, A46:175–186, 1989.

[14] Ciccariello S. The leading asymptotic term of the small-angle intensities scattered by some idealized systems. *J. Appl. Cryst.*, 24:509–515, 1991.

[15] Ciccariello S. Edge contributions to the Kirste-Porod formula: The spherical segment case. *Acta Cryst. A*, 49:750–755, 1993.

[16] Ciccariello S. Integral expressions of the derivatives of the small-angle scattering correlation function. *J. Math. Phys.*, 36:219–246, 1995.

[17] Ciccariello S and Sobry R. The vertex contribution to the Kirste-Porod term. *Acta Cryst.*, A51:60–69, 1995.

[18] Ciccariello S. Particle shape reconstruction from the asymptotic small-angle scattering intensity. *Europhys. Lett.*, 58:823–829, 2002.

[19] Ciccariello S. The intersect distribution of the regular tetrahedron. *J. Appl. Cryst.*, 38:97–106, 2005.

[20] Ciccariello S and Riello P. Small-angle scattering from three-phase samples: Application to coal undergoing an extraction process. *J. Appl. Cryst.*, 40:282–289, 2007.

[21] Ciccariello S. The correlation functions of plane polygons. *Journal of Mathematical Physics*, 50:1035271–20, 2009.

[22] Cooleman R. Random paths through convex bodies. *J. Appl. Prob.*, 6:430–441, 1969.

[23] Cox DR. *Renewal Theory*. John Wiley, New York, 1963.

[24] Damaschun G and Müller JJ. Die Roentgenographische Bestimmung der Packungsdichte im Zweiphasensystem. *Faserforschung u. Textiltechnik*, 18:75–78, 1967.

[25] Damaschun G and Pürschel HV. Berechnung der Form kolloider Teilchen aus Roentgen-Kleinwinkeldiagrammen. *Monatshefte Chemie*, 100:1701–1714, 1969.

[26] Damaschun G, Müller JJ, and Pürschel HV. Roentgenkleinwinkelstreuung von isotropen Proben ohne Fernordnung, Allgemeine Theorie. *Acta Cryst. A*, 27:193–197, 1971.

[27] Debye P and Bueche A. Scattering by an inhomogeneous solid. *J. Appl. Phys.*, 20:518–525, 1949.

[28] Debye P, Anderson JR, and Brumberger H. Scattering by an inhomogeneous solid. II. The correlation function and its application. *J. Appl. Phys.*, 28:679–683, 1957.

[29] Dutkievicz J, Kabisch O, Gille W, Simmich O, Scholz R, and Krol J. Structure changes and precipitation kinetics in melt spun and aged Al-Li-Cu alloy aged AlLiCu-alloys. *Z. Metallkd.*, 11:1247–1252, 2001.

[30] Enke D, Janowski F, Heyer W, Gille W, and Schwieger W. Characterization of colloidal silica in porous glasses by small-angle scattering. Applied Mineralogy, Rammlmair et al. (eds), Balkema, Rotterdam, 2000, 135–138, 2000.

[31] Enke D, Janowski F, Gille W, and Schwieger W. Structure and texture analysis of colloidal silica in porous glasses. *Colloids & Surfaces A: Physicochemical and Engineering Aspects 187-188*, 131–139, 2001.

[32] Enke D, Friedel F, Janowski F, Hahn T, Gille W, Mueller R, and Kaden H. Ultrathin porous glass membranes with controlled texture properties. *Studies in Surface Science and Catalysis*, 144:437–354, 2002.

[33] Enns EG and Ehlers PF. Random paths through a convex region. *Journal of Applied Probability*, 15:144–152, 1978.

[34] Fanter D. *Modelluntersuchungen zur Auswertung von Roentgenstreukurven mittels eines Fourierverfahrens*. Teltow-Seehof, Institute of Polymer Chemistry, 1977.

[35] Fedorova IS and Schmidt PW. A general analytical method for calculating particle-dimension distributions from scattering data. *J. Appl.Cryst.*, 11:405–411, 1978.

[36] Feigin LA and Svergun DI. *Structure Analysis by Small-Angle X-Ray and Neutron Scattering*. Plenum Press, New York, 1987.

[37] Filipescu D, Trandafir R, and Zorilescu D. *Probabilitati Geometrice Si Applicatii (in Roman)*. Editura Dacia, Cluj-Napoca, 1981.

[38] Floriano A, Pipitone G, Caponetti E, and Triolo R. Analysis of SAS from a commercial Al-Li alloy by means of a model incorporating a repulsive step potential. *Phil. Mag. B*, 66:391–404, 1992.

[39] Frisch H and Stillinger F. Contributions to the statistical geometric basis of radiation scattering. *J. Chem. Physics*, 38:2200–2207, 1963.

[40] Fritz G, Bergmann A, and Glatter O. Evaluation of small-angle scattering data of charged particles using the generalized indirect Fourier transformation (GIFT) technique. *J. Chem. Phys.*, 113:9733–9740, 2000.

[41] Gardner RJ. Chord functions of convex bodies. *J. London Math. Soc.*, 36:314–326, 1987.

[42] Gasparyan PM. Pleijel identity and distribution of chord length for plane convex domains. *Teubner Texte zur Mathematik*, 65:91–94, 1984.

[43] Gates J. Some properties of chord length distributions. *Journal of Applied Probability*, 24:863–874, 1987.

[44] Geciauskas E. The distribution function of the distance between two points in a convex domain. *Adv. Appl. Prob.*, 9:427–428, 1977.

[45] Geciauskas E. Geometric parameters of the chord-length distribution of a convex domain. *Lith. Math. J.*, 27:121–123, 1987.

[46] Geciauskas E. Standardization of the distributions of the chord length and distance within oval domains. *Lith. Math. J.*, 39:371–375, 1999.

[47] Gille W, Heyroth W, and Kabisch O. Data evaluation of SAXS-experiments - approaches for the analysis of background scattering and thermal density fluctuations. *Zeitschrift der PH Halle*, 19:27–32, 1981.

[48] Gille W. *Stereologische Charakterisierung von Mikroteilchensystemen mit der Röntgenkleinwinkelstreuung*. Doctoral Thesis, Martin-Luther-Universität Halle-Wittenberg, Halle, 1983.

[49] Gille W. The intercept length distribution of a cylinder of revolution. *Exp. Tech. Phys.*, 35:93–98, 1987.

[50] Gille W. The chord length distribution density of parallelepipeds with their limiting Cases. *Exp. Tech. Phys.*, 36:197–208, 1988.

[51] Gille W. The integrals of small-angle scattering and their calculation. *Journal de Physique IV/I*, 3(C8):503–506, 1994.

[52] Gille W. *Das Konzept der Sehnenverteilung zum Informationsgewinn aus einer Kleinwinkelstreukurve und seine Grenzen*. Habilitationsarbeit, Halle, 1995.

[53] Gille W. Diameter distribution of spherical primary grains in the Boolean model from small-angle scattering. *Part. & Part. Syst. Charact.*, 12:123–131, 1995.

[54] Gille W. *Determination of Volume Fraction and Size Distribution in a Stochastic Model from SAS*. III - rd School of X-ray Investigations of Polymeric Structures, Bielsko Biala, Poland, 5—9 December 1995.

[55] Gille W. *Informationen über Volumenanteil und mittleren Teilchenabstand aus dem 1. Peak einer Kleinwinkelstreukurve*. 4. Konferenz, Strukturuntersuchungen an nichtkristallinen und partiellkristallinen Stoffen, Wolfersdorf-Trockenborn, September, 1996.

[56] Gille W, Kabisch O, and Schmidt U. Small-angle X-ray scattering of a rapidly quenched AlLa30wt.% alloy. *Solidification of Metals and Alloys*, 28:33–40, 1996.

[57] Gille W. Determination of volume fraction and size distribution by use of a stochastic model from small-angle scattering. *Proc. SPIE*,

The International Society for Optical Engineering, 3095:157–162, 1997. Volume 29XR, X-Ray Investigations of Polymer Structures, Edited by Andrzej Wlochowicz, Bielsko-Biala, Poland, June 1996.

[58] Gille W and Kabisch O. SAS studies of an AlLi 9.07at% Sc 0.11at% alloy. In Ciach R, editor, *Advanced Light Alloys and Composites*, pages 319–324. Kluver Academic Publishers, 1998.

[59] Gille W. Local surface curvature of microparticles from scattering experiments. *Comp. Mater. Sci.*, 15:50–62, 1999.

[60] Gille W. The small-angle scattering correlation function of the cuboid. *J. Appl. Cryst.*, 32:1100–1104, 1999.

[61] Gille W. The small-angle scattering correlation function of the hemisphere. *Comp. Mater. Sci.*, 15:449–454, 1999.

[62] Gille W. Properties of the Rayleigh-distribution for particle sizing from SAS experiments. *NanoStructured Materials*, 11:1269–1276, 1999.

[63] Gille W and Handschug H. Chord length distribution density and smallangle scattering correlation function of the right circular cone. *Mathematical & Computer Modelling*, 30:107–130, 1999.

[64] Gille W. Small-Angle Scattering (SAS) and Wide-Angle Scattering (WAS). SAS-laboratory report, Martin-Luther-University Halle-Wittenberg, 2000.

[65] Gille W. Analysis of the chord length distribution of the right circular cone for small chord lengths. *Computers & Mathematics with Applications*, 40:1927–1935, 2000.

[66] Gille W. Small-angle scattering and chord length distributions. *The European Physical Journ. B*, 17:371–383, 2000.

[67] Gille W. Determination of the largest microparticle diameter operating with the correlation function of small-angle scattering. *Comp. Mater. Sci.*, 18:65–75, 2000.

[68] Gille W. Simulation of the Poisson distribution density for small intensities - An instructive example of graphics animation. New–MATH–Wire, Wolfram Research, Illinois, 2000.

[69] Gille W. Direct calculation of the interface distribution function and other structure functions from SAS-curves. *Proc. SPIE*, 4240:14–19, 2000.

[70] Gille W, Koschel B, Schwieger W, and Janowski F. Analysis of the morphology of hectorite by use of small-angle X-ray scattering. *Colloid Polym. Sci.*, 278:805–809, 2000.

[71] Gille W. The function TrigExpand[ ] and definite integration with *Mathematica*. *The Mathematica Journal, Bug report*, Wolfram Research, Illinois, Januar, 2000.

[72] Gille W. The SAS correlation function and the chord length distribution of an infinitely long triangular rod. *Comp. Mat Sci.*, 22:151–154, 2001.

[73] Gille W. Mathematica programs for chord length distributions of selected geometric figures. CDROM Proceedings of the first Scientific and Technical Conference in Science, Technology and Education, *Mathematica*, PrimMath[2001], pp. 117–138, Zagreb, Croatia, September, 2001.

[74] Gille W. Chord length distributions of selected infinitely long geometric figures - connections to the field of small-angle scattering. *Computational Materials Science*, 22:318–332, 2001.

[75] Gille W. Chord length distribution density of an infinitely long circular hollow cylinder. *Mathematical and Computer Modelling*, 34:423–431, 2001.

[76] Gille W. The small-angle scattering correlation function of two infinitely long parallel circular cylinders. *Comp. Mater. Sci.*, 20:181–195, 2001.

[77] Gille W, Enke D, and Janowski F. Order distance estimation in porous glasses via transformed correlation function of small-angle scattering. *Journal of Porous Materials*, 8:111–117, 2001.

[78] Gille W, Enke D, and Janoswki F. Two-phase porous silica: Mesopores inside controlled pore glasses. *Journal of Materials Science*, 36:2349–2357, 2001.

[79] Gille W, Enke D, and Janowski F. Stereological macropore analysis of a controlled pore glass by use of small-angle scattering II. *Journal of Porous Materials*, 8:179–189, 2001.

[80] Gille W. The set covariance of a dead leaves model. *Adv. Appl. Prob.*, 34:11–20, 2002.

[81] Gille W. Chord length distributions of infinitely long geometric figures. *Powder Technology*, 123:192–198, 2002.

[82] Gille W. Linear simulation models for real-space interpretation of small-angle scattering experiments of random two-phase systems. *Waves Random Media*, 12:85–97, 2002.

[83] Gille W, Koschel B, and Schwieger W. The morphology of isomorphous substituted hectorites. *Colloid & Polymer Science*, 280:471–478, 2002.

[84] Gille W, Reichl S, Kabisch O, Enke D, and Janowski F. Characterization of porous glasses via SAS and other methods (the introduction of the so-called new q-parameter). *Microporous and Mesoporous Materials*, 77:612–619, 2002.

[85] Gille W. Volume fraction of random two-phase systems for a certain fixed order range from the SAS correlation function. *Materials Chemistry and Physics*, 77:612–619, 2002.

[86] Gille W, Enke D, and Janowski F. Pore size distribution and chord length distribution of porous VYCOR glass (PVG). *Journal of Porous Materials*, 9:221–230, 2002.

[87] Gille W. Volume fraction of hard particles from small-angle scattering experiments. *Comp. Mat. Sci.*, 25:469–477, 2002.

[88] Gille W. Cross-section structure functions in terms of the three-dimensional structure functions of infinitely long cylinders. *Powder Technology*, 138:124–131, 2003.

[89] Gille W. The SAS-structure functions of a dead leaves tesselation cell. *J. Appl. Cryst.*, 36:1356–1360, 2003.

[90] Gille W. Models of small-angle scattering for the analysis of porous materials. *Fibres and Textiles in Eastern Europe*, 11(5):80–82, 2003.

[91] Gille W. The small-angle scattering structure functions of the single tetrahedron. *J. Appl. Cryst.*, 26:850–853, 2003.

[92] Gille W and Braun A. SAS chord length distribution analysis and porosity estimation of activated and non-activated glassy carbon. *J. of Non-Cryst. Solids*, 321:89–95, 2003.

[93] Gille W, Enke D, Janowski F, and Hahn T. Application of small-angle scattering for the identification of small amounts of platinum supported on porous silica. *Journal of Physics and Chemistry of Solids*, 64:2209–2218, 2003.

[94] Gille W, Enke D, Janowski F, and Hahn T. About the realistic porosity of porous glasses. *Journal of Porous Materials*, 10:179–187, 2003.

[95] Gille W. Analysis of the chord length distribution of a box: Mathematica 5.2 programs for the analysis of large chord lengths. CD-Proceedings of the International *Mathematica* Conference, PrimMath[2003], Zagreb, Croatia, September 2003.

[96] Gille W. Analysis of chord length distributions of cylinders and their cross sections. In Pandalai SG, editor, *Recent Developments in Materials Science*, 4:677–694. Research Signpost, 37/661(2), Fort P. O. Trivandrum, Kerala, India, 2003.

[97] Gille W, Enke D, Janowski F, and Hahn T. Platinum dispersion analysis depending on the pore geometry of the support, *Catalysis Letters*, 93:13–17, 2004.

[98] Gille W. The small-angle scattering correlation function for packages of hard long parallel homogeneous circular cylinders. *Powder Technology*, 149:42–50, 2004.

[99] Gille W. The small-angle scattering curves of two parallel, infinitely long circular cylinders. *Comp. Mat. Sci.*, 32:57–65, 2005.

[100] Gille W. Intersect distributions of a 'Dead Leaves' model with spherical primary grains. *J. Appl. Cryst.*, 38:520–527, 2005.

[101] Gille W. Small-angle scattering analysis of tetrahedral particles. *Fibres & Textiles in Eastern Europe*, 13(5):47–50, 2005.

[102] Gille W, Mazzolo A, and Roesslinger B. Analysis of the initial slope of the small-angle scattering correlation function of a particle. *Part. & Part. Syst. Characterization*, 22:254–260, 2005.

[103] Gille W. Chord length distributions of the hemisphere. *J. Math. Stat.*, 1(1):24–28, 2005.

[104] Gille W. Hemisphere sizing from linear intercept measurement. *J. Math. Stat.*, 1(3):203–211, 2005.

[105] Gille W. Analysis of the chord length distributions along an edge: The wedge case and the quadratic rod case. *J. Math. Stat.*, 1(2):106–112, 2005.

[106] Gille W. Small-angle scattering analysis of the spherical half-shell. *J. Appl. Cryst.*, 40:302–304, 2007.

[107] Gille W. The puzzle-interlayer model: An approach to the analysis of tightly packed arrangements of hard particles. *J. Appl. Cryst.*, 40: 691–695, 2007.

[108] Gille W. Chord length analysis of the fragments of a 'dead leaves' tessellation. *Waves in Random and Complex Media*, 17:121–127, 2007.

[109] Gille W. Interrelations between the CLD moments of random two-phase systems. *Report: Martin Luther University Halle-Wittenberg*, 01:1–22, 2007.

[110] Gille W. Reconstruction of a destroyed 'dead leaves' tessellation from its fragments via the puzzle fitting function. *AIP Conf. Proc., ICCMSE 2007 Corfu*, 963:1208–1211, 2007.

[111] Gille W. Analyse endlicher Reihenterme, die Funktion SequenceLimit leistet Unglaubliches. Vortrag zum 9. Mathematica Tag im WIAS Berlin, November 2007.

[112] Gille W. Analysis of the fragments of a 'dead leaves' tessellation by use of a puzzle fitting function. *Waves in Random and Complex Media*, 18:1–12, 2008.

[113] Gille W. Analysis of randomly shaped puzzle fragments: Punch-matrix/particle puzzles. *AIP Conf. Proc., ICCMSE 2008 Crete*, 1108:174–180, 2009.

[114] Gille W. Analysis of randomly shaped puzzle fragment particles via their chord length distribution. Contribution, ICCMSE 2009, Rhodes/Greece, 2 October 2009.

[115] Gille W. Geometric figures with the same chord length probability density function: Six examples. *Powder Technology*, 192:85–91, 2009.

[116] Gille W. Chord length distribution of pentagonal and hexagonal rods: relation to small-angle scattering. *J. Appl. Cryst.*, 42:326–328, 2009.

[117] Gille W. Was Fit nicht kann: Lineare Regression mit vertauschbaren Koordinatenachsen. Vortrag zum 11. Mathematica Tag im WIAS Berlin, 20.11.2009.

[118] Gille W. Characteristics of the SAS correlation function of long cylinders with oval right section. *J. Appl. Cryst.*, 43:347–349, 2010.

[119] Gille W and Kraus M. Geometric parameters of isotropic ensembles of right cylinders from the small-angle-scattering correlation function. *Acta Cryst. A*, A66:597–601, 2010.

[120] Gille W. Analysis of SAS data: The Fourier transformation of a band-limited scattering pattern. Contribution, 19th SI-HR Crystallographic Meeting, Strunjan/Slovenia, 17.06.2010.

[121] Gille W. Geometric properties of particle ensembles. Contribution, ALGORITMY 2012, 19th Int. Conference on Scientific Computing, Podbanske/Slovakia, September 2012.

[122] Gille W. Geometric properties of particle ensembles in terms of their set covariance. *Conference Proceedings ALGORITLY 2012, 19th Int. Conference on Scientific Computing, Podbanske/Slovakia, September 9-14, 2012*, pp.362–370, 2012.

[123] Gille W. Scattering properties and structure functions of Boolean models. *Computers & Structures*, 89:2309–2315, 2011.

[124] Gille W. Analysis of SAS data: SAS correlation functions for fixed order ranges of the same sample. Contribution, 20th HR-SI Crystallographic Meeting, Baska/Croatia, 18.06.2011.

[125] Gille W. The puzzle-interlayer model (PIM): A universal approach in materials science for describing porous materials. Contribution, 14th YUCOMAT-Conference, Symposium C, Nanostructured Materials, Herceg Novi/Montenegro, 6.09.2011.

[126] Gille W. Geometric properties of particle ensembles in terms of their set covariance. Contribution and Conference Proceedings of the 20th Algoritmy Conference, Podbanske/Slovak Republik, 12.09.2012.

[127] Gille W. Particles puzzles and scattering patterns. Contribution, 21st SI-HR Crystallographic Meeting, Bohinj/Slovenia, 15.06.2012.

[128] Gille W. Analysis of randomly shaped puzzle-fragment-particles via their chord length distribution. *AIP Conference Proceedings*, 1504:737–741, 2012.

[129] Glatter O and Kratky O. *Small-Angle X-Ray Scattering*. Academic Press, London, 1982.

[130] Glatter O. A new method for the evaluation of small-angle scattering data. *J. Appl. Cryst.*, 10:415–421, 1977.

[131] Glatter O. The interpretation of real-space information from small-angle scattering experiments. *J. Appl. Cryst.*, 12:166–175, 1979.

[132] Glatter O. Determination of particle-size distribution functions from small-angle scattering data by means of the indirect transformation method. *J. Appl. Cryst.*, 13:7–11, 1980.

[133] Glatter O. Computation of distance distribution functions and scattering functions of models for small-angle scattering experiments. *Acta Phys. Austriaca*, 52:243–256, 1980.

[134] Glatter O. Evaluation of small-angle scattering data from lamellar and cylindrical particles by the indirect transformation method. *J. Appl. Cryst.*, 13:577–584, 1980.

[135] Glatter O. Convolution square root of band-limited symmetrical functions and its application to small-angle scattering data. *J. Appl. Cryst.*, 14:101–108, 1981.

[136] Glatter O. Section: Interpretation. In Glatter O and Kratky O, editors, *Small Angle X-Ray Scattering*, pages 167–196. Academic Press, London, 1982.

[137] Glatter O. Section: Data Treatment. In Glatter O and Kratky O, editors, *Small Angle X-Ray Scattering*, pages 119–165. Academic Press, London, 1982.

[138] Glatter O. Small-angle techniques: X-ray. *International Tables for Crystallography, Volume C*, 89–105, 1991.

[139] Glatter O. Recent progress in instrumentation, evaluation techniques and applications of small-angle X-ray scattering. In Singh AK, editor, *Advanced X-ray Techniques in Research and Industries*. IOS Press BV, Amsterdam, 2005.

[140] Goodisman JN, and Coppa N. Models for X-Ray scattering from random systems. *Acta Cryst. A*, 37:170–180, 1981.

[141] Gruy F and Jacquier S. The chord length distribution of a two-sphere aggregate. *Comp. Mat. Sci.*, 44:218–223, 2008.

[142] Guinier A and Fournét G. L'etat actuel de la theorie de la diffusion des rayons X aux petits angles. *Le Journal de Physique et le Radium*, 11:516–520, 1950.

[143] Guinier A and Fournét G. *Small-Angle Scattering of X-Rays*. John Wiley & Sons, New York, 1955.

[144] Guinier A. *X-Ray scattering in Crystals, Imperfect Crystals and Amorphous Substances*. John Wiley & Sons, New York, 1966.

[145] Hall P. *Introduction to the Theory of Coverage Processes*. John Wiley & Sons, New York, 1988.

[146] Hansen M, and Zwet E. Nonparametric estimation of the chord length distribution. *MPS-RR 2008-8, Centre for Mathematical Physics and Stochastics, Department of Mathematical Sciences, University of Aarhus, February 2000*.

[147] Hansen M. Determination of chord length distributions by use of the indirect Fourier transformation. *J. Appl. Cryst.*, 36:1190–1196, 2003.

[148] Harutyunyan HS. Chord length distribution function for a regular hexagon. *Uchenye Zapiski Yerevan State Univ.*, 1:17–24, 2007.

[149] Harutyunyan HS and Ohanyan V.K. Chord length distribution function for convex polygons. *International Journal of Mathematical Science Education*, 4:1–15, 2011.

[150] Hermann H. *Stochastic Models of Heterogeneous Materials*. Materials Science Forum, Volume 78. Trans Tech Publications, Brookfield, 1991.

[151] Herrmann C and Gille W. The cuboid correlation function, see www.ordinate.de/wolfram/mma_beisp2.htm. Representation of the cuboid structure functions by a *Mathematica* notebook, 1999.

[152] Hilfer R. Geometric and dielectric characterization of porous media. *Phys. Rev. B*, 44:60–75, 1991.

[153] Hilfer R. Transport and relaxation phenomena in porous media. *Adv. Chem. Phys.*, XCII:299–424, 1996.

[154] Illian J, Penttinen A, Stoyan H, and Stoyan D. *Statistical Analysis and Modelling of Spatial Point Patterns*. Wiley & Sons Ltd, Chichester, 2008.

[155] Innerlohinger J, Wyss HM, and Glatter O. Colloidal systems with attractive interaction: Evaluation of scattering data using the generalized

indirect Fourier transformation method. *J. Phys. Chem. B*, 108:18149-18157, 2004.

[156] Jeulin D. Modelling random media. *Image Anal. Stereol.*, 21:31–40, 2002.

[157] Jeulin D. Morphological models and simulations to predict the physical behaviour of random microstructures. International Conference, Stochastic Geometry and Its Applications, University of Bern, Switzerland, October 2005.

[158] Kabisch O, Gille W, and Krol J. The effect of copper addition on the structure and strength of an Al–Li alloy. *Materials and Design*, 18:385–388, 1998.

[159] Kabisch O, Schmidt U, and Gille W. The SAS behaviour of amorphous AlDyNi samples during the beginning of the crystallization process. *Proc. Int. Conf. on Light Alloys and Composites, Zakopane 1999*, 97–201, 1999.

[160] Kaya H. Scattering from cylinders with globular end-caps. *J. Appl. Cryst.*, 37:223–230, 2004.

[161] Kaya H and Souza N. Scattering from capped cylinders. *J. Appl. Cryst.*, 37:508–509, 2004.

[162] Kellerer AM. Considerations on the random traversal of convex bodies and solutions for general cylinders. *Radiation Res.*, 47:359–376, 1971.

[163] Kinning DJ and Thomas EL. Hard sphere interactions between spherical domains in diblock copolymers. *Macromolecules*, 17:1712–1718, 1984.

[164] Kratky O. *Die Welt der Vernachlaessigten Dimension und die Kleinwinkelstreuung der Roentgen-Strahlen und Neutronen an Biologischen Makromolekuelen.* Nova Acta Leopoldina, number 256, issue 55. Deutsche Akademie der Naturforscher Leopoldina, Halle, 1983.

[165] Levitz P and Tchoubar D. Disordered porous solids: from chord distributions to small-angle scattering. *J. Phys. I France*, 21:771–790, 1992.

[166] Li X, Shew C, Meilleur F, Myles A, Liu E, Zhang Y, Smith S, Herwig W, Pynn R, and Chen W. Scattering functions of Platonic solids. *J. Appl. Cryst.*, 44:545–557, 2011.

[167] Lu B and Torquato S. Lineal-path function for random heterogeneous materials. *Phys. Rev. A*, 45(2):922–929, 1992.

[168] Lu B and Torquato S. Lineal-path function for random heterogeneous materials. II. Effect of polydispersity. *Phys. Rev. A*, 45(10):7292–7301, 1992.

[169] Lu B and Torquato S. Chord-length and free-path distribution functions for many-body systems. *J. Chem. Phys.*, 98(8):6472–6482, 1993.

[170] Mallows CL and Clark JM. Linear intercept distributions do not characterize plane sets. *Journal of Applied Probability*, 7:240–244, 1970.

[171] Mallows CL and Clark JM. Linear intercept distributions do not characterize plane sets (corrections). *Journal of Applied Probability*,8:208–209, 1970.

[172] Mäder U. Chord length distributions for circular cylinders. *Radiat. Res.*, 82:454–466, 1980.

[173] Mason G. Random chord distributions from triangles. *Powder Technology*, 12:277–281, 1975.

[174] Matheron G. *Random Sets and Integral Geometry*. John Wiley, New York, 1975.

[175] Mazzolo A, Roesslinger B, and Gille W. Properties of chord length distributions of nonconvex bodies. *J. Math. Phys.*, 44:6195–6209, 2003.

[176] Mittelbach R and Glatter O. Direct structural analysis of small-angle scattering data from polydisperse colloidal particles. *J. Appl. Cryst.*, 31:600–608, 1998.

[177] Müller JJ, Schmidt PW, Damaschun G, and Walter G. Determination of the largest particle dimension by direct Fourier-cosine-transformation of experimental small-angle X-ray scattering data. *J. Appl. Cryst.*, 13:280–283, 1980.

[178] Müller K and Glatter 0. Practical aspects to the use of indirect Fourier transformation methods. *Makromol. Chem.*, 183:465–479, 1982.

[179] Müller JJ, Glatter O, Zirwer D, and Damaschun G. Calculation of small-angle X-ray and neutron scattering curves and of translational friction coefficients on the common basis of finite elements. *Studia Biophysica*, 93:39–46, 1983.

[180] Müller JJ, Gille W, and Damaschun G. Direct determination of the largest diameter of a particle by a new transformation of X-ray scattering data. *Unpublished report 22 pages, Berlin and Halle*, 1993.

[181] Mocica G. *Probleme de Functii Speciale*. Chapter 5, page 336, Hypergeometric functions (in Romanian language). Edituria Dacia, Bucuresti, 1988.

[182] Méring J and Tchoubar D. Diffusion centrale des ryaons X par les systemes poreux. *J. Appl. Cryst.*, 1:153–165, 1968.

[183] Nagel W. *Das Geometrische Kovariogramm und Verwandte Groessen Zweiter Ordnung*. Habilitationsschrift, Friedrich Schiller-Universität Jena, Mathematische Fakultät, Jena 1992.

[184] Naumovich NV and Kriskovets TI. Influence of the body shape on the profile of chord length distribution. *Acta Stereol.*, 1:51–59, 1982.

[185] Ohser J and Schladitz K. *3D images of Materials Structures - Processing and Analysis*. Wiley-VCH, Weinheim, 2009.

[186] Papoulis L. *The Fourier Integral and Its Applications*. McGraw Hill, New York, 1962.

[187] Patterson A. A direct method for the determination of the components of interatomic distances in crystals. *Z. Krist.*, A90:517–542, 1935.

[188] Percus J and Yevick G. Analysis of classical statistical mechanics by means of collective coordinates. *Phys. Rev.*, 110:1, 1958.

[189] Piefke F. Chord length distribution of the ellipse. *Lithuanian Mathematical Journal*, 19/3:325–333, 1979.

[190] Piefke F. Beziehungen zwischen der Sehnenlaengenverteilung und der Verteilung des Abstandes zweier zufaelliger Punkte im Eikoerper. *Z. f. Wahrscheinlichkeitstheorie*, 43:129–134, 1987.

[191] Porod G. Die Röntgenkleinwinkelstreuung von dichtgepackten kolloiden Systemen. *Kolloid Zeitschrift*, 124:83–114, 1951.

[192] Porod G. Die Röntgenkleinwinkelstreuung von dichtgepackten kolloiden Systemen. *Kolloid Zeitschrift*, 125:51–122, 1952.

[193] Porod G and Mittelbach P. Zur Röntgenkleinwinkelstreuung Kolloider Systeme. *Kolloid Zeitschrift*, 202:40–49, 1965.

[194] Porod G. Section I: The principles of diffraction, general theory. In O. Glatter and O. Kratky, editors, *Small-Angle X-Ray Scattering*, pages 34–40. Academic Press, London, 1982.

[195] Ramlau R and Löffler H. The structure of metastable phases in Al-Zn alloys, HREM investigations on Guinier-Preston-zones. *Crystal Res. Technol.*, 19:1273–1286, 1984.

[196] Ramlau R. Elektronenmikroskopische Untersuchungen zu Struktur und Transformation metastabiler und stabiler Phasen in Al-Zn-Legierungen. Doctoral Thesis, Martin-Luther-Universität Halle-Wittenberg, Halle 1985.

[197] Rosiwal A. Geometrische Gesteinsanalysen gemengter Gesteine. *Verh. K. K. Geol. Reichsanstalt*, 5/6:143–175, 1898.

[198] Roth SV, Burghammer M, Gilles R, Mukherji D, Rösler J and Strunz P. Precipitate scanning in Ni-base-super-alloys. Contribution and private communications, International Conference on Small-Angle Scattering, Venezia 2002.

[199] Roth SV, Burghammer M, Gilles R, Mukherji D, Rösler J and Strunz P. Precipitate scanning in Ni-base-super-alloys. *Nuclear Instruments and Methods in Physics Research B*, 200:255–260, 2003.

[200] Ruland W and Smarsly B. SAXS of self-assembled nanocomposite films with oriented two-dimensional cylinder arrays: an advanced method of evaluation. *J. Appl. Cryst.*, 38:78–86, 2005.

[201] Ryshik I M and Gradstein I S. *Tables of Series, Products and Integrals*. Verlag der Wissenschaften, Berlin, 1957.

[202] Sahian GA. Investigation of the structure of the observed field of an object from the standpoint of integral geometry. *Astro. Zh.*, 68:1036–1045, 1991.

[203] Santaló LA. *Integral Geometry and Geometric Probability*. Addison-Wesley, Reading, MA, 1976.

[204] Schmidt PW. Small-angle studies of disordered, porous and fractal systems. *J. Appl. Cryst.*, 24:414–435, 1991.

[205] Serra J. *Image Analysis and Mathematical Morphology, Volume 1*. Academic Press, London, 1982.

[206] Serra J. *Image Analysis and Mathematical Morphology, Volume 2: Theoretical Advances*. Academic Press, London, 1988.

[207] Sobry R, Ledent J, and Fontaine F. Application of an extended Porod law to the study of the ionic aggregates in telechelic ionomers. *J. Appl. Cryst.*, 24:516–525, 1991.

[208] Sonntag U, Stoyan D, and Hermann H. Random set models in the interpretation of small-angle scattering data. *Phys. Stat. Sol. A*, 68:281–288, 1981.

[209] Stoyan D. On the accuracy of lineal analysis. *Biometrical J.*, 21:439–449, 1979.

[210] Stoyan D, Kendall WS, and Mecke J. *Stochastic Geometry and Its Applications*. Wiley & Sons, Chichester, 1987.

[211] Stoyan D and Stoyan H. *Fraktale Formen Punktefelder*. Akademie Verlag, Berlin, 1992.

[212] Stoyan D, Kendall WS and Mecke J. *Stochastic Geometry and Its Applications*. Wiley & Sons, Chichester, 1995.

[213] Stoyan D. and Schlather M. Random sequential adsorption: relationship to dead leaves and characterization of variability. *J. Statist. Phys.*, 100:969–979, 2000.

[214] Stoyan D. Simulation and characterization of random systems of hard particles. *Image Anal. Stereol.*, 21(suppl 1):41–48. 2002.

[215] Stribeck N and Ruland W. Determination of the interface distribution function of lamellar two-phase systems. *J. Appl. Cryst.*, 11:535–539, 1978.

[216] Sukiasian HS. Three-dimensional Pleijel identity and its applications. *Journal of Contemporary Mathematical Analysis, (Armenian Academy of Sciences)*, 38:79–91, 2003.

[217] Sukiasian HS and Gille W. Relation between the chord length distribution of an infinitely long cylinder and that of its base. *Journal of Mathematical Physics*, 48:053305, 2007.

[218] Svergun DI. Mathematical methods in small-angle scattering data analysis. *J. Appl. Cryst.*, 24:485–492, 1991.

[219] Svergun DI. Determination of the regularization parameter in indirect-transform methods using perceptual criteria. *J. Appl. Cryst.*, 25:495–503, 1992.

[220] Synecek V. Small-angle scattering from dense systems. Contribution at the International Conference on Small-Angle Scattering, Prague, 1983.

[221] Synecek V. Small-angle scattering from dense systems of non-homogeneous particles, parts I and II. Private communications Halle and Prague, 1982 and 1983.

[222] Tchoubar D. *Diffusion centrale des Ryaons X par les Systems Poureux.* Thesis (in French), Paris 1967.

[223] Teichgräber M. *Die quantitative Bestimmung des Ordnungszustandes in partiellkristallinen Polymeren mittels Röntgendiffraktometrie.* Habilitation, Leuna-Merseburg, Technical University, 1971.

[224] Tokunaga TK. Porous media gas diffusivity from a free path distribution model. *J. Chem. Phys.*, 82:5298–5299, 1985.

[225] Torquato S. *Random Heterogeneous Materials: Microstructure and Macroscopic Properties.* Springer-Verlag, New York, 2002.

[226] Torquato S. Stochastic geometry of heterogeneous materials. International Conference, Stochastic Geometry and Its Applications, University of Bern, Switzerland, October 2005.

[227] Truong CT, Kabisch O, Gille W, and Schmidt U. Small angle X-ray scattering and electrical resistivity measurements on an Al-2Li-5Mg-0.1Zr alloy. *Materials Chemistry and Physics*, 73:268–273, 2002.

[228] Vlasov A. Extension of Dirac's chord method to the case of a nonconvex set by use of quasi-probability distributions. *J. Math. Phys.*, 52:053516-14, 2011.

[229] Voss K. Powers of chords for convex sets. *Biom. Journal*, 24:513–516, 1982.

[230] Weibel ER. *Stereological Methods. Volume 2.* Academic Press, London, 1980.

[231] Weil W. Random sets and Boolean models. Manuscript pages 1–50. Mathematisches Institut II, University Karlsruhe, Germany, Course on Stochastic Geometry, Martina Franca, Italy, September 2004.

[232] Wertheim M. Exact solution of the Percus-Yevick integral equation for hard spheres. *Physical Review Letters*, 10:321–323, 1963.

[233] Wertheim M. Analytic solution of the Percus-Yevick equation. *J. Math. Phys.*, 5:643–651, 1964.

[234] Wolfram Research, Inc., Mathematica, versions 4 and 5, Champaign, Illinois 1996/2005.

[235] Wriedt T. A review of elastic light scattering theories. *Part. & Part. Syst. Charact.*, 15:67–74, 1998.

[236] Wu H and Schmidt PW. Intersect distributions and small-angle X-ray scattering theory. *J. Appl. Cryst.*, 6:66–72, 1973.

[237] Wu H and Schmidt PW. The intersect distribution of the ellipsoid. *J. Appl. Cryst.*, 7:131–146, 1974.

Index